Econophysics of Agent-Based Models

New Economic Windows

For further volumes:
www.springer.com/series/6901

Frédéric Abergel • Hideaki Aoyama •
Bikas K. Chakrabarti • Anirban Chakraborti •
Asim Ghosh
Editors

Econophysics of Agent-Based Models

Springer

Editors
Frédéric Abergel
Lab. Mathematics Applied to Systems
École Centrale Paris
Châtenay-Malabry, France

Hideaki Aoyama
Department of Physics
Kyoto University
Kyoto, Japan

Bikas K. Chakrabarti
Saha Institute of Nuclear Physics
Kolkata, India

Anirban Chakraborti
Lab. Mathematics Applied to Systems
École Centrale Paris
Châtenay-Malabry, France

Asim Ghosh
TCMP
Saha Institute of Nuclear Physics
Kolkata, India

ISSN 2039-411X ISSN 2039-4128 (electronic)
New Economic Windows
ISBN 978-3-319-00022-0 ISBN 978-3-319-00023-7 (eBook)
DOI 10.1007/978-3-319-00023-7
Springer Cham Heidelberg New York Dordrecht London

Library of Congress Control Number: 2013941925

Printed on acid-free paper

Springer is part of Springer Science+Business Media (www.springer.com)

Preface

This proceedings volume is based on the conference on ‘Econophysics of Agent-based models’ held at Saha Institute of Nuclear Physics, Kolkata during November 8–12, 2012. Agent-based modeling is one of the most powerful tools, now being widely employed for understanding problems of market dynamics, leading on to very important developments in this area. In most conventional economic models, in order to keep them mathematically tractable, there is usually the ‘representative agent’, who is assumed to be ‘perfectly rational’ and uses the ‘utility maximization’ principle while taking action. There are not many tools available to economists for solving non-linear models of ‘heterogeneous adaptive agents.’ In this respect, the very flexible and diverse multi-agent models, which originate from statistical physics considerations, allow one to go beyond the prototype theories with the ‘representative agent’ in traditional economics.

The Econophys-Kolkata VII conference held last year (2012), the seventh event in this series of international conferences, was dedicated to address and discuss extensively these issues, approaches and the recent developments concerning agent-based models in Econophysics. This event was organized jointly by the École Centrale Paris, the Saha Institute of Nuclear Physics, with the addition of Kyoto University for the first time, and was held at the Saha Institute of Nuclear Physics, Kolkata.

This proceedings volume contains papers by distinguished experts from all over the world, mostly based on the talks and seminars delivered at the meeting, and accepted after refereeing. For completeness, a few articles by the experts who could not participate in the meeting due to unavoidable reasons, were also invited and incorporated in this volume.

These proceedings volume is organized as follows: A first section dedicated to “agent-based models” in the social sciences. A second section on “miscellaneous” presents other on-going studies in related areas on econophysics and sociophysics. We have included in the third section “discussions and commentary”, an extensive note on “evolution of econophysics” which had been intensively discussed during the conference and contributed informally, though significantly, by many formal participants. Two other shorter write-ups—a discussion and a critique on econophysics, arisen out of the various interesting and informal exchanges amongst the participants that took place during the conference, have also been incorporated in this section.

We are grateful to all the participants of the meeting and for all their contributions. We are also grateful to Mauro Gallegati and the Editorial Board of the New Economic Windows series of the Springer-Verlag (Italy) for their support in getting this Proceedings volume published as well, in their esteemed series.[1]

The conveners (editors) also express their thanks to Saha Institute of Nuclear Physics, École Centrale Paris and Kyoto University for their support in organizing this conference.

Châtenay-Malabry, France — Frédéric Abergel
Kyoto, Japan — Hideaki Aoyama
Kolkata, India — Bikas K. Chakrabarti
Châtenay-Malabry, France — Anirban Chakraborti
Kolkata, India — Asim Ghosh
May, 2013

[1]Past volumes:

1. Econophysics of systemic risk and network dynamics, Eds. F. Abergel, B.K. Chakrabarti, A. Chakraborti and A. Ghosh, New Economic Windows, Springer, Milan, 2013.
2. Econophysics of order-driven markets, Eds. F. Abergel, B.K. Chakrabarti, A. Chakraborti, M. Mitra, New Economic Windows, Springer, Milan, 2011.
3. Econophysics & economics of games, social choices and quantitative techniques, Eds. B. Basu, B.K. Chakrabarti, S.R. Chakravarty, K. Gangopadhyay, New Economic Windows, Springer, Milan, 2010.
4. Econophysics of markets and business networks, Eds. A. Chatterjee, B.K. Chakrabarti, New Economic Windows, Springer, Milan, 2007.
5. Econophysics of stock and other markets, Eds. A. Chatterjee, B.K. Chakrabarti, New Economic Windows, Springer, Milan, 2006.
6. Econophysics of wealth distributions, Eds. A. Chatterjee, S. Yarlagadda, B.K. Chakrabarti, New Economic Windows, Springer, Milan, 2005.

Contents

Part I
Agent-Based Models

Chapter 1
Agent-Based Modeling of Zapping Behavior of Viewers, Television Commercial Allocation, and Advertisement Markets

Hiroyuki Kyan and Jun-ichi Inoue

Abstract We propose a simple probabilistic model of zapping behavior of television viewers. Our model might be regarded as a 'theoretical platform' to investigate the human collective behavior in the macroscopic scale through the zapping action of each viewer at the microscopic level. The stochastic process of audience measurements as macroscopic quantities such as television program rating point or the so-called gross rating point (GRP for short) are reconstructed using the microscopic modeling of each viewer's decision making. Assuming that each viewer decides the television station to watch by means of three factors, namely, physical constraints on television controllers, exogenous information such as advertisement of program by television station, and endogenous information given by 'word-of-mouth communication' through the past market history, we shall construct an aggregation probability of Gibbs-Boltzmann-type with the energy function. We discuss the possibility for the ingredients of the model system to exhibit the collective behavior due to not exogenous but endogenous information.

1.1 Introduction

Individual human behaviour is actually an attractive topic for both scientists and engineers, and in particular for psychologists, however, it is still extremely difficult for us to deal with the problem by making use of scientifically reliable investigation. In fact, it seems to be somehow an 'extraordinary material' for exact scientists such as physicists to tackle as their own major. This is because there exists quite large sample-to-sample fluctuation in the observation of individual behaviour. Namely, one cannot overcome the difficulties caused by individual variation to find the universal fact in the behaviour.

H. Kyan · J. Inoue (✉)
Graduate School of Information Science and Technology, Hokkaido University,
N14-W-9, Kita-ku, Sapporo 060-0814, Japan
e-mail: jinoue@cb4.so-net.ne.jp

H. Kyan
e-mail: kyan@complex.ist.hokudai.ac.jp

F. Abergel et al. (eds.), *Econophysics of Agent-Based models*,
New Economic Windows, DOI 10.1007/978-3-319-00023-7_1,

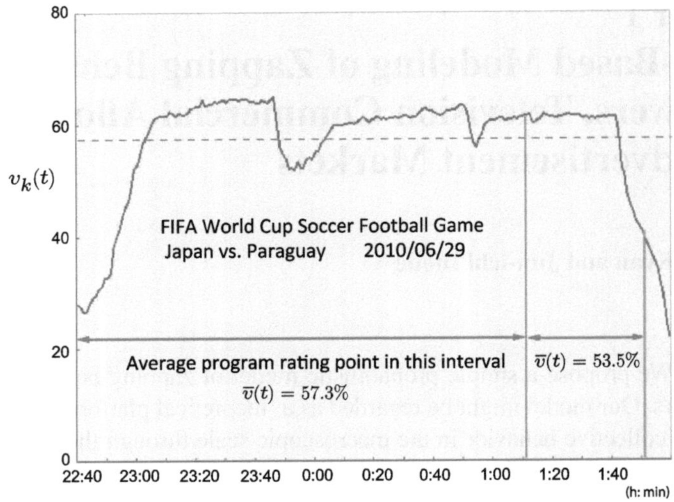

Fig. 1.1 A typical behavior of program rating point. The data set is provided by Video Research Ltd (http://www.videor.co.jp/eng/index.html)

On the other hand, in our human 'collective' behaviour instead of individual, we sometimes observe several universal phenomena which seem to be suitable materials ('many-body systems') for computer scientists to figure out the phenomena through agent-based simulations. In fact, collective behaviour of interacting agents such as flying birds, moving insects or swimming fishes shows highly non-trivial properties. The so-called *BOIDS* (an algorithm for artificial life by simulated flocks) realizes the collective behaviour of animal flocks by taking into account only a few simple rules for each interacting 'intelligent' agent in computer [1–3].

Human collective behavior in macroscopic scale is induced both exogenously and endogenously by the result of decision making of each human being at the microscopic level. To make out the essential mechanism of emergence phenomena, we should describe the system by mathematically tractable models which should be constructed as simple as possible.

It is now manifest that there exist quite a lot of suitable examples around us for such collective behavior emerged by our individual decision making. Among them, the relationship between the zapping actions of viewers and the arrangement of programs or commercials is a remarkably reasonable example.

As an example, in Fig. 1.1, we plot the empirical data for the rating point for the TV program of FIFA World Cup Soccer Football Game, Japan vs. Paraguay which was broadcasted in a Japanese television station on 29th June 2010. From this plot, we clearly observe two large valleys at 23:50 and 0:50. The first valley corresponds to the interval between the 1st half and the 2nd half of the game, whereas the second valley corresponds to the interval between 2nd half of the game and penalty shoot-out. In these intervals, a huge amount of viewers changed the channel to the other stations, and one can naturally assume that the program rating point remarkably dropped in these intervals. Hence, it might be possible to estimate the microscopic viewers' decision makings from the macroscopic behavior of the time series such as the above program rating point.

Commercials usually being broadcasted on the television are now well-established as powerful and effective tools for sponsors to make viewers recognize their commodities or leading brand of the product or service. From the viewpoint of television stations, the commercial is quite important to make a profit as advertising revenue. However, at the same time, each television station has their own wishes to gather viewers of their program without any interruption due to the commercial because the commercial time is also a good chance for the viewers to change the channel to check the other programs which have been broadcasted from the other rival stations. On the other hand, the sponsors seek to maximize the so-called *contact time* with the viewers which has a meaning of duration of their watching the commercials for sponsors' products or survives. To satisfy these two somehow distinct demands for the television station and sponsors, the best possible strategy is to lead the viewers not to zap to the other channels during their program. However, it is very hard requirement because we usually desire to check the other channels in the hope that we might encounter much more attractive programs in that time interval.

As the zapping action of viewers is strongly dependent on the preference of the viewers themselves in the first place, it seems to be very difficult problem for us to understand the phenomena by using exact scientific manner. However, if we consider the 'ensemble' of viewers to figure out the statistical properties of their collective behavior, the agent-based simulation might be an effective tool. Moreover, from the viewpoint of human engineering, there might exist some suitable channel locations for a specific television station in the sense that it is much easier for viewers to zap the channel to arrive as a man-machine interface.

With these mathematical and engineering motivations in mind, here we shall propose a simple mathematical model for zapping process of viewers. Our model system is numerically investigated by means of agent-based simulations. We evaluate several useful quantities such as *television program rating point* or *gross rating point (GRP for short)* from the microscopic description of the decision making by each viewer. Our approach enables us to investigate the television commercial market extensively like financial markets [4].

This paper is organized as follows. In Sect. 1.2, we introduce our mathematical model system and several relevant quantities such as the program rating point or the GRP. In Sect. 1.3, we clearly introduce Ising spin-like variable which denotes the time-dependent microscopic state of a single viewer, a television station for a given arrangement of programs and commercials. In Sect. 1.4, we show that the macroscopic quantities such as program rating point or the GRP are calculated in terms of the microscopic variables which is introduced in the previous Sect. 1.3. In Sect. 1.5, the energy function which specifies the decision making of each viewer is introduced explicitly. The energy function consists of three distinct parts, namely, a physical constraint on the controller, partial energies by exogenous and endogenous information. The exogenous part comes from advertisement of the program by the television station, whereas the endogenous part is regarded as the influence by the average program rating point on the past history of the market. By using the maximum entropy principle under several constraints, we derive the aggregation probability of viewers as a Gibbs-Boltzmann form. In Sect. 1.6, we show our preliminary results obtained by computer simulations. We also consider the 'adaptive location' of commercial advertisements in Sect. 1.7. In this section, we also consider the effects of the so-called *Yamaba CMs*, which are the successive CMs broadcasted intensively at the climax of the program, on the program rating points to the advertisement measurements. The last section is devoted to the concluding remarks.

1.2 The Model System

We first introduce our model system of zapping process and submission procedure of each commercial into the public through the television programs. We will eventually find that these two probabilistic processes turn out to be our effective television commercial markets.

As long as we surveyed carefully, quite a lot of empirical studies on the effect of commercials on consumers' interests have been done, however, up to now there are only a few theoretical studies concerning the present research topic to be addressed. For instance, Siddarth and Chattopadhyay [5] (see also the references therein) introduced a probabilistic model of zapping process, however, they mainly focused on the individual zapping action, and our concept of 'collective behavior' was not taken into account. Ohnishi *et al.* [6] tried to solve the optimal arrangement problem of television commercials for a given set of constraints in the literature of linear programming. Hence, it should be stressed that the goals of their papers are completely different from ours.

1.2.1 Agents and Macroscopic Quantities

To investigate the stochastic process of zapping process and its influence on the television commercial markets, we first introduce two distinct agents, namely, television stations, each of which is specified by the label $k = 1, \dots, K$ and viewers specified by the index $i = 1, \dots, N$. Here it should be noted that $K \ll N$ should hold. The relationship between these two distinct agents is described schematically in Fig. 1.2. In this figure, the thick line segments denote the period of commercial, whereas the thin line segments stand for the program intervals. The set of solid arrows describes a typical trajectory of viewer's zapping process. Then, the *(instant) program rating point* for the station k at time t is given by

$$v_k(t) = \frac{N_k(t)}{N} \tag{1.1}$$

where $N_k(t)$ is the number of viewers who actually watch the television program being broadcasted on the channel (the television station) k at time t.

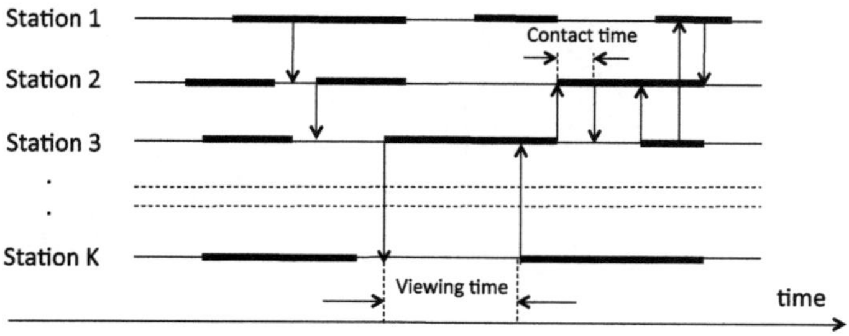

Fig. 1.2 Two-types of agents in our model systems. The *thick line* segments denote the periods of commercial, whereas the thin line segments stand for the program intervals. The set of *solid arrows* describes a typical trajectory of viewer's zapping processes. 'Contact time' and 'viewing time' are clearly defined as intervals for which the viewer watches the commercials and the programs, respectively. The duration and the procedures of casting television programs and CMs would be modeled by Poisson arrival processes. The detail accounts for them will be given in Sects. 1.2.3, 1.2.4 and 1.2.5

On the other hand, the time-slots for commercials are traded between the station and sponsors through the quantity, the so-called *gross rating point (GRP)* which is defined by

$$GRP_k^{(n)} = \frac{\theta_k^{(n)}}{T} \sum_{t=1}^{T} v_k(t) \tag{1.2}$$

where T denotes total observation time, for instance, say $T = 600$ minutes, for evaluating the program rating point. $\theta_k^{(n)}$ stands for the *average contact time* for viewers who are watching the commercial of the sponsor n being broadcasted on the station k during the interval T. Namely, the average GRP is defined by the product of the average program rating point and the average contact time. It should be noted that Eq. (1.2) is defined as the average over the observation time T. Hence, if one seeks for the total GRP of the station k over the observation time T, it should be given as $T \times GRP_k^{(n)}$. Therefore, the average GRP for the sponsor n during the observation interval T is apparently evaluated by the quantity: $GRP^{(n)} = (1/K)\sum_{k=1}^{K} GRP_k^{(n)}$ when one assumes that the sponsor n asked all stations to broadcast their commercials through grand waves.

1.2.2 Zapping as a 'Stochastic Process'

We next model the zapping process of viewers. One should keep in mind that here we consider the controller shown in Fig. 1.3. We should notice that (at least in Japan) there exist two types of channel locations on the controller, namely, 'lattice-type' and 'ring-type' as shown in Fig. 1.3. For the 'lattice-type', each button corresponding to each station is located on the vertex in the two-dimensional square lattice. Therefore, for the case of $K = 9$ stations (channels) for example (see the lower left

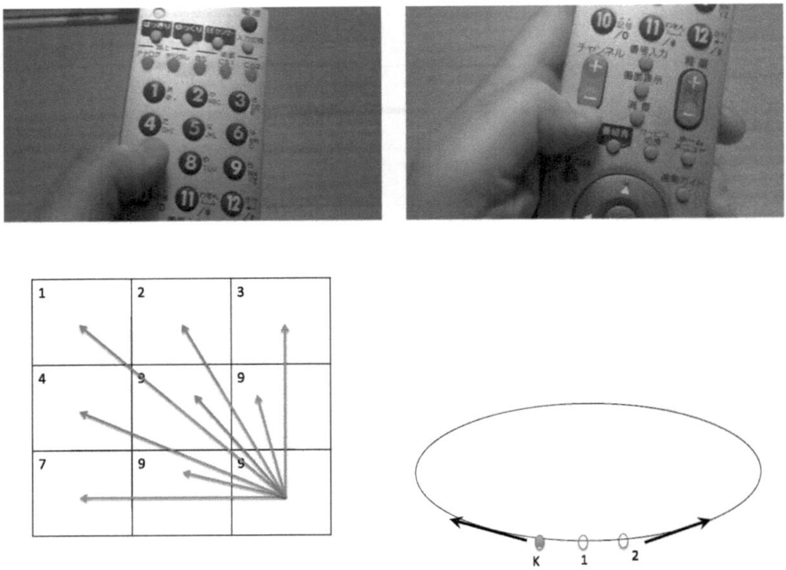

Fig. 1.3 A typical television controller dealt with in this paper. The *left panel* shows ‘lattice-type’ channel, whereas the *right* is a typical channel of ‘ring-type’. Each channel location is modeled as shown in the corresponding cartoons in the lower two panels. Namely, the ‘lattice-type’ (*left*) and the ‘ring-type’ (*right*) for the modeling of the buttons on the television controller. Apparently, there exist a geometrical constraint on the latter type

panel in Fig. 1.3), the viewers can change the channel from an arbitrary station k to the other station $l(\neq k)$, and there is no geometrical constraint for users (viewers). Thus, we naturally define the transition probability $P(l|k)$ which is the probability that a user change the channel from station k to $l(\neq k)$. Taking into account the normalization of the probabilities, we have $\sum_{l\neq k}^{K} P(l|k) = 1, k = 1, \ldots, K$. The simplest choice of the modeling of the transition probability satisfying the above constraint is apparently the uniform one and it is written by

$$P(l|k) = \frac{1}{K-1}, \quad k, l(\neq k) = 1, \ldots, K. \tag{1.3}$$

On the other hand, for the ‘ring-type’ which is shown in the lower right panel of Fig. 1.3, we have a geometrical constraint $P(l = k \pm 1|k) > 0$, $P(l \neq k \pm 1|k) = 0$. From the normalization condition of the probabilities, we also have another type of constraint $P(l = k - 1|k) + P(l = k + 1|k) = 1$ for $k, l(\neq k) = 1, \ldots, K$. The simplest choice of the probability which satisfies the above constraints is given by $P(l = k - 1|k) = P(l = k + 1|k) = 1/2$ for $k, l(\neq k) = 1, \ldots, K$. Namely, the viewers can change from the current channel to the nearest neighboring two stations.

1.2.3 The Duration of Viewer's Stay

After changing the channel stochastically to the other rival stations, the viewer eventually stops to zap and stays the channel to watch the program if he or she is interested in it. Therefore, we should construct the probabilistic model of the length of viewer's stay in both programs and commercials appropriately. In this paper, we assume that the lengths of the viewer i's stay in the programs τ_{on} ('on' is used as an abbreviation for 'on air') and commercials τ_{cm} are given as 'snapshots' from the distributions:

$$P^{(i)}(\tau_{\mathrm{on,cm}}) = a_{\tau_{\mathrm{on,cm}}}^{(i)} \mathrm{e}^{-\tau_{\mathrm{on,cm}}/a_{\tau_{\mathrm{on,cm}}}^{(i)}}, \quad i = 1, \ldots, N \tag{1.4}$$

where $a_{\tau_{\mathrm{on,cm}}}^{(i)}$ stands for the 'relaxation time' of the viewer i for the program and the commercial, respectively. Of course, $\tau_{\mathrm{on,cm}}$ fluctuates from person to person, hence, here we assume that $a_{\tau_{\mathrm{on,cm}}}^{(i)}$ might follow

$$a_{\tau_{\mathrm{on,cm}}}^{(i)} = c_{\mathrm{on,cm}} + \delta, \; P(\delta) = \mathcal{N}(0, \sigma^2). \tag{1.5}$$

Namely, the relaxation times for the program and commercial fluctuate around the typical value $c_{\mathrm{on,cm}}$ by a white Gaussian noise with mean zero and variance σ^2. Obviously, for ordinary viewers, $c_{\mathrm{on}} > c_{\mathrm{cm}}$ should be satisfied. We should notice that the above choice of the length of viewer's stay is independent on the station, program or sponsor. For instance, the length of viewer's stay in commercials might be changed according to the combination of commercials of different kinds of sponsors. However, if one needs, we can modify the model by taking into account the corresponding empirical data.

1.2.4 The Process of Casting Commercials

Here we make a model of casting commercials by television stations. To make a simple model, we specify each sponsor by the label $n = 1, \ldots, M$ and introduce microscopic variables $l_k(t)$ as follows.

$$l_k(t) = \begin{cases} n \in \{1, \ldots, M\} & \text{(The station } k \text{ casts a CM of sponsor } n \text{ at time } t\text{)}, \\ 0 & \text{(The station } k \text{ casts a program at } t\text{)}. \end{cases} \tag{1.6}$$

Hence, if one obtains $l_1(1) = l_1(2) = \cdots = l_1(10) = 3$ and $l_1(11) = 0$, then we conclude that the station $k = 1$ casted the commercial of the sponsor $n = 3$ from $t = 1$ to $t = 10$, and after this commercial period, the station k resumed the program at the next step $t = 11$. On the other hand, if we observe $l_1(1) = l_1(2) = \cdots = l_1(10) = 3$ and $l_1(11) = 2$, we easily recognize that the station $k = 1$ casted the commercial of the sponsor $n = 3$ from $t = 1$ to $t = 10$, and after this commercial period, the same station $k = 1$ casted the commercial of the sponsor $n = 2$ at the next step $t = 11$. Therefore, for a given sequence of variables $l_k(t), k = 1, \ldots, K$ for observation period $t = 1, \ldots, T$, the possible patterns being broadcasted by all television stations are completely determined. Of course, we set $l_k(t), k = 1, \ldots, K$ to the artificial values in our computer simulations, however, empirical evidence might help us to choose them.

1.2.5 Arrangement of Programs and CMs

In following, we shall explain how each television station submits the commercials to appropriate time-slots of their broadcasting. First of all, we set $l_k(0) = 0$ for all stations $k = 1, \ldots, K$. Namely, we assume that all stations start their broadcasting from their own program instead of any commercials of their sponsors. Then, for an arbitrary k-th station, the duration t_{on} between the starting and the ending points of each section in the program is generated by the exponential distribution $\sim \mathrm{e}^{-t_{\mathrm{on}}/L_{\mathrm{on}}}$. Thus, from the definition, we should set $l_k(0) = \cdots = l_k(t_{\mathrm{on}}) = 0$ for the resulting t_{on}. We next choose a sponsor among the M-candidates by sampling from a uniform distribution in $[1, M]$. For the selected sponsor, say, $n \in \{1, M\}$, the duration of their commercial t_{cm} is determined by a snapshot of the exponential distribution $\sim \mathrm{e}^{-t_{\mathrm{cm}}/L_{\mathrm{cm}}}$. Thus, from the definition, we set $l_k(t_{\mathrm{on}} + 1) = \cdots = l_k(t_{\mathrm{on}} + 1 + t_{\mathrm{cm}}) = n$ for the given t_{cm} and t_{on}. We repeat the above procedure from $t = 0$ to $t = T - 1$ for all stations $k = 1, \ldots, K$. Apparently, we should choose these two relaxation times $L_{\mathrm{on}}, L_{\mathrm{cm}}$ so as to satisfy $L_{\mathrm{on}} \gg L_{\mathrm{cm}}$. After this procedure, we obtain the realization of combinations of 'thick' (CMs) and 'thin' (television programs) lines as shown in Fig. 1.2.

1.3 Observation Procedure

In Sect. 1.2, we introduced the model system. To figure out the macroscopic behavior of the system, we should define the observation procedure. For this purpose, we first introduce microscopic binary (Ising spin-like) variables $S_{i,k}^{(l_k(t))}(t) \in \{0, 1\}$ which is defined by

$$S_{i,k}^{(n)}(t) = \begin{cases} 1 & \text{(The viewer } i \text{ watches the CM of sponsor } n \\ & \text{on the station } k \text{ at time } t), \\ 0 & \text{(The viewer } i \text{ does not watch the CM of } n \\ & \text{on the station } k \text{ at time } t). \end{cases} \tag{1.7}$$

We should notice that for the case of $l_k(t) = 0$, the Ising variable $S_{i,k}^{(l_k(t))}(t)$ takes

$$S_{i,k}^{(0)}(t) = \begin{cases} 1 & \text{(The viewer } i \text{ watches the program on the station } k \text{ at time } t), \\ 0 & \text{(The } i \text{ does not watch the program on the station } k \text{ at time } t). \end{cases} \tag{1.8}$$

Thus, the $K \times N$-matrix $\boldsymbol{S}^{(n)}(t)$ written by

$$\boldsymbol{S}^{(n)}(t) = \begin{pmatrix} S_{1,1}^{(n)}(t) & \cdots & S_{1,K}^{(n)}(t) \\ \cdots & \cdots & \cdots \\ \cdots & \cdots & \cdots \\ S_{N,1}^{(n)}(t) & \cdots & S_{N,K}^{(n)}(t) \end{pmatrix} \tag{1.9}$$

becomes a sparsely coded large-size matrix which has only a single non-zero entry in each column. On the other hand, by summing up all elements in each row, the result, say $S_{1,k}^{(n)} + \cdots + S_{N,k}^{(n)}$ denotes the number of viewers who watch the commercial n on the station k at time t. Hence, if the number of sorts of commercials (sponsors) M ($n \in \{1, \ldots, M\}$) is quite large, the element should satisfy $S_{1,k}^{(n)} + \cdots + S_{N,k}^{(n)} \ll N$ (actually, it is a rare event that an extensive number of viewers watch the same commercial on the station k at time t) and this means that the matrix $\boldsymbol{S}^{(n)}(t)$ is a sparse large-size matrix. It should be noted that macroscopic quantities such as program rating point or the GRP are constructed in terms of the Ising variables $S_{i,k}^{(l_k(t))}(t) \in \{0, 1\}$.

For the Ising variables $S_{i,k}^{(l_k(t))}(t) \in \{0, 1\}$, besides we already mentioned above, there might exist several constraints to be satisfied. To begin with, as the system has N-viewers, the condition $\sum_{l_k(t)=0}^{M} \sum_{k=1}^{K} \sum_{i=1}^{N} S_{i,k}^{(l_k(t))}(t) = N$ should be naturally satisfied. On the other hand, assuming that each viewer $i = 1, \ldots, N$ watches the television without any interruption during the observation time T, we immediately have $\sum_{t=1}^{T} \sum_{l_k(t)=0}^{M} \sum_{k=1}^{K} S_{i,k}^{(l_k(t))}(t) = T, i = 1, \ldots, N$. It should bear in mind that the viewer i might watch the program or commercial brought by one of the M-sponsors, hence, we obtain the condition $\sum_{l_k(t)=0}^{M} \sum_{k=1}^{K} S_{i,k}^{(l_k(t))}(t) = 1, i = 1, \ldots, N, t = 1, \ldots, T$.

These conditions might help us to check the validity of programming codes and numerical results.

1.4 Micro-descriptions of Macro-quantities

In this section, we explain how one describes the relevant macroscopic quantities such as average program rating point or the GRP by means of a set of microscopic Ising variables $\{S_{i,k}^{(l_k(t))}(t)\}$ which was introduced in the previous section.

1.4.1 Instant and Average Program Rating Points

Here we should notice that the number of viewers who are watching the program or commercials being broadcasted on the station $k(= 1, \ldots, K)$, namely, $N_k(t)$ is now easily rewritten in terms of the Ising variables at the microscopic level as $N_k(t) =$

$\sum_{i=1}^{N}\sum_{l_k(t)=0}^{M} S_{i,k}^{(l_k(t))}(t)$. Hence, from Eq. (1.1), the instant program rating point of the station k at time t, that is $v_k(t)$, is given explicitly as

$$v_k(t) = \frac{N_k(t)}{N} = \frac{1}{N}\sum_{i=1}^{N}\sum_{l_k(t)=0}^{M} S_{i,k}^{(l_k(t))}(t). \tag{1.10}$$

On the other hand, the average program rating point of the station k, namely, $\overline{v}_k$ is evaluated as $\overline{v}_k = (1/T)\sum_{t=1}^{T} v_k(t)$.

1.4.2 Contact Time and Cumulative GRP

We are confirmed that the contact time which was already introduced in Sect. 1.2 is now calculated in terms of $\{S_{i,k}^{(l_k(t))}(t)\}$ as follows. The contact time of the viewer i with the commercial of the sponsor n is written as

$$\theta_{i,k}^{(n)} = (1/T)\sum_{t=1}^{T}\sum_{l_k(t)=0}^{M} \delta_{l_k(t),n} S_{i,k}^{(l_k(t))}(t) \tag{1.11}$$

where $\delta_{a,b}$ denotes the Kronecker's delta. It should be noted that we scaled the contact time over the observation time T by $1/T$ so as to make the quantity the T-independent value. Hence, the average contact time of all viewers who watch the commercial of sponsor n being broadcasted on the station k is determined by $\theta_k^{(n)} \equiv (1/N)\sum_{i=1}^{N}\theta_{i,k}^{(n)}$. Thus, the cumulative GRP is obtained from the definition (1.2) as

$$\begin{aligned} GRP_k^{(n)} &= \frac{1}{T}\sum_{t=1}^{T} v_k(t) \times \theta_k^{(n)} \\ &= \frac{1}{N^2T^2}\left(\sum_{t=1}^{T}\sum_{i=1}^{N}\sum_{l_k(t)=0}^{M} S_{i,k}^{(l_k(t))}(t)\right) \\ &\quad \times \left(\sum_{i=1}^{N}\sum_{t=1}^{T}\sum_{l_k(t)=0}^{M} \delta_{l_k(t),n} S_{i,k}^{(l_k(t))}(t)\right). \end{aligned} \tag{1.12}$$

From the above argument, we are now confirmed that all relevant quantities in our model system could be calculated in terms of the Ising variables $\{S_{i,k}^{(l_k(t))}(t)\}$ which describe the microscopic state of ingredients in the commercial market.

However, the matrix $\boldsymbol{S}^{(n)}(t)$ itself is determined by the actual stochastic processes of viewer's zapping with arranging the programs and television commercials. Therefore, in the next section, we introduce the energy(cost)-based zapping probability which contains the random selection (1.3) as a special case for the 'lattice-type' controller.

1.5 Energy Function of Zapping Process

As we showed in the literature of our probabilistic labor market [7, 8], it is convenient for us to construct the energy function to quantify the action of each viewer. The main issue to be clarify in this study is the condition on which concentration ('condensation') of viewers to a single television station is occurred due to the endogenous information. The same phenomena refereed to as informational cascade in the financial market is observed by modeling of the price return by means of magnetization in the Ising model [9]. In the financial problem, the interaction J_{ij} between Ising spins S_i and S_j corresponds to endogenous information, whereas the external magnetic field h_i affected on the spin S_i stands for the exogenous information. However, in our commercial market, these two kinds of information would be described by means of a bit different manner. It would be given below.

1.5.1 Physical Constraints on Television Controllers

For this end, let us describe here the location of channel for the station k on the controller as a vertex on the two-dimensional square lattice (grid) as $\mathbf{Z}_k \equiv (x_k, y_k)$. Then, we assume that the channel located on the vertex at which the distance from the channel k is minimized might be more likely to be selected by the viewer who watches the program (or commercial) on the station k at the instance t. In other words, the viewer minimizes the energy function given by $\gamma(\mathbf{Z}_k - \mathbf{Z}_l)^2 (l \neq k)$, where we defined the L_2-norm as the distance $(\mathbf{Z}_k - \mathbf{Z}_l)^2 \equiv (x_k - x_l)^2 + (y_k - y_l)^2$. The justification of the above assumption should be examined from the viewpoint of human-interface engineering.

1.5.2 Exogenous Information

Making the decision of viewers is affected by the exogenous information. For instance, several weeks before World Cup qualifying game, a specific station $\overline{k}$, which will be permitted to broadcast the game, might start to advertise the program of the match. Then, a large fraction of viewers including a soccer football fan might decide to watch the program at the time. Hence, the effect might be taken into account by introducing the energy $-\zeta \prod_{\xi=s}^{e} \delta_{t_\xi,t} \delta_{l,\overline{k}}$ where t_s denotes the time at which the program starts and t_e stands for the time of the end. Therefore, the energy decreases when the viewer watches the match of World Cup qualifying during the time for the program, namely from $t = t_s$ to t_e ($\Delta t \equiv t_e - t_s$: broadcasting hours of the program).

1.5.3 Endogenous Information

The collective behavior might be caused by exogenous information which is corresponding to 'external field' in the literature of statistical physics. However, collective behavior of viewers also could be 'self-organized' by means of endogenous information. To realize the self-organization, we might use the moving average of the instant program rating over the past L-steps ($L \ll T$), namely,

$$\langle v_k(t) \rangle \equiv \frac{1}{L} \sum_{\rho=t-L}^{t-1} v_k(\rho) \tag{1.13}$$

as the endogenous information. Then, we define the 'winner channel' which is more likely to be selected at time t as $\hat{k} = \arg\max_m \langle v_m(t) \rangle$. (One might extend it to a much more general form:

$$\hat{k} = \arg\max_m \langle v_m(r) \rangle \tag{1.14}$$

for a given 'time lag' $r (< t)$). Henceforth, we assume that if the winner channel $\hat{k}$ is selected, a part of total energy $-\beta\delta_{l,\hat{k}}$ decreases. This factor might cause the collective behavior of N-individual viewers. Of course, if one needs, it might be possible for us to recast the representation of the winner channel $\hat{k}$ by means of microscopic Ising variables $\{S_{i,k}^{(l_k(t))}(t)\}$.

Usually, the collective behavior is caused by direct interactions (connections) between agents. However, nowadays, watching television is completely a 'personal action' which is dependent on the personal preference because every person can possess their own television due to the wide-spread drop in the price of the television set. This means that there is no direct interaction between viewers, and the collective behavior we expect here might be caused by some sorts of public information such as program rating point in the previous weeks. In this sense, we are confirmed that the above choice of energy should be naturally accepted.

Therefore, the total energy function at time t is defined by

$$E_k(l) \equiv \gamma(\mathbf{Z}_k - \mathbf{Z}_l)^2 - \zeta \prod_{\xi=s}^{e} \delta_{t_\xi,t}\delta_{l,\overline{k}} - \beta\delta_{l,\hat{k}} \tag{1.15}$$

with $\hat{k} = \arg\max_m \langle v_m(t) \rangle$, where $\beta, \zeta, \gamma \geq 0$ are model parameters to be estimated from the empirical data in order to calibrate our model system. According to the probabilistic labor market which was introduced by one of the present authors, we construct the transition probability $P(l|k)$ as the Gibbs-Boltzmann form by solving the optimization problem of the functional:

$$f\big[P(l|k)\big] \equiv -\sum_{l \neq k} P(l|k) \log P(l|k) - \lambda\Big\{\sum_{l \neq k} P(l|k) - 1\Big\}$$

$$- \lambda' \left\{ \sum_{l \neq k} E_k(l) P(l|k) - E \right\} \quad (1.16)$$

with respect to $P(l|k)$. Then, we immediately obtain the solution of the optimization problem (variational problem) as

$$P(l|k) = \frac{\exp[-E_k(l)]}{\sum_{l \neq k} \exp[-E_k(l)]}$$
$$= \frac{\exp[-\gamma(\mathbf{Z}_k - \mathbf{Z}_l)^2 + \zeta \prod_{\xi=s}^{e} \delta_{t_\xi,t} \delta_{l,\overline{k}} + \beta \delta_{l,\hat{k}}]}{\sum_{l \neq k} \exp[-\gamma(\mathbf{Z}_k - \mathbf{Z}_l)^2 + \zeta \prod_{\xi=s}^{e} \delta_{t_\xi,t} \delta_{l,\overline{k}} + \beta \delta_{l,\hat{k}}]}$$

where we chose one of the Lagrange multipliers in $f[P(l|k)]$ as $\lambda = \log\{\sum_{l \neq k} P(l|k)\} - 1$, and another one $-\lambda'$ is set to 1 for simplicity, which has a physical meaning of 'unit inverse-temperature'. We should notice that in the 'high-temperature limit' $\beta, \zeta, \gamma \to 0$, the above probability becomes identical to that of the random selection (1.3). These system parameters should be calibrated by the empirical evidence.

In the above argument, we focused on the 'lattice-type' controller, however, it is easy for us to modify the energy function to realize the 'ring-type' by replacing $\gamma(\mathbf{Z}_k - \mathbf{Z}_l)^2$ in (1.15) by $\varepsilon_l \equiv -\gamma(\delta_{l,k+1} + \delta_{l,k-1})$, namely, the energy decreases if and only if the viewer who is watching the channel k moves to the television station $k-1$ or $k+1$. This modification immediately leads to

$$P(l|k) = \frac{\exp[-E_k(l)]}{\sum_{l \neq k} \exp[-E_k(l)]} = \frac{\exp[-\varepsilon_l + \zeta \prod_{\xi=s}^{e} \delta_{t_\xi,t} \delta_{l,\overline{k}} + \beta \delta_{l,\hat{k}}]}{\sum_{l \neq k} \exp[-\varepsilon_l + \zeta \prod_{\xi=s}^{e} \delta_{t_\xi,t} \delta_{l,\overline{k}} + \beta \delta_{l,\hat{k}}]}. \quad (1.17)$$

We are easily confirmed that the transition probability for random selection in the 'ring-type' controller is recovered by setting $\zeta = \beta = 0$ as

$$P(k-1|k) = \frac{e^\gamma}{e^\gamma + e^\gamma} = \frac{1}{2} = P(k+1|k)$$

and $P(l|k) = 0$ for $l \neq k \pm 1$.

In the next section, we show the results from our limited contributions by computer simulations.

1.6 A Preliminary: Computer Simulations

In this section, we show our preliminary results. In Fig. 1.4, we plot the typical behavior of instant program rating point $v_k(t)$ for $k = 1, \ldots, K$. We set $K = 9$, $N = 600$, $M = 1$ and $T = 600$ for the case of the simplest choice $\beta = \zeta = \gamma = 0$ leading up to (1.3) ('high-temperature limit'). This case might correspond to the 'unconscious zapping' by viewers. The parameters appearing in the system are chosen as

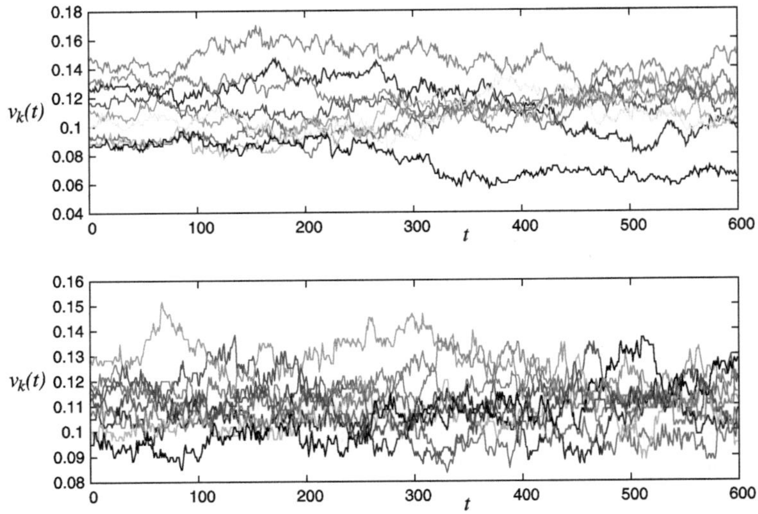

Fig. 1.4 Typical behavior of instant program rating point $v_k(t)$ for $k = 1, \ldots, K$. Here we set $K = 9, M = 1, N = 600$ and $T = 600$. The parameters appearing in the system are chosen as $L_{\rm on} = c_{\rm on} = 12$ and $L_{\rm cm} = c_{\rm cm} = 4$. The *upper panel* is the result of 'lattice-type' with $\beta = \zeta = \gamma = 0$, whereas the *lower panel* shows the case of 'ring-type'. We clearly find that the result of ring-type is less volatile than that of the lattice-type

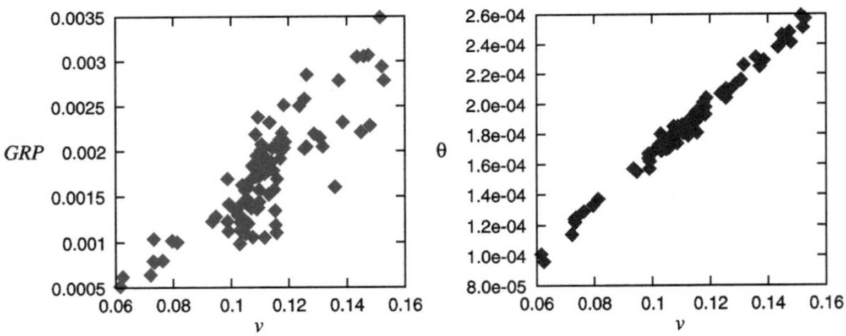

Fig. 1.5 The two-dimensional scattered plot for the average program rating point: $v \equiv (1/K)\sum_{k=1}^{K} \overline{v}_k$ and the average cumulative GRP: $GRP \equiv (1/K)\sum_{k=1}^{K} GRP_k^{(1)}$ for the case of 'lattice-type' channel location (*left*). The *right panel* shows the scatter plot with respect to the v and the effective contact time which is defined by $\theta \equiv (1/KT)\sum_{k=1}^{K} \theta_k^{(1)}$ (we have only a single sponsor). The parameters are set to the same values as in Fig. 1.4

$L_{\rm on} = c_{\rm on} = 12$ and $L_{\rm cm} = c_{\rm cm} = 4$. The upper panel shows the result of 'lattice-type' channel location on the controller, whereas the lower panel is the result of 'ring-type'.

In Fig. 1.5 (left), we display the scattered plot with respect to the GRP and the average program rating point for the case of 'lattice-type' channel location. From this figure, we find that there exists a remarkable positive correlation between these

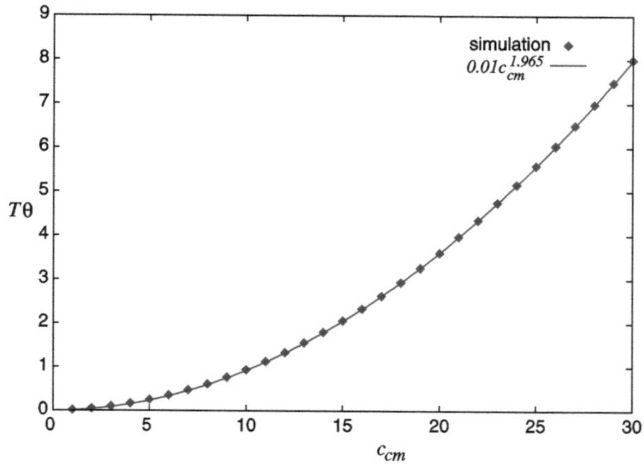

Fig. 1.6 The $c_{\rm cm}$-dependence of the T-scaled effective contact time $T\theta$. We set the observation time $T = 60$ and the other parameters are set to the same values as in Fig. 1.4. We also plot the well-fitting curve $0.01c_{\rm cm}^{1.965}$ for eyes' guide

two quantities (the Pearson coefficient is 0.85). This fact is a justification for us to choose the GRP as a 'market price' for transactions. In the right panel of this figure, the scattered plot with respect to the GRP and the effective contact time defined by $(1/KT)\sum_{k=1}^{K}\theta_k^{(1)}$ is shown. It is clearly found that there also exists a positive correlation with the Pearson coefficient 0.9914.

We also plot the $c_{\rm cm}$-dependence of the T-scaled effective contact time $T\theta$ in Fig. 1.6. This figure tells us that the frequent zapping actions reduce the contact time considerably and it becomes really painful for the sponsors.

1.6.1 Symmetry Breaking Due to Endogenous Information

We next consider the case in which each viewer makes his/her decision according to the market history, namely, we choose $\gamma = 1$, $\zeta = 0$ and set the value of β to $\beta = 0$ and $\beta = 1.8$. We show the numerical results in Fig. 1.7. From this panels, we find that the instant program rating point $v_k(t)$ for a specific television station increases so as to become a 'monopolistic station' when each viewer starts to select the station according to the market history, namely, $\beta > 0$. In other words, the symmetry of the system with respect to the program rating point is broken as the parameter β increases.

To measure the degree of the 'symmetry breaking' in the behavior of the instant program rating points more explicitly, we introduce the following order parameter:

$$B(t) \equiv \frac{1}{K}\sum_{k=1}^{K}\left|v_k(t) - \frac{1}{K}\right| \tag{1.18}$$

which is defined as the cumulative difference between v_k and the value for the 'perfect equality' $1/K$. We plot the $B(t)$ for the case of $\beta = 0, 1$ and 2. For finite β, the symmetry is apparently broken around $t = 100$ and the system changes from

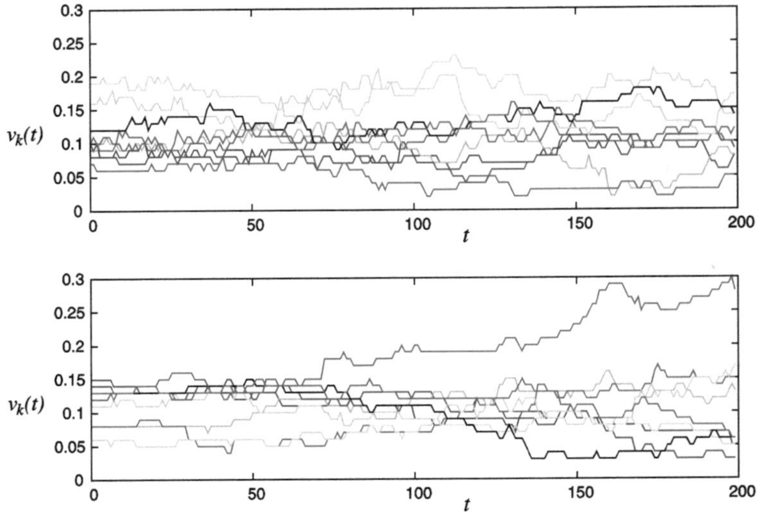

Fig. 1.7 Typical behavior of instant program rating point $v_k(t)$ for $k = 1, \ldots, K$. Here we set $K = 9, M = 1, N = 600$ and $T = 200$. The parameters which specify the energy function of 'lattice-type' controller are chosen as $(\gamma, \zeta, \beta) = (1, 0, 0)$ (the *upper panel*) and $(\gamma, \zeta, \beta) = (1, 0, 1.8)$ (the *lower panel*) with the history length $L = 5$. The parameters appearing in the system are chosen as the same values as in Fig. 1.4

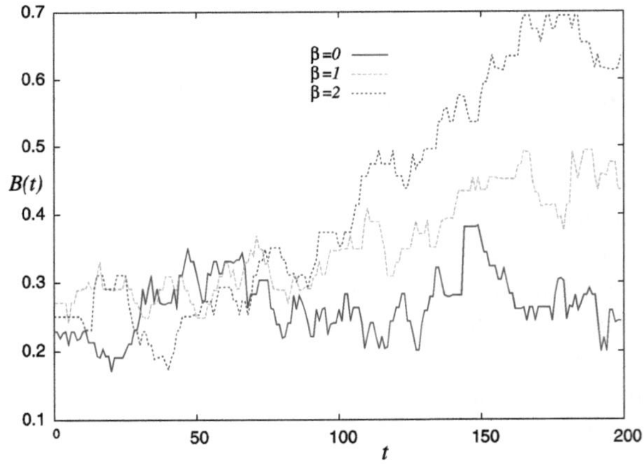

Fig. 1.8 The behavior of order parameter $B(t)$ which measures the degree of symmetry breaking in the v_k. The symmetry is apparently broken for $\beta = 1, 2$ around $t = 100$ and the system changes from symmetric phase (small $B(t)$) to the symmetry breaking phase (large $B(t)$)

symmetric phase (small $B(t)$) to the *symmetry breaking phase* (large $B(t)$) (see Fig. 1.8). We next evaluate the degree of the symmetry breaking by means of the following Shannon's entropy:

$$H(t) = -\sum_{k=1}^{K} v_k(t) \log v_k(t) \tag{1.19}$$

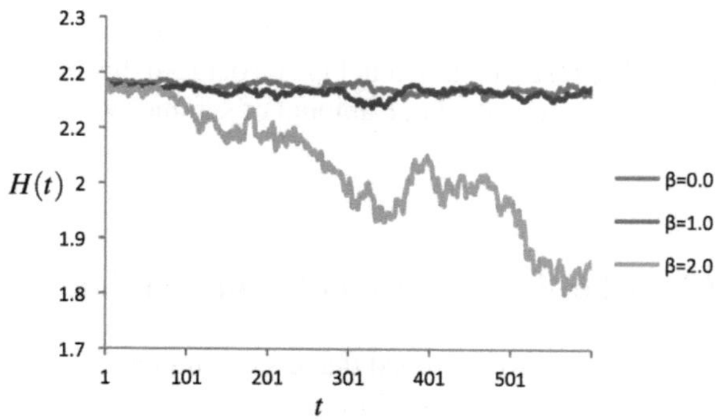

Fig. 1.9 A typical behavior of the Shannon's entropy $H(t)$

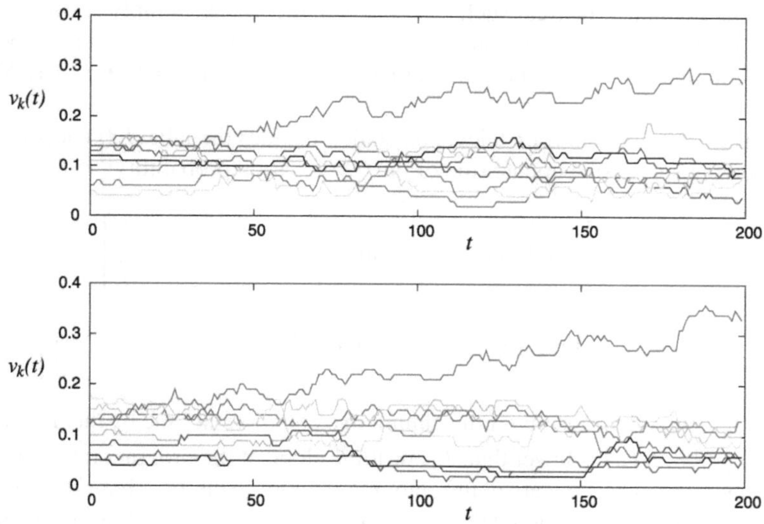

Fig. 1.10 Typical behavior of instant program rating point $v_k(t)$ for $k = 1, \ldots, K$ for the choice of the time lag $r = 10$ (the *upper panel*) and $r = 40$ (the *lower panel*). Here we set $K = 9$, $M = 1$, $N = 600$ and $T = 200$. The parameters which specify the energy function of 'lattice-type' controller are chosen as $(\gamma, \zeta, \beta) = (1, 0, 1.5)$ with history length $L = 5$. The parameters appearing in the system are chosen as the same values as in Fig. 1.4

where the above $H(t)$ takes the maximum for the symmetric solution $v_k(t) = 1/K$ as

$$H(t) = -K \times \frac{1}{K} \log(1/K) = \log K \tag{1.20}$$

whereas the minimum $H(t) = 0$ is achieved for $v_k = 1$ and $v_{k' \neq k} = 0$, which is apparently corresponding to the symmetry breaking phase. In Fig. 1.9, we plot the $H(t)$ for several choices of β as $\beta = 0, 1$ and 2. From this figure, we find that for finite β, the system gradually moves from the symmetric phase to the symmetry breaking phase due to the endogenous information (*e.g.* word-of-mouth communication).

Finally, we consider the time lag r-dependence (see Eq. (1.14)) of the resulting $v_k(t)$. The result is shown in Fig. 1.10. From this figure, we clearly find that the large time lag r causes the large amount of symmetry breaking in the program rating point $v_k(t)$.

1.7 Adaptive Location of Commercials

In Sect. 1.2.4, we assumed that each commercial advertisement is posted according to the Poisson process. However, it is rather artificial and we should consider the case in which each television station decides the location of the commercials using the adaptive manner. To treat such case mathematically, we simply set $K = 2$ and $M = 1$, namely, only two stations cast the same commercial of a single sponsor. Thus, we should notice that one can define $l_k(t) = 0$ (on air) or $l_k(t) = 1$ (CM) for $k = 0, 1$.

Then, we assume that each television station decides the label $l_k(t)$ according to the following successive update rule of the CM location probability:

$$P\big(l_k(t)\big) = \frac{1}{4}\left\{1 + \big(2l_k(t) - 1\big)\tanh \Omega\left(L_c - \sum_{\rho=t-1}^{t-\mathcal{L}} l_k(\rho)\right)\right\} \times \left\{1 + \big(2l_k(t) - 1\big)\tanh \Omega\left(\frac{v_0(t-\tau) - v_0(t-1)}{\tau}\right)\right\} \quad (1.21)$$

for $k = 0, 1$, which means that if the cumulative commercial time by the duration $\mathcal{L}$, that is, $\sum_{\rho=t-1}^{t-\mathcal{L}} l_k(\rho)$ is lower than L_c, or if the slope of the program rating point v_k during the interval τ is negative, the station k is more likely to submit the commercial at time t. It should be noted that in the limit of $\Omega \to \infty$, the above probabilistic location becomes the following deterministic location model (see also Fig. 1.11)

$$l_k(t) = \Theta\left(L_c - \sum_{\rho=t-1}^{t-\mathcal{L}} l_k(\rho)\right)\Theta\left(\frac{v_k(t-\tau) - v_k(t-1)}{\tau}\right) \quad (1.22)$$

for $k = 0, 1$. From the nature of two television stations, $v_0(t-1) + v_1(t-1) = 1$ and $v_0(t-\tau) + v_1(t-\tau) = 1$ should be satisfied. Thus, the possible combinations of $(l_0(t), l_1(t))$ are now restricted to $(l_0(t), l_1(t)) = (0, 0), (0, 1), (1, 0)$, and $(l_0(t), l_1(t)) = (1, 1)$ is not allowed to be realized. In other words, for the deterministic location model described by (1.22), there is no chance for the stations $k = 0, 1$ to cast the same CM advertisement at the same time.

On the other hand, the viewer also might select the station according to the length of the commercial times in the past history. Taking into account the assumption, we define the station $\tilde{k}$ by

$$\tilde{k} = \frac{1}{2}\big\{1 - \mathrm{sgn}\big(\langle l_1(t)\rangle - \langle l_0(t)\rangle\big)\big\} \quad (1.23)$$

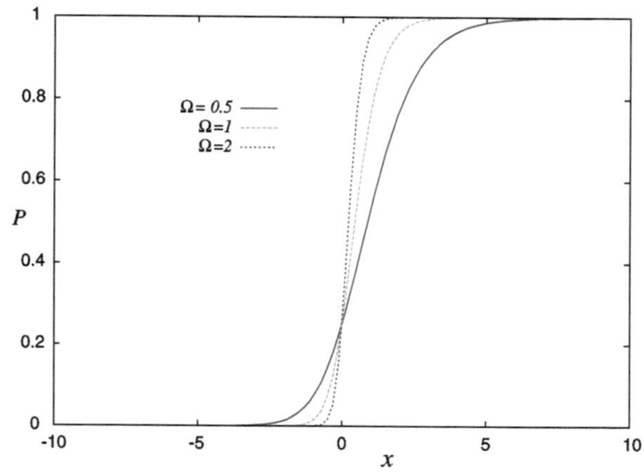

Fig. 1.11 The behavior of $P(x) = (1+\tanh(\Omega x)) \times (1+\tanh(\Omega x))/4$

with

$$\langle l_k(t)\rangle \equiv (1/\mathcal{L}) \sum_{\rho=t-1}^{t-\mathcal{L}} l_k(t), \quad k=0,1. \tag{1.24}$$

Then, the $\tilde{k}$ denotes the station which casted shorter commercial times during the past time steps $\mathcal{L}$ than the other. Hence, we rewrite the energy function for the two stations model in terms of the $\tilde{k}$ as follows.

$$E_k(l) \equiv -\zeta \prod_{\xi=s}^{e} \delta_{t_\xi,t}\delta_{l,\overline{k}} - \beta\delta_{l,\hat{k}} - \xi\delta_{l,\tilde{k}} \tag{1.25}$$

where we omitted the term $\gamma(\mathbf{Z}_k - \mathbf{Z}_l)^2$ due to the symmetry $(\mathbf{Z}_0 - \mathbf{Z}_1)^2 = (\mathbf{Z}_1 - \mathbf{Z}_0)^2$. As the result, the transition probability is rewritten as follows.

$$P(l|k) = \frac{\exp[\zeta \prod_{\xi=s}^{e} \delta_{t_\xi,t}\delta_{l,\overline{k}} + \beta\delta_{l,\hat{k}} + \xi\delta_{l,\tilde{k}}]}{\sum_{l\neq k} \exp[\zeta \prod_{\xi=s}^{e} \delta_{t_\xi,t}\delta_{l,\overline{k}} + \beta\delta_{l,\hat{k}} + \xi\delta_{l,\tilde{k}}]}. \tag{1.26}$$

The case without the exogenous information, that is, $\zeta = 0$, we have the following simple transition probability for two stations.

$$P(1|0) = \frac{\exp(\beta\delta_{1,\hat{k}} + \xi\delta_{1,\tilde{k}})}{\exp(\beta\delta_{0,\hat{k}} + \xi\delta_{0,\tilde{k}}) + \exp(\beta\delta_{1,\hat{k}} + \xi\delta_{1,\tilde{k}})}, \tag{1.27}$$

$$P(0|1) = \frac{\exp(\beta\delta_{0,\hat{k}} + \xi\delta_{0,\tilde{k}})}{\exp(\beta\delta_{0,\hat{k}} + \xi\delta_{0,\tilde{k}}) + \exp(\beta\delta_{1,\hat{k}} + \xi\delta_{1,\tilde{k}})} \tag{1.28}$$

with $P(0|0) = 1 - P(1|0)$ and $P(1|1) = 1 - P(0|1)$.

We simulate the CM advertisement market described by (1.27), (1.28) and (1.22) and show the limited result in Fig. 1.12. From this figure, we find that for the case of

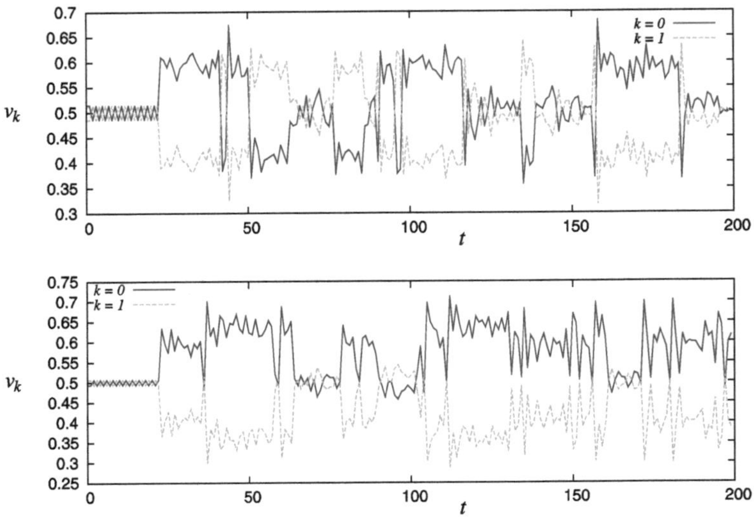

Fig. 1.12 Typical behavior of instant program rating points $v_k(t)$ for the two stations $k = 0, 1$ which are generated by the adaptive CM locations (1.22) and zapping probabilities (1.28) and (1.27). We chose $\beta = 0$ (the *upper panel*) and $\beta = 1$ (the *lower panel*). Here we set $M = 1, N = 600$ and $T = 200$. The parameters which specify the energy function and CM locations are chosen as $\zeta = 1, L_c = 2, L = \mathcal{L} = \tau = 20(= T/10)$. The other parameters appearing in the system are chosen as the same values as in Fig. 1.4

$\beta = 0$, the superiority of two stations changes frequently, however, the superiority is almost 'frozen', namely, the superiority does not change in time when we add the endogenous information to the system by setting $\beta = 1$. For both cases ($\beta = 0, 1$), the behavior of the instant program rating as a 'macroscopic quantity' seems to be 'chaotic'. The detail analysis of this issue should be addressed as one of our future studies.

1.7.1 Frequent CM Locations at the Climax of Program

The results given in the previous sections partially have been reported by the present authors in the reference [10]. Here we consider a slightly different aspect of the television commercial markets.

Recently in Japan, we sometimes have encountered the situation in which a television station broadcasts their CMs frequently at the climax of the program. Especially, in a quiz program, a question master speaks with an air of importance to open the answer and the successive CMs start before the answer comes out. Even after the program restarts, the master puts on airs and he never gives the answer and the program is again interrupted by the CMs. This kind of CMs is now refereed to as *Yamaba CM* ('Yamaba' has a meaning of 'climax' in Japanese). To investigate the psychological effects on viewers' mind, Sakaki [11] carried out a questionnaire

Table 1.1 A questionnaire survey for viewers' impression on the so-called *Yamaba CM* [11]

Question	Yes	I do not know	No
Is *Yamaba CM* unpleasant?	86 %	7 %	7 %
Is *Yamaba CM* not favorable?	84 %	14 %	2 %
Do you purchase the product advertised by *Yamaba CM*?	66 %	37 %	97 %

survey and the result is given in Table 1.1. From this table, we find that more than eighty percent of viewers might feel that the *Yamaba CM* is unpleasant and not favorable. With this empirical fact in mind, in following, we shall carry out computer simulations in which the CMs broadcasted by a specific television station are located intensively at the climax of the program.

1.7.1.1 Effects on the Program Rating Points

We first consider the effects of the *Yamaba CMs* on the program rating points. The results are shown in Fig. 1.13 as a typical behavior of the program rating points $v_k(t), k = 1, \dots, 9$. In this simulation, we fix the total length of CMs in a program so as to be less than eighteen percent of the total broadcasting time T of the whole program including CMs. In the upper panel, we distribute the CMs of all television stations randomly, whereas in the lower panel, the CMs of a specific station (the line in the panel is distinguished from the other eight stations by a purple thick line) are located intensively at the climax (the end of the program) and for the other eight stations, the CMs are located randomly. The other conditions in the simulations are selected as the same as in Fig. 1.4. From this figure, we find that the program rating point for the station $k = 1$ which broadcasts *Yamaba CMs* intensively at the climax apparently decreases at the climax in comparison with the other stations.

To check the effect of the relaxation time L_{cm} on the results, we carry out the simulation by changing the value as $L_{\mathrm{cm}} = 1, 2, 3$ and 5. The results are shown in Fig. 1.14. From this figure, we clearly find that the program rating point for the station $k = 1$ which broadcasts *Yamaba CMs* intensively at the climax apparently decreases around $t = t_{\mathrm{c}} = 400$ and the t_{c} is independent of the length of L_{cm}.

1.7.1.2 Effects on the Advertisement Measurements

We next evaluate of the effects of the so-called *Yamaba CMs* on the advertisement measurements such as the GRP or average contact time of the CMs by viewers. To quantify the effects, we consider the GRP_k-$\overline{v}_k$ diagram for $K = 9$ stations, where $GRP_k \equiv GRP_k^{(1)}$ in the definition of (1.12) because now we consider the case of $M = 1$ for simplicity. We plot the result in Fig. 1.15 (left). In this panel, there is no station broadcasting the *Yamaba CMs*. When we define the advertisement efficiency η for the sponsor by the slope of these points, the efficiency for this unbiased case is $\eta \simeq$

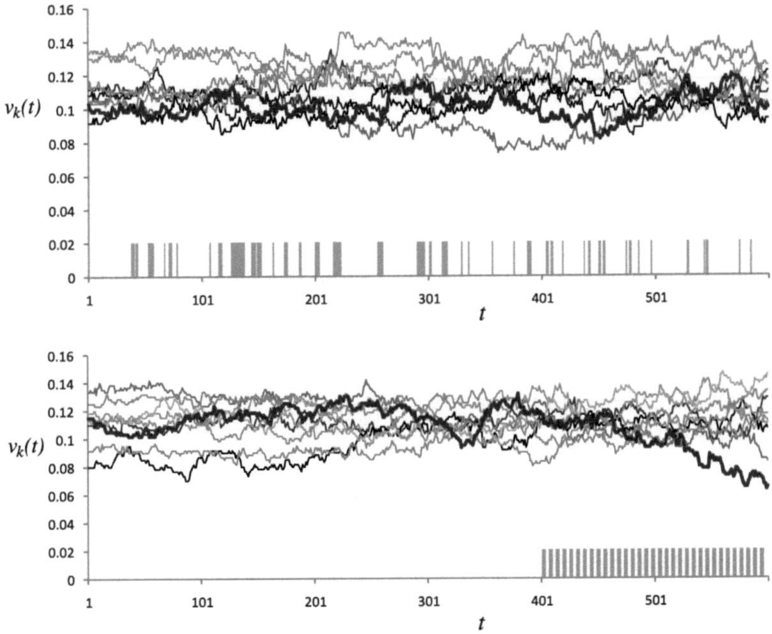

Fig. 1.13 The resulting program rating points for random locations of CMs (the *upper panel*) and for biased locations of CMs for a specific television station (say, $k = 1$, the *line* in the panel is distinguished from the other eight stations by a *purple thick line*). The *perpendicular purple lines* stand for the period of CMs for the specific station $k = 1$

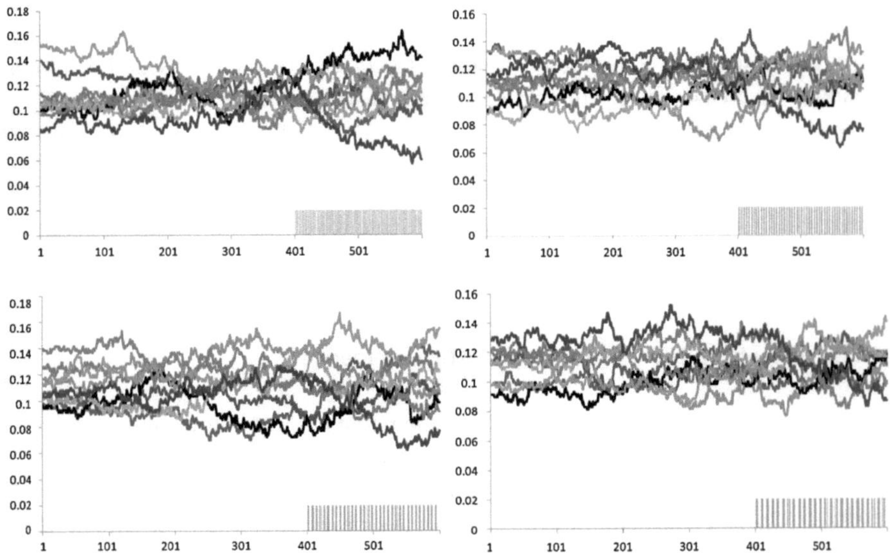

Fig. 1.14 The L_{cm}-dependence of the program rating points $v_k(t), k = 1, \ldots, 9$ shown in Fig. 1.13. From the upper left to the lower right, $L_{\mathrm{cm}} = 1, 2, 3$ and $L_{\mathrm{cm}} = 5$

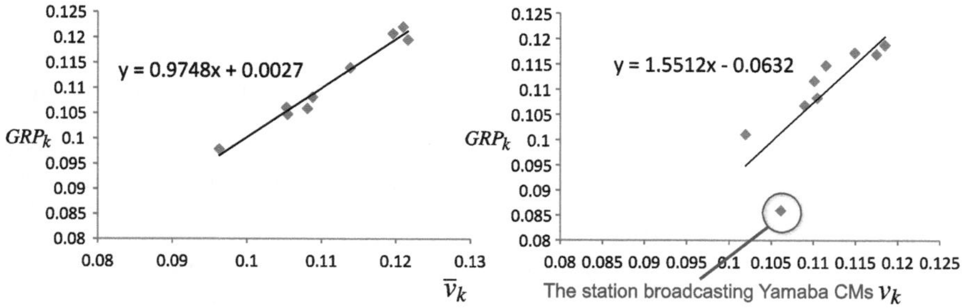

Fig. 1.15 The GRP_k-$\overline{v}_k$ diagram for $K = 9$ stations. Each point corresponds to each station. In the *left panel*, there is no station broadcasting the so-called *Yamaba CMs*, whereas only a specific station, say $k = 1$ is broadcasting the *Yamaba CMs* in the *right panel*. We set $M = 1$, $N = T = 600$

0.9748. We next consider the case in which a specific station, say, $k = 1$ broadcasts the *Yamaba CMs*. The results are shown in the right panel of Fig. 1.15. From this panel, we are confirmed that the both $GRP_{k=1}$ and $\overline{v}_{k=1}$ apparently decrease in comparison with the other eight stations. Hence, the slope η calculated by the eight stations (except for $k = 1$) increases up to $\eta \simeq 1.5512$ because viewers who was watching the program of the station $k = 1$ moved (changed the channel) to the other eight stations and it might increase the $GRP_{k \neq 1}$ and $\overline{v}_{k \neq 1}$ extensively.

From the results given in this section, we might conclude that the *Yamaba CMs* (biased CM locations) are not effective from the view points of viewers, sponsors and television stations although our simulations were carried out for limited artificial situations.

1.8 Concluding Remarks

We proposed a 'theoretical platform' to investigate the human collective behavior in the macroscopic scale through viewers' zapping actions at the microscopic level. We just showed a very preliminary result without any comparison with empirical data. However, several issues, in particular, much more mathematically rigorous argument based on the queueing theory [12, 13], data visualization via the MDS [14], portfolio optimization [15] and a mathematical relationship between our system and the so-called *regime-switching processes* [16] should be addressed as our future studies.

Acknowledgements This work was financially supported by Grant-in-Aid for Scientific Research (C) of Japan Society for the Promotion of Science, No. 22500195. One of the present authors (JI) thanks Nivedita Deo, Sanjay Jain, Sudhir Shah for fruitful discussion at the workshop: *Exploring an Interface Between Economics & Physics* at University of Delhi. He also thanks George Yin at Department of Mathematics, Wayne State University, USA, for drawing our attention to the reference [16]. We thank organizers of *Econophysics-Kolkata VII*, in particular, Frederic Abergel, Anirban Chakraborti, Asim K. Ghosh, Bikas K. Chakrabarti and Hideaki Aoyama.

References

1. Reynolds CW (1987) Flocks, herds, and schools: a distributed behavioral model. Comput Graph 21:25
2. Makiguchi M, Inoue J (2010) Numerical study on the emergence of anisotropy in artificial flocks: a BOIDS modelling and simulations of empirical findings. In: Proceedings of the operational research society simulation workshop 2010 (SW10). pp 96–102, CD-ROM. The preprint version: arxiv:1004.3837
3. Makiguchi M, Inoue J (2010) Emergence of anisotropy in flock simulations and its computational analysis. Trans Soc Instrum Control Eng 46(11):666–675 (in Japanese)
4. Ibuki T, Inoue J (2011) Response of double-auction markets to instantaneous selling-buying signals with stochastic bid-ask spread. J Econ Interact Coord 6(2):93–120
5. Siddarth S, Chattopadhyay A (1997) To zap or not to zap: a mixture model of channel switching during commercials. Working paper, series no MKTG97.091
6. Ohnishi H, Ishida K, Aoyama H, Saruwatari Y, Ikai M (2005) J Oper Res Soc Jpn 50(3):151–158 (in Japanese)
7. Chen H, Inoue J (2012) Dynamics of probabilistic labor markets: statistical physics perspective. In: Managing market complexity. Lecture notes in economics and mathematical systems, vol 662. Springer, Berlin, pp 53–64
8. Chen H, Inoue J (2013) Statistical mechanics of labor markets, econophysics of systemic risk and network dynamics. In: New economic windows. Springer, Milan, pp 157–171
9. Ibuki T, Higano S, Suzuki S, Inoue J (2012) Hierarchical information cascade: visualization and prediction of human collective behaviour at financial crisis by using stock-correlation. ASE Hum J 1(2):74–87
10. Kyan H, Inoue J (2012) Modeling television commercial advertisement markets: to zap or not to zap, that is the question for sponsors of TV programs. In: Proceedings of IEEE/SICE symposium SII2012. pp 816–823, CD-ROM
11. Sakaki H (2011) Annu Rep Nikkei Advert Res Inst 255:19–26
12. Sazuka N, Inoue J, Scalas E (2009) The distribution of first-passage times and durations in FOREX and future markets. Physica A 388(14):2839–2853
13. Inoue J, Sazuka N (2010) Queueing theoretical analysis of foreign currency exchange rates. Quant Finance 10(10):121–130
14. Ibuki T, Suzuki S, Inoue J (2013) Cluster analysis and Gaussian mixture estimation of correlated time-series by means of multi-dimensional scaling, econophysics of systemic risk and network dynamics. In: New economic windows. Springer, Milan, pp 239–259
15. Livan G, Inoue J, Scalas E (2012) On the non-stationarity of financial time series: impact on optimal portfolio selection. J Stat Mech P07025
16. Nguyen SL, Yin G (2010) Weak convergence of Markov-modulated random sequences. Stoch Int J Probab Stoch Process 82(6):521–552

Chapter 2
Agent-Based Modeling of Housing Asset Bubble: A Simple Utility Function Based Investigation

Kausik Gangopadhyay and Kousik Guhathakurta

Abstract The housing asset bubble and mortgage crisis of 2007–2008 in the US market poses a challenge to understanding of market and hypotheses related to market efficiency. The contribution of our paper is bifold. First, we present a survey of the existing literature which explains the housing asset bubble. We have emphasized on agent based modeling approaches in this context. The second part of the paper frames an economic model to demonstrate the power of "irrational exuberance" hypothesis, a term coined by Robert J. Shiller. Using a felicity function based framework, this shows the power of irrational expectation in bringing about an artificial and unintended boost in demand for investment of housing asset.

2.1 Introduction

The world at large was at a loss to explain the magnitude as well as nature of calamity that hit the US market in 2007. Economic thoughts are being re-organised and re-structured even now in search of a definite analytical framework to explain the failure of what was thought to be a fail-proof wealth generating system. This unprecedented crisis in the financial market engendered theories on financial markets which eventually adds to refinement of economic thinking. The qualitative way of thinking can point out to factors relating to human behaviour and its departure from presumed economic rationality regarding decision-making. However, any qualitative story should be supplemented with sufficient quantitative illustration for general acceptance. A quantitative model of an underlying qualitative story provides us with the power of the story to explain this phenomenon.

K. Gangopadhyay (✉) · K. Guhathakurta
Indian Institute of Management Kozhikode, IIMK P.O., Kunnamangalam, Kerala 673570, India
e-mail: kausik@iimk.ac.in

K. Guhathakurta
e-mail: kousikg@iimk.ac.in

F. Abergel et al. (eds.), *Econophysics of Agent-Based models*,
New Economic Windows, DOI 10.1007/978-3-319-00023-7_2,

What is the root underlying cause of this financial crisis? Is it engendered by "bad" but otherwise improbable run of flawed decisions, or is there any fundamental flaw in the financial system? Hellwig [2] argues that the cause of this crisis was not embedded in some flawed decisions rather in financial system architecture. The International Monetary Fund (IMF) forecasted a total loss of 750 billion dollars [3] in US residential real-estate lending, by October 2008. This amount, though a large sum in the housing market, is relatively low if analysed in the context of the size of global financial system. Moreover, the decline in the value of financial securities is far more compared to the fall in housing prices. These indicate deeper flaw in the financial system than a mere coincidental fall in housing prices.

The first part of Hellwig's work [2] discusses the "financial system architecture". More specifically, the mechanisms of risk management are discussed. One example is the process of mortgage securitization. It acts based on the principle of diversification to mitigate risks from interest rate fluctuations. The problems in the financial system architecture was traced back to basic theories in economics of information—"moral hazard" and "adverse selection". The paper cites one specific example to illustrate that sellers used to display higher prices in the contract and the additional money used to be given back to buyers as advance payment. This system takes risk away from buyers, one of the stakeholders in a risky investment of purchasing a house, to create moral hazard problem. The other related failures from different stakeholders such as, rating agency, internal correction, market discipline, has also been discussed.

The contribution of incidence of systemic risk is analysed in the second part of Hellwig's work [2]. There could be various ways to understand the incidence of systematic risk and its augmentation. The perennial problem is transformation of long term investment to short term investment which is done through conduits and structured-investment vehicles (SIVs). There is some systemic risk involved in this transformation. At the onset of this financial crisis in August 2007, an excessive amount of assets are transformed in this manner compared to historical average. This excessive supply actually plummeted down security prices considerably. At the instance of public recognition of delinquencies and defaults, there is a host of factors which led to market breakdown. These factors are identified as lack of fair value accounting and the insufficiency of equity capital at financial institutions.

Many theoretical and empirical works have come out to explain the phenomenon particularly the housing asset bubble that started it all. We surveyed the present literature in sufficient detail to present a thorough understanding to reader. The present literature may not be conclusive enough. We draw key insights from this literature, and then present our model based on our insights. The organization of this paper follows here. Section 2.2 presents an elaborate survey of the present literature. Section 2.3 states our economic environment and its contribution to demonstrate the power of "irrational exuberance" in defining expectation. Section 2.4 discusses our contribution to this literature.

2.2 Existing Literature: Sub Prime Crisis and Agent Based Modelling

Wray [4] (also, [5]) use Hyman P. Minsky's approach to analyze the international financial crisis initiated by problems in the US real estate market. They examine the role played by each of the key players which includes brokers, appraisers, borrowers, securitizers, insurers, and regulators in creating the crisis. In a 1987 manuscript, Minsky had already recognized the importance of the trend toward securitization of home mortgages. This paper identifies the causes and consequences of the financial innovations that created the real estate boom and bust and proposes short-run solutions to the current crisis, as well as longer-run policy measures to prevent a debt deflation from happening again.

Goodman and Thibodeau [6], takes a call on the housing asset bubble in the US market from the perspective of economic fundamentals. They investigate how much of the price rise was caused by the demand supply dynamics of the housing economy. The exorbitant rise of housing price index in the early years of the first decade of the new millennium led many economists to postulate theories of speculative bubbles and herd behaviour. Such a conclusion can only be confirmed if the rise in price cannot be explained by the fundamentals of the governing economy. The authors worked on this premise and examined the demand supply dynamics of the housing market in the US. They examined what was the relative contribution of the fundamentals and the speculative phenomenon to the price rise. They have approached this problem in a two pronged manner. At first, they have used a simulation based model of the housing price behaviour in the long run. After that they have done empirical investigation of 133 metropolitan areas across the USA testing for the elasticity of supply in the housing market in these areas.

While working on the simulation, they [6] examine the 'shift in aggregate demand' necessary to result in a 10.3 % rise in home ownership. They also verify whether the price rise resulted from a shift in equilibrium. They have simulated the demand for various elasticities of supply against a constant elasticity of demand. Their findings indicate that the price rise is extremely sensitive to the elasticity of supply. For their empirical work, the authors work with a "long-run equilibrium housing market model" which is able to explain the price variation of housing assets across the areas. Their analysis has resulted in positive elasticity of supply in 84 Metropolitan Statistical Areas (MSA) during the no-bubble period of 1990–2000. Then, their analysis has been extended to the bubble period. They have accounted for the change in fundamentals and estimated the expected price rise from the model. Results show that the speculative increase in housing asset prices was an extremely localised phenomenon as opposed the general impression of a market-wide speculative bubble. As a benchmark, the authors have considered 30 % over the expected increase as a housing bubble threshold.

Coleman et al. [7] show that the subprime crisis was more of a systemic issue arising out of complex interaction of multiple agents of the financial system making it rather a "joint product" of the institutional, political and regulatory framework prevailing at the time of the crisis. This is an alternative perspective to the one that prescribes the speculative pricing of housing assets as the main cause of the crisis. Their study reveals that the crisis cannot be ascribed to the existence of sub prime market alone. They use a simple model which is based on the change in loan intensity being leading indicator of future home prices. They have considered quantity of housing demanded in period *t* as the dependent variable. They model the movement of this dependent variable using several independent variables such as, housing prices at time *t*, vector of loan type intensity lagged, vector of macroeconomic, demographic, and financial controls, cost of capital, quantity of housing supplied in period *t*, housing market supply regulation and cost to supply housing. Subsequently, the authors impose the demand supply equilibrium conditions allowing the market imperfections to get corrected over time. A pooled cross-sectional time series was constructed which included 20 metropolitan areas for 36 quarters during the period 1998–2006.

One of the most striking findings of their initial model is that there was almost no significance of the proportion of sub prime loans among all loan in explaining the future housing price. These results also show that the increase in proportion of sub prime loans on the other hand had a positive correlation with past returns on housing prices indicating a strong dependence. This was a reinforcement of their argument that the intensity of sub prime loans did not play a causal role in the run-up of housing prices. The main macroeconomic variables which were identified as the drivers of housing price movement are aggregate level of mortgage lending, population growth, and the unemployment rate. Using "supply constraint" index as a proxy for regulatory policy, they have found a positive relationship between this variable-in-question and the housing price. The supply price was found to be significant in explaining the price movement only in case of high price assets while its significance was not present in case of middle and low value assets. The authors [7] use a second model to investigate the effect of shift of the role of dominant players from the Government agencies like Freddie Mac and Fannie Mae to the private players on the dynamics of housing market. For the data set belonging to the period before the shift, the macroeconomic fundamentals could explain the price significantly, while their explanatory power was not significant in case of the data set belonging to post-shift period. Overall, it may be concluded that this study absolved the sub prime loan products of their role as the prime accused in the crisis. The whole affair was phenomenological in nature with the entire multi body system contributing to the problem in a complex manner.

Crouhyet al. [8], examine the various factors that have led to the subprime mortgage credit crisis. They identify the following factors as the main causes for the failure: yield enhancement, investment management, agency problems, lax underwriting standards, rating agency incentive problems, poor risk management by financial institutions, the lack of market transparency, the limitation of extant valuation models, the complexity of financial instruments, and the failure of regulators to understand the implications of the changing environment for the financial system. Looking at the chronological development of the problem the authors have examined the evolution of the crisis and analysed the factors that the crisis may be attributed to. The final picture that emerges in their analysis is a multi-agent complex phenomenon. The several causes that the authors point out are described here to justify the above statement. Firstly, the interest rate being low there was a need for "yield enhancement". Then there was an automatic demand for asset pooling to take benefit of financial engineering by dumping the high yield assets into the collateral pool. For this an automatic choice was sub prime loans along with auto loan and credit cards. Securitisation meant that mortgage originators were out of "default risk" and therefore they had no reason to perform with due diligence. This was coupled with relaxed regulatory activities and fraudulence. Banks also joined in the fray in search of reduction of capital requirements. The complex multi layered derivatives ensured that the same risky assets may be a part of myriad instruments and structures leading to systemic risk and cascading effect. The rating agencies relying on past data did not use models that reflected the true risk of the underlying assets in terms of the probability of default, recovery rates and default dependence. They were lax in recognizing the ascending risk in the sub prime sector. The business of rating agencies depended on the volume of transactions generated by their client, who were the originators of the engineered products. The volume was positively related to upward ratings. This in itself constituted a classic problem of conflict of interests. Performance incentives in financial institutes were designed to promote short run profitability. Lack of transparency also added to the layer of problems already mounting. In sum, the crisis may be viewed as an end result of interaction of systemic, economic and regulatory issues.

Francis A. Longstaff [9], does an empirical analysis of the pricing of subprime asset-backed collateralized debt obligations (CDOs) and looks into the cascading effects of the same on the other sections of the financial market. The results indicate the sub prime market witnessed "significant price discovery" during the crisis. The crisis of 2007 which was initiated in a specific section of a specific asset (housing) market and later pervaded to entire financial market of the US and major parts of the globe, provides an ideal platform for studying "contagion effect". Using data from the 'ABX' index for the sub prime markets, the author examines whether the effect cascaded across the markets. One of the most commonly used definitions of contagion in the literature is "significant temporary increase in cross-market linkages after a major distress event". With this definition as the theoretical support, he uses a Vector Auto Regression (VAR) framework to distinguish the pre-crisis correlation between the ABX market and other markets from the corresponding figure during post-crisis phrase.

The results indicate that the cross-market linkages have increased significantly post crisis. Before the subprime crisis the information contained in ABX market did not have significant explanatory or predictive power in explaining returns from other markets. In the post-crisis period, however, the ABX indices showed significant predictive power for treasury bond yields, corporate yield spreads, stock market returns and changes in the VIX volatility index. As noted by the author, the ABX indices, generally speaking, show the ability to forecast treasury yields, corporate yield spreads, stock market returns and changes in the VIX up to three weeks ahead, with strikingly high R^2 values. Such results are strong evidence in favour of a "contagion effect" having spread across markets after the crisis.

One of the major contributions of this work is that with these results it is possible to distinguish between the various existing models in the literature on contagion.

One could clearly see from the length of the forecast horizon (which often ran to as long as three weeks), the view that contagion is spread via the correlated information channel, may not hold much water. The author argues that if that was the case then the price discovery would occur much more rapidly in liquid stock, bond and similar markets because of the faster spread of information. On the other hand, the fact that the results showed the ability of the ABX market to predict the trading pattern for both the liquid stock and bond markets and the market for engineered products, reinforces the model based on spread of contagion via a liquidity/financing channel.

Apart from the analytical and investigative works on the crisis, several researchers have tried to model the bubble and the market failure as well. Glaeser et al. [10] construct a simple model of housing bubbles that predicts that "places with more elastic housing supply have fewer and shorter bubbles, with smaller price increases". The objective of their work is to find out how much of the housing asset bubble could be predicted by a rational model and is related to the fundamentals, and if and when the irrational exuberance does play a role. For this purpose they construct a simple continuous time model with an underlying assumption that housing asset prices are formed by demand supply interaction, dynamically. They introduce scarcity of resources in production resulting in linear monotonic increase of prices with production. Using this model, they show that if housing supply is elastic, with a finite number of potential buyers, there is no equilibrium with the number of houses being offered exceeding the number of buyers available. Thus a rational bubble can exist only with inelastic supply.

In the next phase they turn their attention to the "irrational bubbles". They model the same as temporary spike in the buyers' expected price of the housing assets ascribing the same to "irrational exuberance" as described by Shiller [1]. This rise in expectation is a purely exogenous factor with a fixed life, the buyers having no knowledge of the influence of the same. The authors then propose that during bubble the exogenous factor has a multiplying effect in increasing the prices while after the factor the effect is inversed and results in multiplying the decline in prices. They also show that the interaction between the exogenous bubble factor and the supply inelasticity is similar to that of supply inelasticity and shifts in demand. The bubbles persist more in case of inelastic supply whereas in case of elastic supply they pop up much faster. The authors then proceed to analyse the data on housing prices, construction and supply elasticity during the periods of price boom and bust. Empirical analysis supports the propositions of their model.

One important aspect of any speculative bubble like the one that might lead to the housing asset crisis is the behavioural dynamics of the economic agents participating in the market. Earl et al. [11], inspect this aspect by bringing in the concept of decision cascade (as against the information cascade) and later on combine the same with Minsky's financial fragility analysis, and evolutionary economics to provide a theoretical platform for analysing the behavioural dynamics of such bubbles. The authors introduce the concept of decision cascades clearly distinguishing the same from information cascade as highlighted by Shiller [12]. The authors argue that the information cascade theory which hangs on the non-availability of information may not hold water in terms of explaining the stock and financial market phenomena as the efficient market hypothesis seems to be a reasonable approximation in light of today's media presence. The inefficiency may creep more out of the interpretation of the information than the availability of information, an issue that is not captured in the information cascade idea. This is exactly what brings in the idea of decision cascades. Decision cascade refer to the interaction effect of the decision rules as they change during the mutual transfer of information among the agents. The cascade essentially means the probability of domination of a set of decision rules over others because of social interaction norms. Any new information is thus processed according to the dominant rule set. In a speculative market such biases will ultimately lead to herding and synchronisation.

Next, the concept of the rule degradation is brought in. As decision rules cascade from agent to agent there is a possibility of "Chinese whisper" effect on the same. In effect, it means that the decision rules that ultimately prevail in sections of the market may actually be a high order derivative of the original set. Literature has shown that degradation of decision rules will happen from 'opportunism' and "tacit knowledge". Because of failure of strategies based on some decision rules, there will exist always a need for new rules in the market. People will be attracted towards newer set of strategies that are producing wealth and there will be herding towards such decision rules. Because of the way our societies have evolved, decision cascades have a great impact on our decisions, especially the speculative ones. There has indeed been numerous studies which establish the social decision making process as a cascading one.

The authors then suggest the use of agent based modelling in an evolutionary economic framework to simulate this decision cascade process. They arrive at the idea that "the degeneration of decision rules is easily modelled in a multi-agent replicator setting through imposing some kind of entropy condition on the adoption process and with replicator dynamic pay-offs of rules in relation to the population of others playing the same strategy." The evolutionary framework is an extension of Minsky [13]. The paper ultimately shows that successful modelling of decision cascade using an evolutionary framework for rule degradation can help us identify and analyse the process of speculative bubbles in a robust manner.

2.2.1 Agent Based Modelling in Financial Markets

We now shift our attention to existing literature on agent based modelling in general to understand the state of the art. We particularly focus on few works that concentrate on financial markets and speculative behaviour. Roszczynska et al. [14] has presented a technique based on agent based simulation. This technique is able to generate a robust measure of detachment of trading choices created by feedback, and forecast the timing of speculative bubbles in experiments with human subjects. Their work is a combination of laboratory experiments with human subjects and agent based simulations. Such a unique framework helps reveal the behavioural aspects which are crucial to formation of bubbles and whose identification ultimately may lead to preventive measures. For the experiment they used the Minority Game [15] to study the price discovery mechanism in financial markets. While already a body of literature exists on agent based simulations which model the Minority Game, the researchers shed new light on the same by incorporating experiments with human subjects. Repeating the experiment with various initial conditions, they studied the process of synchronisation into a bubble state and observed the trading strategies of the investors the interaction of which resulted in such synchronisation. To explain the process of synchronisation they take help of the "decoupling process" as explained in [16]. Decoupling essentially means trading strategies that are independent of trends. In other words the strategy is not coupled with the price movement at a particular moment of time. They used this knowledge to investigate whether decoupling plays a role in Minority game subjects and they found out that decoupling indeed played a role in synchronising the prices to reach a bubble state. They also used the additional test of false feedback post-bubble stage and found that the results corroborate. Using the subject behaviour as the cue, they perform redesigned simulations on the computer incorporating the decoupling moments in their design. The results are very encouraging with almost 87 % success rate in predicting the bubble states. The results are also invariant with size which indicates that the model is robust and is capable of capturing the complexities associated with higher size. This work also highlights the importance of factoring the decision biases in the agent based simulation model to be able to capture the market dynamics properly.

Rabertoa et al. [17] simulate an agent-based artificial financial market which they call Genoa market, where trading is done on one single asset by heterogeneous traders. The price discovery happens through a trading mechanism which mimics the real trading rules reasonably well. The objective of their work is to represent the trading complexities as much as possible while focusing on the finiteness of available resources. The programme allows for the pricing to be determined by demand supply interaction. Initial conditions allow the agents to start with a finite amount of cash and a portfolio of finite investment opportunities. The process does not allow creating money, there is a law of conservation of total cash in operation. In each subsequent epoch stochastic buy and sell orders of the agents are simulated. The decisions are bounded by the resource constraints and are dependent on prior period volatility also generating clustering effect. The uniqueness of the model lies in its ability to reproduce the fat tails in the probability distribution of the log returns of the assets as well as the phenomenon of volatility clustering.

Using the state of the art programming and object oriented technology, the Genoa market may be used as a platform to perform various degrees of experiment to address both research problems as well as practical issues. The authors do agree, however, that the model suffers from some lacunae which need to be addressed to increase its usability. To start with, the volatility clustering is sensitive to the size of the market. As the number of traders increases, the clustering gives way to pure stochastic volatility. Next, the model fails to capture all the stylized facts that have been empirically established about financial market behaviour. It is a well established fact from several empirical works that volatility exhibits power law decay in financial time series. The model, however, results in exponential decay. It may be added, though, that the power law decay has not been perfectly modelled as yet. Even with GARCH and ARCH models exponential decay of volatility is exhibited with correlated time steps.

The sub prime crisis was studied using a Systems Dynamics model by An et al. [18], looking at the whole problem as a multi agent complex system with the crisis evolving from the interactions between the agents. Taking cue from various literature on the recent crisis they build their system as an interaction of three subsystems namely, an aggregate banking system, an aggregate housing market and an economic environment. They followed the principle underlying the system dynamics model of Jay Forrester [19] which represents the system as a stock flow diagram. The asset-liability management decisions of the banking system are affected by the housing market as well as the economic system determining the price characteristics of the assets involved. Five classes of assets are considered: cash, short-term securities, the mortgage-backed securities, the bank-owned houses and the non-mortgage backed securities. The assets flows are modelled as linear equations taking into consideration the interest rates and asset returns associated with each class. The purchasing power of investors/homeowners which is the prime mover of the housing asset market is in turn governed by the lending policy and capacity of the banking system. This determines the stock flow of the housing market. Three kinds of housing assets are considered: houses that are currently occupied, houses owned by banks and houses not owned by banks but available for purchase. The flow model is built consistent with the banking system model taking into account the unemployment rate, mortgage interest rate, subprime loan availability and average family lifespan. The housing price is a function of the ratio of supply over demand. The economic subsystem is a dynamics between business credit and banking system liquidity. The model parameters include gross production output per unit time-period, household income per unit time-period and unemployment per unit time-period. The aggregate demand per period is a function of consumption and business environment.

The model is then simulated with given initial conditions followed by shocks thereafter. The system is perturbed by increasing the availability of subprime loan. The system observations reveal that as a result of the shock the building rates increase leading to higher expected mortgage payment per period, ultimately reaching a stage when overall mortgage payments due overshoot the affordability causing defaults to start. The model is then subject to economic stimulus in the form of government aids or stimulus money that adds to the aggregate demand. The mortgage payments return to stability provided that the stimulus is above a threshold. This in turn means that the government deficit shoots up. Overall, this model is a simple interactive system which can be used to investigate some key aspects of the crisis.

The utility of agent based models in analysing a systemic failure is brought out very clearly by Thurner [20]. Economic crisis, the author argues, is a systemic phenomenon, involving complex interaction of institutions, markets, businesses, individuals and the state. The global economic structure of today lend further complexity to the system by bringing in interaction of multiple sovereign states. The complexities associated with such interactions are not possible to capture using the standard steady state economic models which thrive on general equilibrium. Hence the need for a model, which is adaptable, allows for correlation of parameters and shocks of states and does not need general equilibrium conditions. Indeed, agent based models fulfill all these criteria. The author describes in detail the process of building a model for the financial system. One main advantage is the use of 'non-representative' agents which allow the model to have varied levels of tolerance with respect to the decision making parameters. The differences between the agents can lead to results that have "macro effects". In the model there are three classes of investors: investors who allocate funds based on analysis of given information, investors who are not informed but intuitively place orders randomly and investors who place their funds in the custody of financial institutions. All these different types of investors place buy and sell orders in a market dealing with a single non-dividend paying asset. The other agents involved are the banks and regulators which govern the liquidity and leverage available in the market. The interactions between all these agents are observed as the initial conditions vary and the parameter values are allowed to evolve dynamically.

The simulation reveals that there are two scenarios which may lead to crash under leverage pressure. Firstly, the random shocks in demand caused by the uninformed investor can pull down the price of the asset much below its intrinsic value. Secondly, investment funds can take excessively large positions in the market causing concentration of risk. Both these factors may combine to form a major crisis. One of the major findings of the simulations was that in absence of regulation, both banks and investors find it attractive to increase leverage. Regulations have strange effect of producing separating and pooling equilibria under different leverage conditions. During moderate leverage they seem to work well while they seem to amplify the synchronisation during high leverage. There is an agent interaction effect which disturbs the price signal. This error emanating from interaction of agents (mutual influence) is not captured in standard models. It was also observed that during times of high leverage, even value adding strategies may act counter-intuitively adding to the crisis. One more observation about leverage was that it has a stabilizing effect when the levels are moderate. Finally, repeated simulations helped to identify the onset of crashes.

Katalin Boer-Sorbán [21] takes a detailed look at how different models can be adopted in agent based simulation framework to capture the financial market dynamics by developing an agent based simulation that would capture the behavioural characteristics of markets and investors. To develop such a system Boer-Sorbán has conducted studies on the systemic issues and general behaviour of the financial markets that are in place and pinpointed the relevant aspects of the stock market which are to be considered for modelling. They fall under two major categories: organisational and behavioural. There are delineation of six different organisational attributes that may be mapped—traded instruments, orders and quotes, market participants, trading sessions, execution systems and market rules. Apart from these well-defined observable 'hard' variables, identification of 'soft' behavioural aspects of the agents (investors, brokers and market makers) are done. The key variable for the investors is identified as "the order generation mechanism". For the brokers four distinctive variables were mapped: "order selection mechanisms, order execution mechanisms, negotiation strategies, strategies to determine transaction prices". The market makers are distinguished by "order execution mechanisms, determination and timing of quotes and handling the limit order book". With these variables Boer-Sorbán builds the framework describing the real market.

With the key variables for understanding the real market established, Boer-Sorbán studies the existing literature on artificial markets. The purpose of the study was to throw light on the relative success/ failure of various available agent based simulation models to capture the various key aspects of the real market. In the process, Boer-Sorbán has generated a conceptual framework for a taxonomy of agent based simulation models that results in extension of the conceptual framework for description of stock markets with design and implementation aspects. The analysis shows that continuous trading sessions need to be studied and incorporated in a model. The survey also reveals that most models focus mainly on the investors while ignoring the brokers and often markets as well. This has led Boer-Sorbán to propose the ABSTRACTE model of trading which takes care of the real market aspects and the best of existing ABS models.

2.3 A Basic Economic Model

We have built up a basic economic model to quantitatively demonstrate the role of irrational perception in investment. Rationality has its appeal in economics partly because of the fact that it is rather possible to model a rational agent. On the other hand, image of a completely directionless irrational agent is blurred in our mind. To resolve this dilemma, a thumb rule is to model a rational agent and use some departures in his behaviour from the rationality. This is called bounded rational agent which is often used when we talk about limits in rationality. This is how we can quantify departure from rationality in an agent and measure the impact of his behaviour as a function of departure from rationality. Since this crisis is about investment in housing market, we require a dynamic model to address issues associated with this crisis. In our model, agents have a simple utility function given below.

$$U(\{c_t\}) = \sum_{t=0}^{\infty} \beta^t u(c_t). \tag{2.1}$$

The agents earn an income of y_t at time period t. The agents can either consume the numeraire good or can invest in an asset. The total amount of spending is equal to total income. Therefore,

$$c_t + s_{t+1} = y_t + R_t \cdot s_t \tag{2.2}$$

where s_t is the amount of savings which is invested into an asset of housing. R_t indicates the return on an asset at time period t which is a random variable following a Gaussian process with mean μ and standard deviation σ. The shocks are autocorrelated with correlation coefficient ρ. The agent maximizes utility function (Eq. (2.1)) subject to the budget constraint (2.2). We also impose additional no debt constraint that $s_t \geq 0$ for all t.

This utility function is quite common in the economics literature, so we use it to illustrate a baseline scenario. We add one element of bounded rationality in the model. Bounded rational agents compute the expectation and the standard deviation of the interest rate fluctuation based on some few periods. The supply of housing is fixed. Therefore, price of housing is proportional to demand for investment.

What is the condition for utility maximization for an agent? The first order condition or the Euler equation, as it is popularly known, is:

$$u'(c_t) = E\big[R_t \cdot u'(c_{t+1})\big]. \tag{2.3}$$

By solving this equation, one can derive the decision of an agent, $\{c_t, s_{t+1}\}_{t=0}^{\infty}$, based on levels of his savings and the state of the economy, demonstrated through the shock in the interest rate.

$u'(\cdot)$ is chosen from the class of constant elasticity of substitution functions,

$$u(c) = \begin{cases} \frac{c^{1-\frac{1}{\eta}}-1}{1-\frac{1}{\eta}} & \text{for } \eta \neq 1, \\ \log(c) & \text{for } \eta = 1 \end{cases} \tag{2.4}$$

where η is the elasticity of substitution. Typically, η is great than or equal to one.

Our framework is not novel but this is used first by Aiyagari [22] to analyse savings in a heterogenous agents model. The shocks to agents are idiosyncratic in nature. The reason for this assumption is quite straightforward. The emphasis of our modelling is not for the purpose of analysing macroeconomic fluctuations but systemic fluctuations and agents expectations. Therefore, our exclusion of economy-wide shocks is rather justified.

2.3.1 Algorithm for Computation

We may not be able to have any closed form solution for our optimization problem. We have an alternative of computing the numerical solution. We use the following algorithm for computing our equilibrium.

1. Discretization: We discretize the shocks using the method described by Tauchen [23]. In doing so, this methods allows us to construct a transition matrix between various shocks. We also construct a discrete grid of savings around the income of agents.
2. Initial Guess: We construct an initial guess for savings choice of agents. The consumption of an agent is the minimum of value of income and savings.
3. We compute right hand side of Eq. (2.3) from the prior estimate of consumption in different states of nature. We use linear interpolation to estimate this value as and when required.

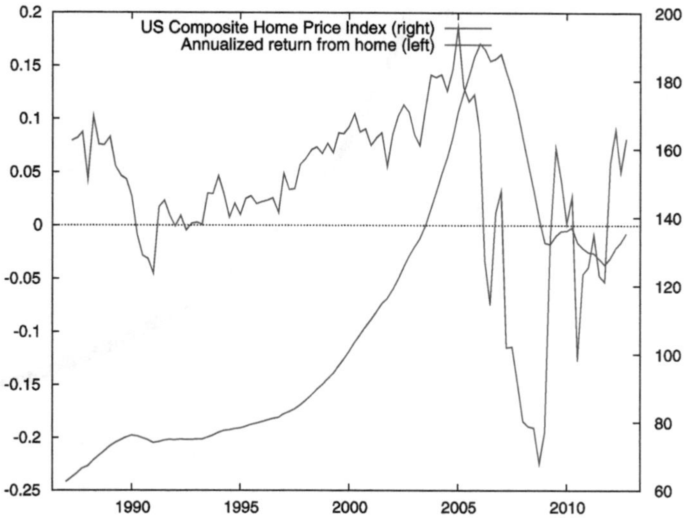

Fig. 2.1 National composite home price index and annualized returns constructed (quarterly data, US) (Color figure online)

4. For various values of contemporary consumption, the left hand side of Eq. (2.3) assumes different values. We solve for (2.3) using Newton-Raphson method.
5. We update the values for the consumption grid.
6. If the difference between previous values and updated values is sufficiently small (using L^∞ norm), we stop. Otherwise, we go back to step 3.

2.3.2 Calibration for Demonstrating Savings Mechanism

We experiment with some numerical values to illustrate the mechanism of our model to the readers. Risk-averseness of an agent is a monotone function of the elasticity of substitution parameter, η. We choose a conservative value of η at unity. As far as, mean, variance and autocorrelation coefficient of shocks are concerned, we directly estimate it from the United States data. Since our focus is housing price, we look into the returns in the housing sector. National Composite Home Price Index for the United States is a series maintained by Standard & Poor. The quarterly data runs between 1987:Q1 to 2012:Q4. The mean annualized return is 3 % and standard deviation is 0.08 % (see Fig. 2.1). The autocorrelation coefficient is computed as 0.67. We take discount factor as reciprocal of the mean returns.

We have discretized the shock in the return into seven discrete states between $\mu - 3\sigma$ and $\mu + 3\sigma$. As autocorrelation is positive, it implies that when the return is comparatively lower, the expectation of future return is also lower. The lower expectation of future return dictates the contemporary consumption to be comparatively higher, absolute value wise. On the other hand, an agent will have less re-

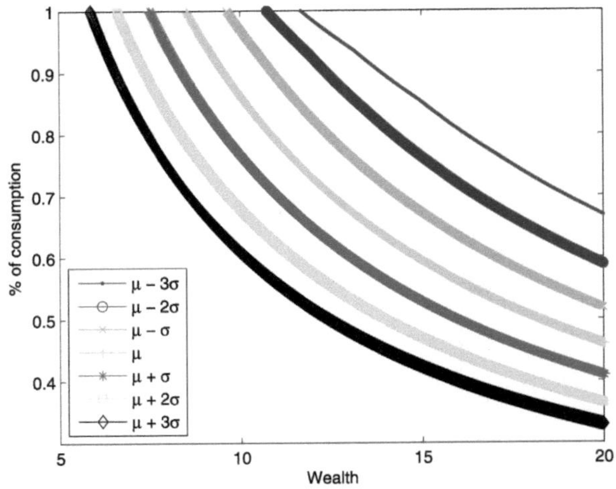

Fig. 2.2 Consumption as a percentage of wealth for various shocks in the return. Periodic income, without any loss of generality, is fixed at an arbitrary value of 10. The other parameters are calibrated (Color figure online)

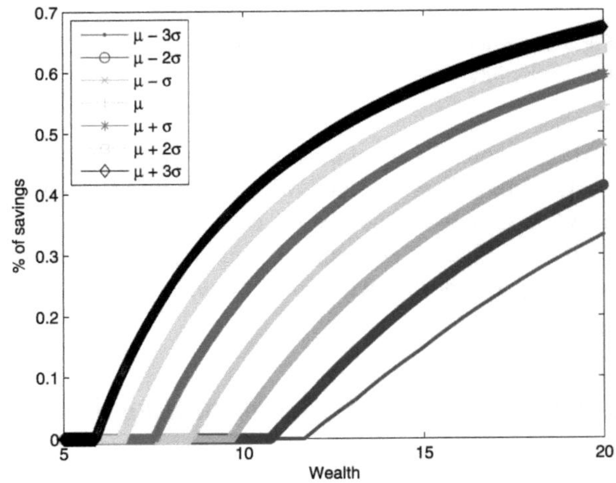

Fig. 2.3 Savings as a percentage of wealth for various shocks in the return. Periodic income, without any loss of generality, is fixed at an arbitrary value of 10. The other parameters are calibrated (Color figure online)

sources available when returns are low and this motivates him to spend comparatively lower fraction of income. Overall, the contemporary consumption is dominated by the former factor compared to the latter one. Figure 2.2 illustrates proportion of consumption against wealth for different values of returns shock whereas Fig. 2.3 portrays proportion of consumption against wealth for different values of returns shock.

2.3.3 Wealth Distribution and Numerical Experiments

The distribution of wealth in our model is matched with the corresponding figure of the united states. Essentially we follow an empirical approach here. Prior

Table 2.1 Investment in various scenarios: results from a numerical experiment conducted. Percentage increase in investment from the baseline case for one percent change in the parameter value is reported in the parenthesis

	Mean	Median
Baseline case	0.8010	0.0070
Increase in average return by 10 %	1.6467	0.9998
	(10.6 %)	(1418.3 %)
Increase in standard deviation in return by 10 %	0.8073	0.0094
	(0.08 %)	(3.4 %)
Increase in autocorrelation coefficient by 10 %	0.8063	0.0105
	(0.07 %)	(5.0 %)

research [24] suggests that the wealth distribution fits the log-normal distribution at the lower tail and Pareto distribution at the upper end. The Pareto tail is restricted [25] to 10 % of the population, at the most. We require to calculate mean and standard deviation of these distributions. For that purpose, we use the data [26] published by the US Census Bureau. We find that the median income of a family was approximately 60,000 USD in 2008 whereas the median family wealth was around 120,000 USD. Therefore, the median of the log-normal distribution is double the periodic income. Table 717 in [26] elaborates that approximately 10 % of the population hold 1.5 million USD, which is 25 times the periodic income, in wealth. This may be one indication of the extent of power law in wealth distribution. The mean wealth is approximately 556,000 USD, which is more than 9 times the periodic income.

In our analysis, we set the periodic income to unity, without any loss of generality. The median of the log-normal distribution is set at 2 accordingly. The mean of the log-normal distribution is 9. We know that if a random variable $X \sim N(\mu, \sigma^2)$ then $\exp(X)$ follows a log-normal distribution with median $\exp(\mu)$ and mean $\exp(\mu + 0.5 \cdot \sigma^2)$. Since, we know the mean and median of the log-normal distribution, we can calculate the mean and standard deviation of the underlying normal distribution. Thereby, we derive μ as $\log(2)$ and σ as $\sqrt{2 \cdot (\log 9 - \log 2)}$. We simulate wealth of agents from this distribution ignoring the power tail for the sake of convenience.

We have enforced the idea of "irrational exuberance" through three channels: (a) increase in perceived mean, μ, (b) increase in perceived standard deviation, σ, and (c) increase in perceived autocorrelation coefficient, ρ. In the baseline case, we note the average investment of all agents when they perceive the parameters correctly. In the three numerical experiments, we implement agents' perception and note the increase in investment in each case. We increase the parameters values by 10 % and note the change in investment per one percentage point change in parameters in Table 2.1.

2.4 Discussion

The results indicate that a perception of shift in mean has an explosive effect in drawing investment. A mere one percent increase in the mean returns boosts average investment tremendously by more than ten percent points. The effect of increase in mean return on agents who are at the lower range of wealth, is even more gigantic. One can safely say that a perception of change in mean will boost investment tremendously. This additional investment will definitely augment the price of house to a considerable extent. This is even truer when we consider that supply of housing is rather inelastic in the short run. This artificial boost in price will plummet down tremendously, once agents' perception falls back to the exact level causing a sharp decline in the housing prices. This is Shiller's idea of irrational exuberance. Shiller [1], for example ascribe it to "irrational exuberance" that drove the stock market bubble in the 1990s and the housing market bubble between 2000 and 2007. The speculative bubbles may be caused by "information cascades" or "decision cascades" which means that individuals in a group disregard their individually collected information because they feel that everyone else can not be wrong. He also shows how bubbles led to dangerous overextension of credit and finally to the global credit crunch. We see the same story being repeated, albeit less dramatically, when agents' perception of standard deviation in return or autocorrelation coefficient for returns changes to goad them to invest more.

We have not discussed any systemic failure in our story but provided a simple narrative when a collective perception which is different from reality may cause irrational exuberance for agents. This simple narrative may not stand the test of time if systems of financial engineering are placed to prevent and thwart any mishaps in proper time. This is where the extent of systemic flaws becomes important. A small loophole may not exacerbate a problem to a great extent but will be rectified at a higher level before becoming endemic, whereas a flawed architecture will encourage and snowball even a small problem to the level of catastrophe.

One way of looking into the dynamics of the system to analyse the nature and causes of failure is to view the entire economic process associated with the housing asset bubble as a multi-agent interaction process. In that case, the extent of failure of various parts could be measured using numerical experiment. As Farmer and Foley [27] points out that both the "econometric" as well as "dynamic stochastic models" are inadequate to map the dynamics of the crisis of this magnitude. A better alternative is the use of agent-based models. This is a computerised simulation of the decision making process of a large number of entities (agents) which may be individuals, institutions and other market participants and regulators. It is more of an evolutionary process rather than a prescriptive model. The dynamics of the complex system that led to the housing asset bubble and the subsequent economic crisis can be captured with the help of agent based simulation and multiple tiers of agents. We illustrate our case with an example.

A model can be posed in four tiers involving the buyer seller interaction in the housing market at the lowest level. A trade between a buyer and a seller takes place with the attendant instrument of price. In the second level, the buyers approach the mortgage banks for credit to purchase the house mortgaging the property. The contract between a borrowers and a lender happens at this stage. At the third level, the interaction of the large investment banks with the mortgage banks occurs. Consequently comes the creation of Special Purpose Vehicles for asset securitisation. The interaction of the investment banks with the economic system happens at the highest level. We have shown the magnitude of a bounded rational perception when there is no interaction from upper tiers. In others words, with a simple utility function the level have been modelled in this article. We have demonstrated that a small departure from reality could be a source of enormous over-investment leading to economic crisis. One may question our assumption of a wrong perception pervasive among all agents of the economy. Our results will remain essentially the same even when there are some rational and some bounded rational agents present in the economy so long as proportion of bounded rational agents are significant in number. The magnitude of the impact of a hyped expectation has to be adjusted accordingly.

References

1. Shiller RJ (2008) The subprime solution. Addison-Wesley, Cambridge
2. Hellwig M. Working paper. Preprints of the Max Planck Institute for Research on Collective Goods, no 2008/43
3. International Monetary Fund. Financial stress and delivering: macro-financial implications and policy, global financial stability report, October 2008
4. Wray LR. Working papers. The Levy Economics Institute, no 522
5. Wray LR (2009) Camb J Econ 33:807–828
6. Goodman AC, Thibodeau TG (2008) J Hous Econ 17:117–137
7. Coleman M IV, LaCour-Little M, Vandell KD (2008) J Hous Econ 17:272–290
8. Crouhy MG, Jarrow RA, Turnbull SM (2008) J Deriv 16:81–110
9. Longstaff FA (2010) J Financ Econ 97:436–450
10. Glaeser EL, Gyourkob J, Saizb A (2008) J Urban Econ 64:198–217
11. Earl PE, Peng T-C, Potts J (2007) J Econ Psychol 28:351–364
12. Shiller R (2000) Irrational exuberance. Princeton University Press, Princeton
13. Minsky HP (1975) John Maynard Keynes. Columbia University Press, New York
14. Roszczynska M, Nowak A, Kamieniarz D, Solomon S, Andersen JV (2008) arXiv: 0806.2124v1 [q-fin.TR]
15. Challet D, Marsili M, Zhang Y-C (2004) Minority games: interacting agents in financial market, Oxford University Press, Oxford
16. Andersen JV, Sornette D (2005) Europhys Lett 70:5
17. Rabertoa M, Cincottia S, Focardib SM, Marchesi M (2001) Physica A 299:319–327
18. An L, Subramanian D, King A (2009) In: The 27th international conference of the system dynamics society, Albuquerque, NM, USA, July 2009
19. Forrester JW (1961) Industry dynamics. MIT Press, Cambridge
20. Thurner S (2011) OECD/IFP project on future global shocks
21. Boer-Sorbán K (2008) PhD thesis, Erasmus University Rotterdam
22. Aiyagari SR (1994) Uninsured idiosyncratic risk and aggregate saving. Q J Econ 109.3:659–684
23. Tauchen G (1986) Finite state Markov-chain approximations to univariate and vector autoregressions. Econ Lett 20(2):177–181

24. Davies JB, Shorrocks AF (2000) Handbook of income distribution, vol 1, pp 605–675
25. Dragulescu A, Yakovenko VM (2001) Phys A, Stat Mech Appl 299.1:213–221
26. US Census Bureau (2012) Statistical abstract of the United States (131st edition), Washington, DC; accessed at http://www.census.gov/compendia/statab/ on March 30, 2013
27. Farmer JD, Foley D (2009) Nature 460(7256):685–686

Chapter 3
Urn Model-Based Adaptive Multi-arm Clinical Trials: A Stochastic Approximation Approach

Sophie Laruelle and Gilles Pagès

Abstract This paper presents the link between stochastic approximation and multi-arm clinical trials based on randomized urn models investigated in Bai et al. (J. Multivar. Anal. 81(1):1–18, 2002) where the urn updating depends on the past performances of the treatments. We reformulate the dynamics of the urn composition, the assigned treatments and the successes of assigned treatments as standard stochastic approximation (*SA*) algorithms with remainder. Then, we derive the *a.s.* convergence of the normalized procedure under less stringent assumptions by calling upon the *ODE* and a new asymptotic normality result (Central Limit Theorem *CLT*) by calling upon the *SDE* methods.

3.1 Introduction

The aim of this paper is to illustrate the efficiency of Stochastic Approximation (*SA*) Theory by revisiting recent results on randomized urn models known as "multi-arm clinical test" (introduced in [6]), where the urn updating which produces the adaptive design is based on statistical estimators of the past efficiency of the assigned treatments.

Clinical trials essentially deal with the asymptotic behavior of the patient allocation to several treatments during the procedure. This adaptive approach relies on the cumulative information provided by the responses to treatments of previous patients in order to adjust treatment allocation to the new patients. To this end, many urn models have been suggested in the literature (see [11, 16, 20, 22, 23]). The most widespread random adaptive model is the Generalized Friedman Urn (*GFU*) (see [2] and more recently [15, 19]), also called Generalized Pólya Urn (*GPU*).

S. Laruelle (✉)
Chaire de Finance Quantitative, Laboratoire de Mathématiques Appliquées aux Systèmes, École Centrale Paris, Grande Voie des Vignes, 92290 Châtenay-Malabry, France
e-mail: sophie.laruelle@ecp.fr

G. Pagès
Laboratoire de Probabilités et Modèles aléatoires, UMR 7599, Université Paris 6, case 188, 4, pl. Jussieu, 75252 Paris Cedex 5, France
e-mail: gilles.pages@upmc.fr

F. Abergel et al. (eds.), *Econophysics of Agent-Based models*,
New Economic Windows, DOI 10.1007/978-3-319-00023-7_3,

The idea of this modeling is that the urn contains balls of d different types representative of the treatments. All random variables involved in the model are supposed to be defined on the same probability space $(\Omega, \mathcal{A}, \mathbb{P})$. Denote $Y_0 = (Y_0^i)_{i=1,\dots,d} \in \mathbb{R}_+^d \setminus \{0\}$ the initial composition of the urn, where Y_0^i denotes the number of balls of type i, $i = 1, \dots, d$ (of course a more realistic though not mandatory assumption would be $Y_0 \in \mathbb{N}^d \setminus \{0\}$). The allocation of the treatments is sequential and the urn composition at draw n is denoted by $Y_n = (Y_n^i)_{i=1,\dots,d}$. When the nth patient presents, one draws randomly (*i.e.* uniformly) a ball from the urn with instant replacement. If the ball is of type j, then the treatment j is assigned to the nth patient, $j = 1, \dots, d$, $n \geq 1$. The urn composition is updated by taking into account the response of the nth patient to the treatment j, or the responses of all patients up to the nth one (*i.e.* the efficiency of the assigned treatment), namely by adding D_n^{ij} balls of type i, $i = 1, \dots, d$. The procedure is iterated as long as patients present. Consequently the larger the number of balls of a given type is, the more efficient the treatment is. The urn composition at stage n, modeled by an $\mathbb{R}^d$-valued vector Y_n, satisfies the following recursive procedure:

$$Y_n = Y_{n-1} + D_n X_n, \quad n \geq 1, \quad Y_0 \in \mathbb{R}_+^d \setminus \{0\}, \tag{3.1}$$

with $D_n = (D_n^{ij})_{1 \leq i,j \leq d}$ is the addition rule matrix and X_n is the result of the nth draw and $X_n : (\Omega, \mathcal{A}, \mathbb{P}) \to \{e^1, \dots, e^d\}$ models the selected treatment ($\{e^1, \dots, e^d\}$ denotes the canonical basis of $\mathbb{R}^d$ and e^j stands for treatment j). We assume that there is no extinction *i.e.* $Y_n \in \mathbb{R}_+^d \setminus \{0\}$ *a.s.* for every $n \geq 1$: so is the case if all the entries D_n^{ij} are *a.s.* nonnegative. We model the drawing in the urn by setting

$$X_n = \sum_{j=1}^{d} \mathbb{1}_{\left\{\frac{\sum_{\ell=1}^{j-1} Y_{n-1}^\ell}{\sum_{\ell=1}^{d} Y_{n-1}^\ell} < U_n \leq \frac{\sum_{\ell=1}^{j} Y_{n-1}^\ell}{\sum_{\ell=1}^{d} Y_{n-1}^\ell}\right\}} e^j, \quad n \geq 1, \tag{3.2}$$

where $(U_n)_{n\geq 1}$ is i.i.d. with distribution $U_1 \stackrel{\mathcal{L}}{\sim} \mathcal{U}_{[0,1]}$.

Let $\mathcal{F}_n = \sigma(Y_0, U_k, D_k, 1 \leq k \leq n)$ be the filtration of the procedure. The *generating matrices* are defined as the $\mathcal{F}_n$-compensator of the additions rule sequence *i.e.*

$$H_n = \left(\mathbb{E}\left[D_n^{ij} \,|\, \mathcal{F}_{n-1}\right]\right)_{1\leq i,j\leq d}, \quad n \geq 1.$$

The first designs under consideration were the homogeneous *GFU* models where the addition rules D_n are i.i.d. and the generating matrices $H_n = H = \mathbb{E}D_n$ are identical, non-random, with nonnegative entries and irreducible (see [2, 3, 13, 14]). For practical matters, inhomogeneous *GFU* models have been introduced (see [4]) in which H_n are not random but converge to a deterministic limit H, under the assumption that the total number of balls added at each stage is constant; then homogeneous Extended Pólya Urn (*EPU*) models have been introduced in [21] in which only the balance is constant. Finally, in [5] the authors proposed a nonhomogeneous *EPU* model because in applications, the addition rule D_n depends on the past history of previous trials (see [1]), so that the general generating matrix H_n is usually random. Thus the entries of H may not be all nonnegative (*e.g.*, when there is no replacement after the draw diagonal terms may become negative), and they assume that the matrix H has a unique maximal eigenvalue λ with associated (right) eigenvector $v^* = (v^{*,i})_{i=1,\dots,d}$ with $\sum_{i=1}^d v^{*,i} = 1$. Furthermore the conditional expectation of the total number of balls added at each stage was constant.

The adaptive design applied to multi-arm clinical trials that we study has already been introduced in [6] with first consistency results. This kind of models is clearly the most interesting for practitioners since it takes into account the past results of the assigned treatments in the addition rule matrices, denoted S_n at time n (S_n^i denotes the number of cured patients by treatment i among the N_n^i treated ones with $N_n := \sum_{k=1}^n X_k$). First we prove the *a.s.* convergence of the generating matrices H_n towards an irreductible matrix with positive entries. Then, by considering an appropriate recursive procedure for the normalized urn composition derived from (3.1) we prove by the *ODE* method its *a.s.* convergence toward v^*. The *a.s.* convergence of the treatment allocation frequency N_n/n toward the same v^* follows from the previous one and the *a.s.* convergence of the treatment successes frequency S_n/n follows from the one of N_n/n and the one of H_n. As concerns asymptotic normality, we show that the triplet $(Y_n/n, N_n/n, S_n/n)$ can be written as a recursive *SA* algorithm with remainder satisfying a *CLT*. Thus we illustrate on this example that *SA* Theory is a powerful tool to investigate this kind of adaptive design problem. The main difficulty is to exhibit the appropriate form for the recursion by making *a priori* the balance between significant asymptotic terms and remainder terms.

The paper is organized as follows. In Sect. 3.2, we present the *GPU* model introduced in [6] and prove the convergence of the generating matrix towards an irreducible limit. Then we rewrite the dynamics (3.1) of the urn composition as a stochastic approximation procedure with state variable for $\widetilde{Y}_n := Y_n/n$ in Sect. 3.3 and establish the *a.s.* convergence of $\widetilde{Y}_n$, $\widetilde{N}_n := N_n/n$ and $\widetilde{S}_n := S_n/n$ by using the *ODE* method of *SA*. The rate of convergence is investigated in Sect. 3.4: we obtain a new *CLT* for this model, when the generating matrix H_n satisfies itself a *CLT*, which relies again on Stochastic Approximation techniques.

Notations $\forall u = (u^i)_{i=1,\dots,d} \in \mathbb{R}^d$, $\|u\|$ denotes the canonical Euclidean norm of the column vector u on $\mathbb{R}^d$, $w(u) = \sum_{k=1}^d u^k$ denotes its "weight", u^t denotes its transpose; $|||A|||$ denotes the operator norm of the matrix $A \in \mathcal{M}_{d,q}(\mathbb{R})$ with d rows and q columns with respect to canonical Euclidean norms. When $d = q$, $\mathrm{Sp}(A)$ denotes the set of eigenvalues of A. $\mathbf{1} = (1 \cdots 1)^t$ denotes the unit column vector in $\mathbb{R}^d$, I_d denotes the $d \times d$ identity matrix and $\mathrm{diag}(u) = [\delta_{ij} u_i]_{1 \le i,j \le d}$, where δ_{ij} is the Kronecker symbol.

3.2 Presentation of Bai-Hu-Shen *GFU* Model

We consider here the model introduced in [6], where balls are added depending on the success probabilities of each treatment. Define an *efficiency indicator* as follows:

let $(T_n^i)_{n\geq 1}$, $1\leq i\leq d$, be d independent sequences of $[0,1]$-valued i.i.d. random variables, independent of the i.i.d. *sampling* sequence $(U_n)_{n\geq 1}$ so that

$$\mathbb{E}\big[T_n^i\big]=p^i,\quad 0<p^i<1,\ 1\leq i\leq d. \tag{3.3}$$

Remark If $(T_n^i)_{n\geq 1}$, $1\leq i\leq d$, is simply a *success indicator*, namely d independent sequences of i.i.d. $\{0,1\}$-valued Bernoulli trials with respective parameter p^i, then the convention is to set $T_n^i=1$ to indicate that the response of the ith treatment in the nth trial is a success and $T_n^i=0$ otherwise.

In this framework one considers the filtration $\mathcal{F}_n=\sigma(Y_0,U_k,T_k,1\leq k\leq n)$, $n\geq 0$. Let $N_n=(N_n^1,\dots,N_n^d)^t$ and $S_n=(S_n^1,\dots,S_n^d)^t$, where $N_n^i=N_{n-1}^i+X_n^i$, $n\geq 1$, still denotes the number of times the ith treatment is selected among the first n stages and

$$S_n^i=S_{n-1}^i+T_n^iX_n^i,\quad n\geq 1,$$

denotes the *number of successes* of the ith treatment among these N_n^i trials, $i=1,\dots,d$. However, to avoid degeneracy of the procedure, we will make the following initialization assumption

$$N_0^i=1,\qquad S_0^i=1,\quad i=1,\dots,d$$

(which makes the above interpretation of these quantities correct "up to one unit").

Define $\Pi_n=(\Pi_n^1,\dots,\Pi_n^d)^t$, where $\Pi_n^i=\frac{S_n^i}{N_n^i}$, $i=1,\dots,d$. In [6] the authors consider the following addition rule matrices,

$$D_{n+1}=\begin{pmatrix} T_{n+1}^1 & \frac{\Pi_n^1(1-T_{n+1}^2)}{\sum_{j\neq 2}\Pi_n^j} & \cdots & \frac{\Pi_n^1(1-T_{n+1}^d)}{\sum_{j\neq d}\Pi_n^j}\\ \frac{\Pi_n^2(1-T_{n+1}^1)}{\sum_{j\neq 1}\Pi_n^j} & T_{n+1}^2 & \cdots & \frac{\Pi_n^2(1-T_{n+1}^d)}{\sum_{j\neq d}\Pi_n^j}\\ \vdots & \vdots & \ddots & \vdots\\ \frac{\Pi_n^d(1-T_{n+1}^1)}{\sum_{j\neq 1}^d\Pi_n^j} & \frac{\Pi_n^d(1-T_{n+1}^2)}{\sum_{j\neq 2}^d\Pi_n^j} & \cdots & T_{n+1}^d \end{pmatrix},$$

i.e. at stage $n+1$, if the response of the jth treatment is a success, then one ball of type j is added in the urn. Otherwise, $\frac{\Pi_n^i}{\sum_{k\neq j}\Pi_n^k}$ (virtual) balls of type i, $i\neq j$, are added. Furthermore we have that for every $n\geq 1$, the matrix D_n *a.s.* has nonnegative entries, is conditionally independent of the drawing procedure X_n given $\mathcal{F}_{n-1}$ and satisfies

$$\forall 1\leq j\leq d,\quad \sup_{n\geq 1}\mathbb{E}\big[\big\|D_n^{\cdot j}\big\|^2\,\big|\,\mathcal{F}_{n-1}\big]<+\infty\quad a.s. \tag{3.4}$$

where $D_n^{\cdot j} = (D_n^{ij})_{i=1,\dots,d}$. Then, one easily checks that the generating matrices are given by

$$H_{n+1} = \mathbb{E}[D_{n+1} \mid \mathcal{F}_n] = \begin{pmatrix} p^1 & \frac{\Pi_n^1(1-p^2)}{\sum_{j\neq 2}\Pi_n^j} & \cdots & \frac{\Pi_n^1(1-p^d)}{\sum_{j\neq d}\Pi_n^j} \\ \frac{\Pi_n^2(1-p^1)}{\sum_{j\neq 1}\Pi_n^j} & p^2 & \cdots & \frac{\Pi_n^2(1-p^d)}{\sum_{j\neq d}\Pi_n^j} \\ \vdots & \vdots & \ddots & \vdots \\ \frac{\Pi_n^d(1-p^1)}{\sum_{j\neq 1}\Pi_n^j} & \frac{\Pi_n^d(1-p^2)}{\sum_{j\neq 2}\Pi_n^j} & \cdots & p^d \end{pmatrix}$$

and satisfy

$$\forall n \geq 1,\ \forall j \in \{1,\dots,d\},\quad \sum_{i=1}^{d} H_n^{ij} = 1 \quad a.s. \tag{3.5}$$

Lemma 3.1 *If the assumption* (3.1) *holds and* $Y_0 \in \mathbb{R}_+^d \setminus \{0\}$, *then* $\Pi_n \xrightarrow[n\to\infty]{a.s.} p = (p^1,\dots,p^d)$, *so that* $H_n \xrightarrow[n\to\infty]{a.s.} H$ *where*

$$H = \begin{pmatrix} p^1 & \frac{p^1(1-p^2)}{\sum_{j\neq 2}p^j} & \cdots & \frac{p^1(1-p^d)}{\sum_{j\neq d}p^j} \\ \frac{p^2(1-p^1)}{\sum_{j\neq 1}p^j} & p^2 & \cdots & \frac{p^2(1-p^d)}{\sum_{j\neq d}p^j} \\ \vdots & \vdots & \ddots & \vdots \\ \frac{p^d(1-p^1)}{\sum_{j\neq 1}p^j} & \frac{p^d(1-p^2)}{\sum_{j\neq 2}p^j} & \cdots & p^d \end{pmatrix}$$

is clearly irreducible since $0 < p^i < 1$, $1 \leq i \leq d$.

Remark If we assume that $Y_0^i > 0$, $1 \leq i \leq d$, then we can prove that $\lim_n N_n^i = +\infty$ *a.s.*, $1 \leq i \leq d$, faster than in the proof of this result by using that $Y_n^i \geq Y_0^i$, $1 \leq i \leq d$, $n \geq 1$. The proof considers the more general case where $Y_0 \in \mathbb{R}_+^d \setminus \{0\}$ and can be found in [18].

The combination of (3.5) and Lemma 3.1 guarantees that H satisfies the assumptions of the Perron-Frobenius Theorem (see [9]) so that 1 is the eigenvalue of H with the highest norm (maximal eigenvalue) and that the components of its right eigenvector υ can be chosen all positive. Therefore, we may normalize this vector υ^* such that $w(\upsilon^*) = 1$. It is given by

$$\upsilon^{*i} = \frac{\frac{p^i}{1-p^i}\sum_{k\neq i}p^k}{\sum_{1\leq j\leq d}\frac{p^j}{1-p^j}\sum_{k\neq j}p^k}, \qquad i = 1,\dots,d.$$

Note that if $p^i > p^j$, $\frac{p^i}{p^j}\frac{\sum_{k\neq i} p^k}{\sum_{k\neq j} p^k} > 1$ and $\frac{1-p^j}{1-p^i} > 1$ so that $v^{*i} > v^{*j}$. Hence the entries v^{*i} are ordered according to the increasing efficiency p^i of the treatments.

3.3 Asymptotic Consistency for Multi-arm Clinical Trials for the BHS GFU Model

Theorem 3.1 *Assume that $Y_0 \in \mathbb{R}_+^d \setminus \{0\}$. Then*

$$\frac{w(Y_n)}{n} \xrightarrow[n\to\infty]{a.s.} 1, \qquad \frac{Y_n}{w(Y_n)} \xrightarrow[n\to\infty]{a.s.} v^*,$$

$$\widetilde{N}_n := \frac{N_n}{n} \xrightarrow[n\to\infty]{a.s.} v^* \quad \textit{and} \quad \widetilde{S}_n := \frac{S_n}{n} \xrightarrow[n\to\infty]{a.s.} \operatorname{diag}(p)v^*.$$

Proof Step 1 (Convergence of $\frac{Y_n}{w(Y_n)}$). Our aim is to reformulate the dynamics (3.1)–(3.2) into a recursive stochastic algorithm. Then we aim at applying the most powerful tools of *SA*, namely the "*ODE*" method to elucidate the *a.s.* convergence of the urn composition. We start from (3.1) with $Y_0 \in \mathbb{R}_+^d \setminus \{0\}$. For $n \geq 1$, using the definition of the generating matrix H_n, we have

$$Y_{n+1} = Y_n + D_{n+1}X_{n+1} = Y_n + \mathbb{E}[D_{n+1}X_{n+1} \,|\, \mathcal{F}_n] + \Delta M_{n+1}, \tag{3.6}$$

where

$$\Delta M_{n+1} := D_{n+1}X_{n+1} - \mathbb{E}[D_{n+1}X_{n+1} \,|\, \mathcal{F}_n] = D_{n+1}X_{n+1} - H_{n+1}\frac{Y_n}{w(Y_n)}$$

is an $\mathcal{F}_n$-martingale increment. Then, by setting $\widetilde{Y}_n = \frac{Y_n}{n}$, $n \geq 1$, we obtain a canonical recursive stochastic approximation procedure

$$\widetilde{Y}_{n+1} = \widetilde{Y}_n - \frac{1}{n+1}(I_d - H)\widetilde{Y}_n + \frac{1}{n+1}(\Delta M_{n+1} + r_{n+1}) \tag{3.7}$$

with step $\gamma_n = \frac{1}{n}$ and a remainder term given by

$$r_{n+1} := \left(\frac{n}{w(Y_n)} - 1\right)H_{n+1}\widetilde{Y}_n + (H_{n+1} - H)\widetilde{Y}_n. \tag{3.8}$$

Furthermore, in order to establish the *a.s.* boundedness of $(\widetilde{Y}_n)_{n\geq 1}$, we will rely on the following recursive equation satisfied by $w(Y_n)$ obtained by using the properties of the generating matrix H_{n+1}

$$w(Y_{n+1}) = w(Y_n) + 1 + w(\Delta M_{n+1}). \tag{3.9}$$

By (3.4), we have that $\sup_{n\geq 1}\mathbb{E}[\|\Delta M_{n+1}\|^2\,|\,\mathcal{F}_n]<+\infty$ *a.s.* Therefore thanks to the strong law of large numbers for conditionally L^2-bounded martingale increments, we have $\frac{M_n}{n}\xrightarrow[n\to\infty]{a.s.}0$. Consequently it follows from (3.9) that

$$\frac{w(Y_n)}{n}=1+\frac{w(Y_0)-1}{n}+\frac{w(M_n)}{n}\xrightarrow[n\to\infty]{a.s.}1. \tag{3.10}$$

Since the components of $\widetilde{Y}_n=\frac{Y_n}{n}$ are nonnegative and $w(\widetilde{Y}_n)=\frac{w(Y_n)}{n}\xrightarrow[n\to\infty]{a.s.}1$, it is clear that $(\widetilde{Y}_n)_{n\geq 1}$ is *a.s.* bounded and that *a.s.* the set $\mathcal{Y}_\infty$ of all its limiting value is contained in

$$\mathcal{V}=w^{-1}\{1\}=\big\{u\in\mathbb{R}^d_+\,\big|\,w(u)=1\big\}.$$

So we may try applying the *ODE* method (see Appendix Theorem 3.3). Since $\widetilde{Y}_n$ and $H_{n+1}\widetilde{Y}_n$ are *a.s.* bounded, (3.10) and Lemma 3.1 imply that $r_n\xrightarrow[n\to\infty]{a.s.}0$.

The *ODE* associated to the recursive procedure reads

$$ODE_{I_d-H}\equiv\dot{y}=-(I_d-H)y.$$

Owing to Lemma 3.1, I_d-H admits v^* as unique zero in $\mathcal{V}$. The restriction of ODE_{I_d-H} to the affine hyperplane $\mathcal{V}$ is the linear system $\dot{z}=-(I_d-H)z$, where $z=y-v^*$ takes values in $\mathcal{V}_0=\{u\in\mathbb{R}^d\,|\,w(u)=0\}$. Since $\mathrm{Sp}((I_d-H)_{|\mathcal{V}_0})\subset\{\lambda\in\mathbb{C},\Re e(\lambda)>0\}$ by Lemma 3.1, v^* is an uniformly stable equilibrium for the restriction of ODE_{I_d-H} to $\mathcal{V}$, the whole hyperplane, as an attracting area. The fundamental result derived from the *ODE* method (see Theorem 3.3 in Appendix and the notations therein, in particular the remainder r_n) yields the expected result

$$\widetilde{Y}_n\xrightarrow[n\to\infty]{a.s.}v^*.$$

Step 2 (Convergence of $\widetilde{N}_n$). We will prove that $\frac{Y_n}{w(Y_n)}\xrightarrow[n\to\infty]{a.s.}v^*$ implies that $\frac{N_n}{n}\xrightarrow[n\to\infty]{a.s.}v^*$. We have

$$\mathbb{E}[X_n\,|\,\mathcal{F}_{n-1}]=\sum_{i=1}^{d}\frac{Y^i_{n-1}}{w(Y_{n-1})}e^i=\frac{Y_{n-1}}{w(Y_{n-1})}$$

and, by construction $\|X_n\|^2=1$ so that $\mathbb{E}[\|X_n\|^2\,|\,\mathcal{F}_{n-1}]=1$. Hence the martingale

$$\widetilde{M}_n=\sum_{k=1}^{n}\frac{X_k-\mathbb{E}[X_k\,|\,\mathcal{F}_{k-1}]}{k}\xrightarrow[n\to\infty]{a.s.\&L^2}\widetilde{M}_\infty\in L^2,$$

and by the Kronecker Lemma we obtain

$$\frac{1}{n}\sum_{k=1}^{n} X_k - \frac{1}{n}\sum_{k=1}^{n}\frac{Y_{k-1}}{w(Y_{k-1})} \xrightarrow[n\to\infty]{a.s.} 0.$$

This yields the announced implication owing to the Cesaro Lemma.

Step 3 (Convergence of $\widetilde{S}_n$). We have that $\mathrm{diag}(\widetilde{S}_n) = \mathrm{diag}(\Pi_n)\widetilde{N}_n$. Then by using Lemma 3.1 and *Step 2*, we obtain that $\widetilde{S}_n \xrightarrow[n\to\infty]{a.s.} \mathrm{diag}(p)v^*$. □

3.4 Asymptotic Normality for Multi-arm Clinical Trials for the BHS GFU Model

In [6] in order to derive a *CLT*, not with the bias $\mathbb{E}Y_n$ but with nv^*, from their own general asymptotic normality result they need to fulfill the following convergence rate assumption for H_n

$$\sum_{n\geq 1}\frac{\|H_n - H\|_\infty}{\sqrt{n}} < +\infty \tag{3.11}$$

where $\|\cdot\|_\infty$ is the norm on $L^\infty_{\mathbb{R}^{d\times d}}(\mathbb{P})$. In [6], the *a.s.* rate of decay $|||H_n - H|||_\infty = o(n^{-\frac{1}{4}})$ is shown which is clearly not fast enough to fulfill (3.11).

However, by enlarging the dimension of the structure process of the procedure by considering the $3d$-dimensional random sequence $\theta_n = (\widetilde{Y}_n, \widetilde{N}_n, \widetilde{S}_n)^t$, $n \geq 1$, we will establish that a *CLT* does hold for the BHS *GFU* model.

The first step is to notice that the generating matrix H_{n+1} can may be written as a function depending on $\widetilde{S}_n$ and $\widetilde{N}_n$, *i.e.* $H_{n+1} = \Phi(\widetilde{S}_n, \widetilde{N}_n)$, where $\Phi : \mathbb{R}^d_+ \times (0,\infty)^d \to \mathcal{M}_d(\mathbb{R})$ is a differentiable function defined by

$$\Phi(s,\nu) = \big(\Phi^{ij}(s,\nu)\big)_{1\leq i,j\leq d} \quad \text{where}$$

$$\begin{cases} \Phi^{ii}(s,\nu) = p^i, & 1\leq i\leq d,\\ \Phi^{ij}(s,\nu) = \dfrac{s^i/\nu^i}{\sum_{k\neq j} s^k/\nu^k}q^j, & 1\leq i,j\leq d,\ i\neq j. \end{cases}$$

Then the following strong consistency and *CLT* hold for $(\theta_n)_{n\geq 1}$.

Theorem 3.2 *Assume that $Y_0 \in \mathbb{R}^d_+ \setminus \{0\}$. If $\Re e(\mathrm{Sp}(H)\setminus\{1\}) < \frac{1}{2}$, then*

$$\sqrt{n}\big(\theta_n - \theta^*\big) \xrightarrow[n\to\infty]{\mathcal{L}} \mathcal{N}(0,\Sigma),$$

where

$$\theta^* := \big(v^*, v^*, \mathrm{diag}(p)v^*\big)^t, \qquad \Sigma = \int_0^{+\infty} e^{u(Dh(\theta^*)-\frac{I}{2})}\,\Gamma\, e^{u(Dh(\theta^*)-\frac{I}{2})^t}\,du$$

with

$$\Gamma = \begin{pmatrix} \sum_{k=1}^{d} v^{*k} C^k - v^*(v^*)^t & H(\operatorname{diag}(v^*) - v^*(v^*)^t) & (\operatorname{diag}(v^*) - v^*(v^*)^t)\operatorname{diag}(p) \\ (\operatorname{diag}(v^*) - v^*(v^*)^t)^t H^t & \operatorname{diag}(v^*) - v^*(v^*)^t & (\operatorname{diag}(v^*) - v^*(v^*)^t)\operatorname{diag}(p) \\ \operatorname{diag}(p)(\operatorname{diag}(v^*) - v^*(v^*)^t)^t & \operatorname{diag}(p)(\operatorname{diag}(v^*) - v^*(v^*)^t)^t & \operatorname{diag}(p)(v^* - v^* v^{*t}\operatorname{diag}(p)) \end{pmatrix}$$

where $C^k = (C^k_{ij})_{1\le i,j\le d}$, $1 \le k \le d$, *are* $d \times d$ *positive definite matrices with*

$$C^k_{ij} = \frac{p^i p^j (1-p^k)}{(\sum_{\ell \neq k} p^\ell)^2} \mathbb{1}_{\{i,j\neq k\}} + p^k \mathbb{1}_{\{i=j=k\}},$$

and

$$Dh(\theta^*) = \begin{pmatrix} I_d - H + v^*\mathbf{1}^t & -\frac{\partial}{\partial v}(\Phi(s,\nu)y)_{|\widetilde{\theta}=\widetilde{\theta}^*} & -\frac{\partial}{\partial s}(\Phi(s,\nu)y)_{|\widetilde{\theta}=\widetilde{\theta}^*} \\ v^*\mathbf{1}^t - I_d & I_d & 0_{\mathcal{M}_d(\mathbb{R})} \\ \operatorname{diag}(p)(v^*\mathbf{1}^t - I_d) & 0_{\mathcal{M}_d(\mathbb{R})} & I_d \end{pmatrix}$$

which is invertible.

Proof We will show that $(\theta_n)_{n\ge1}$ satisfies an appropriate recursion to apply Theorem 3.4 (*CLT*). First, we write a recursive procedure for $\widetilde{N}_n$ and $\widetilde{S}_n$. Having in mind that $N_n = 1 + \sum_{1\le k\le n} X_k$ and $S_n = 1 + \sum_{1\le k\le n} \operatorname{diag}(T_k)X_k$, we get, for $n \ge 1$,

$$\begin{aligned} \widetilde{N}_{n+1} &= \widetilde{N}_n - \frac{1}{n+1}(\widetilde{N}_n - X_{n+1}) = \widetilde{N}_n - \frac{1}{n+1}\left(\widetilde{N}_n - \frac{\widetilde{Y}_n}{w(\widetilde{Y}_n)}\right) + \frac{1}{n+1}\Delta\widetilde{M}_{n+1} \\ &= \widetilde{N}_n - \frac{1}{n+1}\big(\widetilde{N}_n - \big(2 - w(\widetilde{Y}_n)\big)\widetilde{Y}_n\big) + \frac{1}{n+1}(\Delta\widetilde{M}_{n+1} + \widetilde{r}_{n+1}) \end{aligned} \tag{3.12}$$

where $\Delta\widetilde{M}_{n+1} := X_{n+1} - \frac{Y_n}{w(Y_n)}$ is an $\mathcal{F}_n$-martingale increment and $\widetilde{r}_{n+1} := \frac{(w(\widetilde{Y}_n)-1)^2}{w(\widetilde{Y}_n)}\widetilde{Y}_n$ and likewise,

$$\widetilde{S}_{n+1} = \widetilde{S}_n - \frac{1}{n+1}\big(\widetilde{S}_n - \operatorname{diag}(p)\big(2 - w(\widetilde{Y}_n)\big)\widetilde{Y}_n\big) + \frac{1}{n+1}(\Delta\widehat{M}_{n+1} + \widehat{r}_{n+1}) \tag{3.13}$$

where $\Delta\widehat{M}_{n+1} := \operatorname{diag}(T_{n+1})X_{n+1} - \mathbb{E}[\operatorname{diag}(T_{n+1})X_{n+1} \mid \mathcal{F}_n] = \operatorname{diag}(T_{n+1})X_{n+1} - \operatorname{diag}(p)\frac{\widetilde{Y}_n}{w(\widetilde{Y}_n)}$ is an $\mathcal{F}_n$-martingale increment and $\widehat{r}_{n+1} = \operatorname{diag}(p)\frac{(w(\widetilde{Y}_n)-1)^2}{w(\widetilde{Y}_n)}\widetilde{Y}_n$. Then we rewrite the dynamics satisfied by $\widetilde{Y}_n$ as follows

$$\widetilde{Y}_{n+1} = \widetilde{Y}_n - \frac{1}{n+1}\big(I_d - \big(2 - w(\widetilde{Y}_n)\big)H_{n+1}\big)\widetilde{Y}_n + \frac{1}{n+1}(\Delta M_{n+1} + \mathbf{r}_{n+1}), \tag{3.14}$$

where $\mathbf{r}_{n+1} := \frac{(w(\widetilde{Y}_n)-1)^2}{w(\widetilde{Y}_n)} H_{n+1}\widetilde{Y}_n$. Finally, we get the following recursive procedure for θ_n

$$\theta_{n+1} = \theta_n - \frac{1}{n+1}h(\theta_n) + \frac{1}{n+1}(\Delta\mathbf{M}_{n+1} + R_{n+1}), \quad n \geq 1,$$

where, for every $\theta = (y, v, s)^t \in \mathbb{R}_+^{3d}$,

$$h(\theta) := \begin{pmatrix} (I_d - (2 - w(y))\Phi(s, v))y \\ v - (2 - w(y))y \\ s - (2 - w(y))\,\mathrm{diag}(p)y \end{pmatrix},$$

$$\Delta\mathbf{M}_{n+1} := \begin{pmatrix} \Delta M_{n+1} \\ \Delta\widetilde{M}_{n+1} \\ \Delta\widehat{M}_{n+1} \end{pmatrix} \quad \text{and} \quad R_{n+1} := \begin{pmatrix} \mathbf{r}_{n+1} \\ \widetilde{r}_{n+1} \\ \widehat{r}_{n+1} \end{pmatrix}.$$

Let us check that the addition rule matrices satisfy (3.17). For every $j \in \{1, \ldots, d\}$, let set $C_n^j = \mathbb{E}[D_{n+1}^{\cdot j}(D_{n+1}^{\cdot j})^t|\mathcal{F}_n]$. We have that

$$\begin{aligned}(C_n^j)_{ii'} &= \mathbb{E}\big[D_{n+1}^{ij}(D_{n+1}^{i'j})^t|\mathcal{F}_n\big] \\ &= \frac{Q_n^i Q_n^{i'}}{(\sum_{k\neq j} Q_n^k)^2}\mathbb{E}\big[(1 - T_{n+1}^j)^2|\mathcal{F}_n\big]\mathbb{1}_{\{i,i'\neq j\}} + \mathbb{E}\big[(T_{n+1}^j)^2|\mathcal{F}_n\big]\mathbb{1}_{\{i=i'=j\}}\end{aligned}$$

because $T_{n+1}^j(1 - T_{n+1}^j) = 0$. Then owing to Lemma 3.1, $C_n^j \xrightarrow[n\to+\infty]{a.s.} C^j$ with

$$C_{ii'}^j = \frac{p^i p^{i'}(1 - p^j)}{(\sum_{k\neq j} p^k)^2}\mathbb{1}_{\{i,i'\neq j\}} + p^j \mathbb{1}_{\{i=i'=j\}}.$$

We can check that C^j is a positive definite matrix. Consequently (3.17) holds.

The function Φ being differentiable at the equilibrium point θ^*, we have

$$Dh(\theta^*) = \begin{pmatrix} I_d - H + v^*\mathbf{1}^t & -\frac{\partial}{\partial v}(\Phi(s, v)y)_{|\theta=\theta^*} & -\frac{\partial}{\partial s}(\Phi(s, v)y)_{|\theta=\theta^*} \\ v^*\mathbf{1}^t - I_d & I_d & 0_{\mathcal{M}_d(\mathbb{R})} \\ \mathrm{diag}(p)(v^*\mathbf{1}^t - I_d) & 0_{\mathcal{M}_d(\mathbb{R})} & I_d \end{pmatrix}$$

which is invertible since by Schur complement we have $\det(Dh(\theta^*)) = \det(I_d - H + v^*\mathbf{1}^t)$ thanks to $\frac{\partial}{\partial v}(\Phi(s, v)y)_{|\theta=\theta^*} = -\,\mathrm{diag}(p)\frac{\partial}{\partial s}(\Phi(s, v)y)_{|\theta=\theta^*}$.

Finally, it remains to check that the remainder sequence $(R_n)_{n\geq 1}$ satisfies (3.18) for an $\epsilon > 0$:

$$\mathbb{E}\big[(n + 1)\|R_{n+1}\|^2\mathbb{1}_{\{\|\theta_n-\theta^*\|\leq\epsilon\}}\big] \xrightarrow[n\to\infty]{} 0. \tag{3.15}$$

We note that $\|R_{n+1}\|^2 = \|\mathbf{r}_{n+1}\|^2 + \|\widetilde{r}_{n+1}\|^2 + \|\widehat{r}_{n+1}\|^2$. It follows from the definition of $\mathbf{r}_{n+1}$ and the elementary facts $\|\widetilde{Y}_n - v^*\| \leq \|\theta_n - \theta^*\|$ and $w(\widetilde{Y}_n) \geq \|\widetilde{Y}_n\|$

that

$$\|\mathbf{r}_{n+1}\|^2 \mathbb{1}_{\{\|\theta_n-\theta^*\|\le \frac{\|v^*\|}{2}\}} \le 6\big(w(\widetilde{Y}_n)-1\big)^4 \mathbb{1}_{\{\|\theta_n-\theta^*\|\le \frac{\|v^*\|}{2}\}}.$$

But $w(\widetilde{Y}_n)-1 = \frac{w(\Delta M_n)}{n}$ where $\sup_{n\ge 0}\mathbb{E}[|w(\Delta M_{n+1})|^{2+\delta}|\mathcal{F}_n] \le C'$, $\delta > 0$. Now using that $|w(y)| \le C_d\|y\|$,

$$\begin{aligned}\mathbb{E}\big[n\big|w(\widetilde{Y}_n)-1\big|^4 \mathbb{1}_{\{\|\theta_n-\theta^*\|\le \frac{\|v^*\|}{2}\}}\big] &\le C^*_\delta n\mathbb{E}\big[\big|w(\widetilde{Y}_n)-1\big|^{2+\delta}\big]\\ &= \frac{C_d}{n^{1+\delta}}\mathbb{E}\big[\big|w(\Delta M_n)\big|^{2+\delta}\big] \le \frac{C'_d}{n^{1+\delta}},\end{aligned}$$

where $C^*_\delta > 0$ is a real constant. Consequently

$$\mathbb{E}\big[\|\bar{r}_{n+1}\|^2 \mathbb{1}_{\{\|\theta_n-\theta^*\|\le \frac{\|v^*\|}{2}\}}\big] = o\left(\frac{1}{n}\right).$$

The same argument yields

$$\mathbb{E}\big[\|\widetilde{r}_{n+1}\|^2 \mathbb{1}_{\{\|\theta_n-\theta^*\|\le \frac{\|v^*\|}{2}\}}\big] = o\left(\frac{1}{n}\right), \qquad \mathbb{E}\big[\|\widehat{r}_{n+1}\|^2 \mathbb{1}_{\{\|\theta_n-\theta^*\|\le \frac{\|v^*\|}{2}\}}\big] = o\left(\frac{1}{n}\right),$$

therefore (3.15) is satisfied. We refer to [18] for the computation of the matrix Γ. The three results of convergence rate follows from Theorem 3.4 in the Appendix. The details are left to the reader. □

Corollary 3.1 *Under the assumptions of Theorem* 3.2,

$$\sqrt{n}(H_n - H) \xrightarrow[n\to\infty]{\mathcal{L}} \mathcal{N}(0, \Gamma_H)$$

where Γ_H *is a* $d^2 \times d^2$ *matrix given by*

$$\Gamma_H = D\Phi\big(u^*, v^*\big)[\Sigma_{i+d,j+d}]_{1\le i,j\le 2d} D\Phi\big(u^*, v^*\big)^t.$$

Proof This is an easy consequence of the so-called Δ-method since

$$\begin{aligned}H_n = \Phi(\widetilde{S}_n, \widetilde{N}_n) &= \Phi\big(u^*, v^*\big) + D\Phi\big(u^*, v^*\big).\big(\widetilde{S}_n - u^*, \widetilde{N}_n - v^*\big)\\ &+ \big\|\big(\widetilde{S}_n - u^*, \widetilde{N}_n - v^*\big)\big\|\varepsilon(\widetilde{S}_n, \widetilde{N}_n)\end{aligned}$$

with $\lim_{y\to(u^*,v^*)}\varepsilon(y) = 0$. Consequently

$$\sqrt{n}(H_n - H) = D\Phi\big(u^*, v^*\big).\big(\sqrt{n}\big(\widetilde{S}_n - u^*\big), \sqrt{n}\big(\widetilde{N}_n - v^*\big)\big) + \varepsilon_{\mathbb{P}}(n)$$

where $\varepsilon_{\mathbb{P}}(n)$ goes to 0 in probability (as the product of a tight sequence and an *a.s.* convergent sequence). This concludes the proof. □

Numerical Example: BHS Model We consider the case $d = 2$, where $n = 2 \times 10^3$, $p^1 = 0.5$, $p^2 = 0.7$, so $v^{*i} = \frac{1-p^i}{2-p^1-p^2}$, $i = 1, 2$. Figure 3.1 shows the *a.s.* convergence of the normalized urn composition, the frequency of drawing of each treatment and their frequency of success. Figure 3.2 illustrates the *a.s.* convergence of the estimator of the success probability of each treatment.

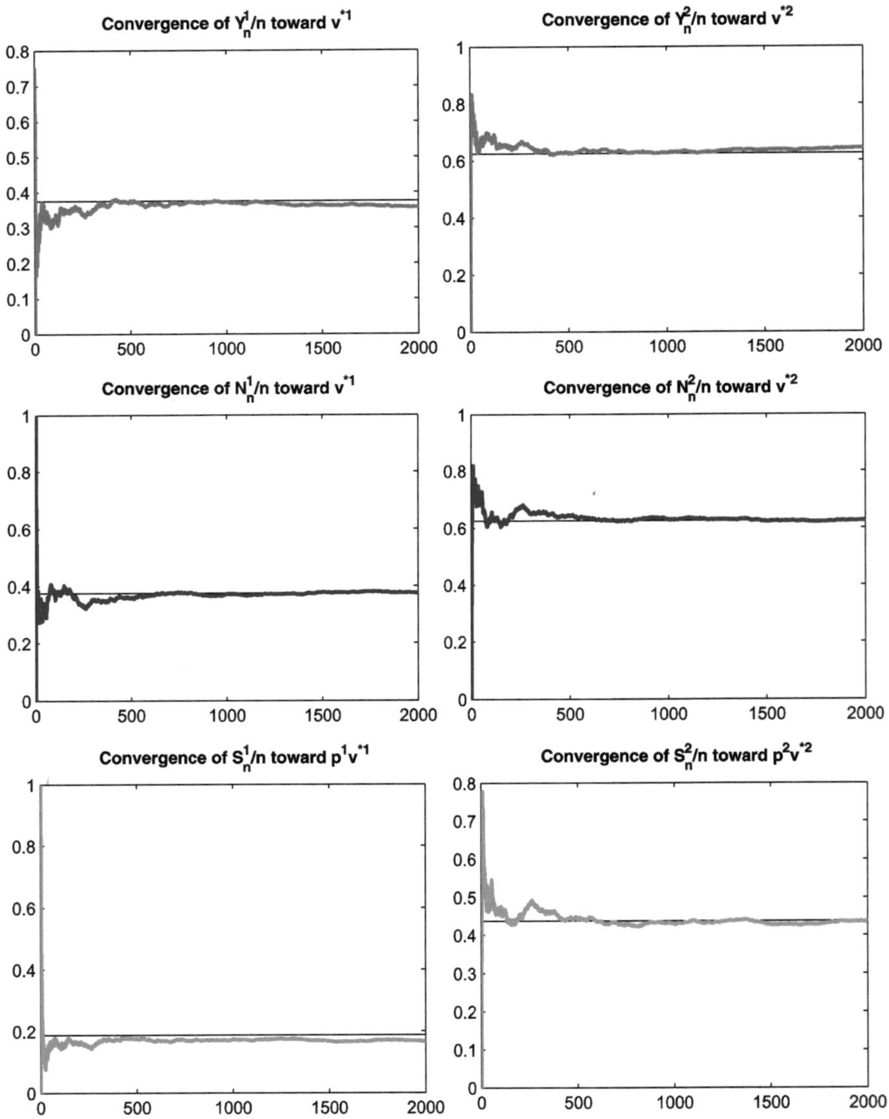

Fig. 3.1 Convergences of $\frac{Y_n}{n}$ toward v^* (*up-windows*), of $\frac{N_n}{n}$ toward v^* (*middle-windows*) and of $\frac{S_n}{n}$ toward diag$(p)v^*$ (*down-windows*): $d = 2$, $n = 2.10^3$, $p^1 = 0.5$, $p^2 = 0.7$, $Y_0 = (0.5, 0.5)^t$, $N_0 = (1, 1)^t$ and $S_0 = (1, 1)^t$

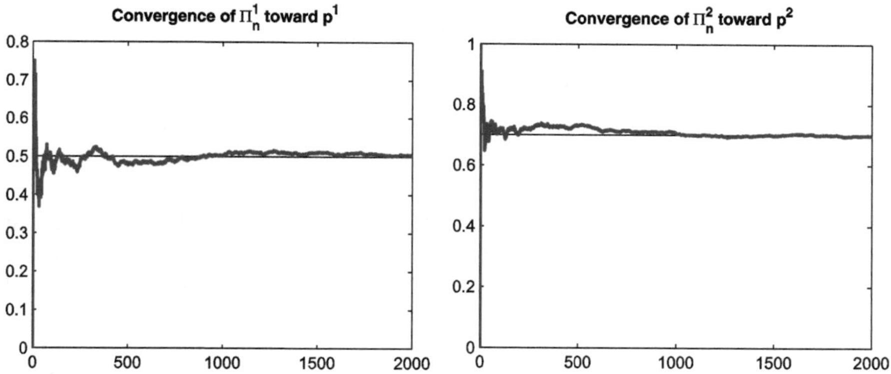

Fig. 3.2 Convergences of Π_n toward p: $d=2, n=2\times 10^3$, $p^1=0.5$, $p^2=0.7$, $Y_0=(0.5,0.5)^t$, $N_0=(1,1)^t$ and $S_0=(1,1)^t$

Appendix: Basic Tools of Stochastic Approximation

Consider the following recursive procedure defined on a filtered probability space $(\Omega, \mathcal{A}, (\mathcal{F}_n)_{n\geq 0}, \mathbb{P})$

$$\forall n \geq n_0, \quad \theta_{n+1} = \theta_n - \gamma_{n+1} h(\theta_n) + \gamma_{n+1}(\Delta M_{n+1} + r_{n+1}), \tag{3.16}$$

where $h: \mathbb{R}^d \to \mathbb{R}^d$ is a locally Lipschitz continuous function, θ_{n_0} an $\mathcal{F}_{n_0}$-measurable finite random vector and, for every $n \geq n_0$, ΔM_{n+1} is an $\mathcal{F}_n$-martingale increment and r_n is an $\mathcal{F}_n$-adapted remainder term.

Theorem 3.3 (*A.s.* Convergence with *ODE* Method, see *e.g.* [7, 8, 10, 12, 17]) *Assume that h is locally Lipschitz, that*

$$r_n \xrightarrow[n\to\infty]{a.s.} 0 \quad \text{and} \quad \sup_{n\geq n_0} \mathbb{E}\big[\|\Delta M_{n+1}\|^2 \,\big|\, \mathcal{F}_n\big] < +\infty \quad a.s.,$$

and that $(\gamma_n)_{n\geq 1}$ *is a positive sequence satisfying*

$$\sum_{n\geq 1} \gamma_n = +\infty \quad \text{and} \quad \sum_{n\geq 1} \gamma_n^2 < +\infty.$$

Then the set Θ^∞ *of its limiting values as* $n \to +\infty$ *is a.s. a compact connected set, stable by the flow of*

$$ODE_h \equiv \dot{\theta} = -h(\theta).$$

Furthermore if $\theta^* \in \Theta^\infty$ *is a uniformly stable equilibrium on* Θ^∞ *of* ODE_h, *then*

$$\theta_n \xrightarrow[n\to\infty]{a.s.} \theta^*.$$

Comments By uniformly stable we mean that

$$\sup_{\theta\in\Theta^\infty} \left|\theta(\theta_0,t)-\theta^*\right| \longrightarrow 0 \quad \text{as } t\to+\infty,$$

where $\theta(\theta_0,t)_{\theta_0\in\Theta^\infty, t\in\mathbb{R}_+}$ is the flow of ODE_h on Θ^∞.

Theorem 3.4 (Rate of Convergence see [10], Theorem 3.III.14, p.131 (for *CLT* see also *e.g.* [8, 17])) *Let θ^* be an equilibrium point of $\{h=0\}$. Assume that the function h is differentiable at θ^* and all the eigenvalues of $Dh(\theta^*)$ have positive real parts. Assume that for some $\delta>0$,*

$$\sup_{n\geq n_0} \mathbb{E}\big[\|\Delta M_{n+1}\|^{2+\delta}\,\big|\,\mathcal{F}_n\big]<+\infty \quad a.s., \quad \mathbb{E}\big[\Delta M_{n+1}\Delta M_{n+1}^t\,\big|\,\mathcal{F}_n\big] \xrightarrow[n\to\infty]{a.s.} \Gamma, \tag{3.17}$$

where Γ is a deterministic symmetric definite positive matrix and for an $\epsilon>0$,

$$\mathbb{E}\big[(n+1)\|r_{n+1}\|^2 \mathbb{1}_{\{\|\theta_n-\theta^*\|\leq\epsilon\}}\big] \xrightarrow[n\to\infty]{} 0. \tag{3.18}$$

Specify the gain parameter sequence as follows: for every $n\geq 1$, $\gamma_n=\frac{1}{n}$. If $\Lambda:=\Re e(\lambda_{\min})>\frac{1}{2}$, where $\lambda_{\min}$ denotes the eigenvalue of $Dh(\theta^*)$ with the lowest real part, then, the above a.s. convergence is ruled on the convergence set $\{\theta_n\to\theta^*\}$ by the following Central Limit Theorem

$$\sqrt{n}\big(\theta_n-\theta^*\big) \xrightarrow[n\to\infty]{\mathcal{L}} \mathcal{N}\left(0, \frac{\Sigma}{2\Lambda-1}\right)$$

with

$$\Sigma := \int_0^{+\infty} \big(e^{-(Dh(\theta^*)-\frac{I_d}{2})u}\big)^t\, \Gamma\, e^{-(Dh(\theta^*)-\frac{I_d}{2})u}\,du.$$

References

1. Andersena J, Fariesa D, Ramuraa R (1994) A randomized play-the-winner design for multi-arm clinical trials. Commun Stat, Theory Methods 23:309–323
2. Athreya KB, Karlin S (1967) Limit theorems for the split times of branching processes. J Math Mech 17:257–277
3. Athreya KB, Karlin S (1968) Embedding of urn schemes into continuous time Markov branching processes and related limit theorems. Ann Math Stat 39:1801–1817
4. Bai Z-D, Hu F (1999) Asymptotic theorems for urn models with nonhomogeneous generating matrices. Stoch Process Appl 80(1):87–101
5. Bai Z-D, Hu F (2005) Asymptotics in randomized urn models. Ann Appl Probab 15(1B):914–940
6. Bai Z-D, Hu F, Shen L (2002) An adaptive design for multi-arm clinical trials. J Multivar Anal 81(1):1–18
7. Benaïm M (1999) Dynamics of stochastic approximation algorithms. In: Séminaire de probabilités, XXXIII. Lecture notes in mathematics, vol 1709. Springer, Berlin, pp 1–68

8. Benveniste A, Métivier M, Priouret P (1990) Adaptive algorithms and stochastic approximations. Applications of mathematics (New York), vol 22. Springer, Berlin. Translated from the French by Stephen S. Wilson
9. Berman A, Plemmons RJ (1994) Nonnegative matrices in the mathematical sciences. Classics in applied mathematics, vol 9. SIAM, Philadelphia. Revised reprint of the 1979 original
10. Duflo M (1997) Random iterative models. Applications of mathematics (New York), vol 34. Springer, Berlin. Translated from the 1990 French original by Stephen S. Wilson and revised by the author
11. Flournoy N, Rosenberger WF (eds) (1995) Adaptive designs. Institute of mathematical statistics lecture notes—monograph series, vol 25. Institute of Mathematical Statistics, Hayward. Papers from the Joint AMS-IMS-SIAM Summer conference held at Mt, Holyoke College, South Hadley, MA, 1992
12. Fort J-C, Pagès G (1996) Convergence of stochastic algorithms: from the Kushner-Clark theorem to the Lyapounov functional method. Adv Appl Probab 28(4):1072–1094
13. Freedman DA (1965) Bernard Friedman's urn. Ann Math Stat 36:956–970
14. Gouet R (1993) Martingale functional central limit theorems for a generalized Pólya urn. Ann Probab 21(3):1624–1639
15. Janson S (2004) Functional limit theorems for multitype branching processes and generalized Pólya urns. Stoch Process Appl 110(2):177–245
16. Johnson NL, Kotz S (1977) Urn models and their application. Wiley, New York
17. Kushner HJ, Yin GG (2003) Stochastic approximation and recursive algorithms and applications 2nd edn. Applications of Mathematics (New York), vol 35. Springer, New York
18. Laruelle S, Pagès G (2013) Randomized urn models revisited using stochastic approximation. Ann Appl Probab 23(4):1409–1436. doi:10.1214/12-AAP875. Permanent link to this document: http://projecteuclid.org/euclid.aoap/1371834033
19. Pouyanne N (2008) An algebraic approach to Pólya processes. Ann Inst Henri Poincaré 44(2):293–323
20. Rosenberger WF (1996) New directions in adaptive designs. Stat Sci 11(2):137–149
21. Smythe RT (1996) Central limit theorems for urn models. Stoch Process Appl 65(1):115–137
22. Wei LJ (1979) The generalized Polya's urn design for sequential medical trials. Ann Stat 7(2):291–296
23. Zelen M (1969) Play the winner rule and the controlled clinical trial. J Am Stat Assoc 64:131–146

Chapter 4
Logistic Modeling of a Religious Sect Cult and Financial Features

Marcel Ausloos

Abstract The financial characteristics of sects are challenging topics. The present paper concerns the Antoinist Cult community (ACC), which has appeared at the end of the 19-th century in Belgium, have had quite an expansion, and is now decaying. The historical perspective is described in an Appendix. Although surely of marginal importance in religious history, the numerical and analytic description of the ACC growth AND decay evolution *per se* should hopefully permit generalizations toward behaviors of other sects, with either longer life time, i.e. so called religions or churches, or to others with shorter life time. Due to the specific aims and rules of the community, in particular the lack of proselytism, and strict acceptance of only anonymous financial gifts, an indirect measure of their member number evolution can only be studied. This is done here first through the time dependence of new temple inaugurations, between 1910 and 1940. Besides, the community yearly financial reports can be analyzed. They are legally known between 1920 and 2000.

Interestingly, several regimes are seen, with different time spans. The agent based model chosen to describe both temple number and finance evolutions is the Verhulst logistic function taking into account the limited resources of the population. Such a function remarkably fits the number of temple evolution, taking into account a no construction time gap, historically explained. The empirical Gompertz law can also be used for fitting this number of temple evolution data, as shown in an Appendix. It is thereby concluded that strong social forces have been acting both in the growth and decay phases.

M. Ausloos (✉)
Beauvallon, r. Belle Jardinière 483, 4031 Liège, Belgium
e-mail: marcel.ausloos@ulg.ac.be

F. Abergel et al. (eds.), *Econophysics of Agent-Based models*,
New Economic Windows, DOI 10.1007/978-3-319-00023-7_4,

4.1 Introduction

In recent years, some description of various communities, in particular by physicists [14, 15, 25, 36, 48], has been of interest quite outside sociology,[1] since the communities are made of *agents*, defined by several *degrees of freedom*, like sex, age, race, citizenship, wealth, intellectual quotient, love for music groups, sport activity, language, religion, etc.

Interest in religious movements has much led scholars to consider those through field observations rather than through surveys. Studies on Bruderhof [56], Jehovah's Witnesses [11], Satanist groups [10], the Unification Church [45], Scientology [54], the Divine Light Mission [21], and other movements [26] have focussed more on the importance of symbolism, ritual, and discourse in the construction of "religious meanings" [12, 49, 55] than on the motivation within social, financial or general sectary aspects.

Of course, Marx and Engels [34] did ask to what extent religion serves as an opiate that stifles social change, but this political view emphasizes exogenous goals and means. Durkheim [22] did explore the "degree" to which a cult, later on a religion, through a church, entices a source of cohesion that stimulates agent collective endogenous actions. One should claim that such a "degree" contains some measure of a (at least, moral) satisfaction. Still, too much consideration on the various goals of religious organizations, i.e. saving the body OR (my emphasis) the soul, might obliterate an objective/quantitative approach. Thus, not disregarding the need for a qualitative understanding of the complex role of religions in social service delivery performance, a more quantitative approach, along agent based modeling considerations, might shine some light on successes and failures of some cult or sect.

In fact, within Comte ideas [17], one has often attempted to describe economic and sociological features within some analytic equations, involving agents, (i) with degrees of freedom, thus interacting with "external fields", like a spin with a magnetic field or a charge with an electric field, *and* (ii) interacting with each other within some cluster. External fields can be mapped into so called *social forces* [38] through a change of vocabulary. For simulating the numerical evolution in size of a religious movement, many available models of opinion dynamics are also available, several taking into account preferential attachment,—seen in [2, 7–9] as one of the fundamental dynamical causes of the evolution of such religious movements.

Beside such considerations, one may wonder about financial and/or economic aspects of religious communities, and how they evolve in some so called market. There is a huge literature, going back to [29], and much intense work on the economics of religious adepts and their hierarchy; both at the micro- and macro-levels. Many interesting considerations exist and are worth to be read, but quantitative modeling

[1] At this level, one should pay some respect to earlier work, i.e. in 1974, Montroll and Badger [39] have introduced, into social phenomena, quantitative approaches, as physicists should do along mechanics lines. Indeed communities can be considered as in a thermodynamic state,—as Boltzmann discovered [13] after Comte [17].

is somewhat absent, as well as some search for empirical laws, and for explaining them, say mathematically.

In view of the above, I was interested in finding whether one could get some economic or financial data on the evolution of a community made of agents having a well defined so called "degree of freedom", like their religious adhesion. For various reasons, this is not so easy to find: such communities do not want to be appreciated as rich, or as poor. Financial data are rarely released. There are known psychological difficulties in merging considerations about money and religion [29]. One may wonder why!

One crucial request, for a study within the so defined framework, is to find some community for which growth and decay have been observed, but have not been influenced by too many, violent or not, competitive aspects with other cults. Therefore, the time life of an interesting community should be rather short, yet long enough to have some meaningful data.

A community like the Antoinists, here below called the Antoinist Cult Community (ACC), exists for about more than a century, has markedly grown, [18, 20] and is now apparently decaying. For the reader information, some comment on the community origin and roles is left for Appendix A. In France, the religious association community is considered as a sect [32] but it is a *Etablissement d'utilité publique* (Organization of Public Utility)[2] in Belgium, since 1922. For short, the ACC will be called a religious sect, nevertheless.

It would have been of interest to have some quantitative measure of the true number of adepts, in order to relate such a data with previous studies, as in [2, 7–9]. However, one remarkable and quite respectable characteristics of the ACC, beside *not* to have any proselytism action, unlike the more financially active religious movements, is not to keep anything, like a financial gift, which would induce some private information. Thus, information on adept number evolution can only come from indirect measures. The presently relevant, thereafter studied, data is limited to: (i) Sect. 4.2.1, the number of temples, built in Belgium, before World War II, and (ii) Sect. 4.2.2, the financial activity of the community, i.e. yearly income and expenses, for over about 80 years. In Sect. 4.2.3, the most commonly accepted kinetic growth law for describing a population evolution, i.e. Verhulst law [50], is introduced as the analytic solution of a possibly relevant agent based model (ABM).

Interestingly, the growth in the number of temples will be found to follow such a Verhulst law [50], in Sect. 4.3.1. In estimating acceleration and deceleration processes in the inauguration of temples, one will observe and quantify some social force effect, in the sense of Montroll [38], in Sect. 4.3.2. Some ABM interpretation is given, in Sect. 4.3.3.

In Sect. 4.4, the evolution of financial data is also studied starting from fits according to the Verhulst logistic function [50]. A succession of three logistic regimes is found, in Sect. 4.4.1, both for income and expenses data.

---

[2]Such a legal association status seems to have been invented for the Antoinist Cult in Belgium, though no formal proof of the latter can be found in notes of the parliament related to the 1921 law elaboration process. Notice that the Minister in charge of the application is the Minister of Justice,—apparently due to suspicion going on with such unfamiliar religious/charity matters.

Since there is some natural interest in considering both growth and decay processes of a community, the decay of income and expenses after 1980 is also examined, in Sect. 4.4.2. A trivially simple behavior is *not* found though. The non symmetric time dependence of growth and decay regimes leads to an open question.

Section 4.5 serves as a conclusion: the complexity of quantitatively studying a religious community through its social history will be emphasized. Indeed several growth regimes are found. On the other hand, the recent decay process is hardly mapped into a simple analytical form. This is in marked contrast with ecological or laboratory based data on population evolutions. Thus, *as should most likely be really expected*, it is concluded that social phenomena are very complex processes offering much challenge for future physics modelling, suggesting challenging investigations through agent based model simulations.

Note that the Verhulst mapping is sometimes criticized as an unrealistic, too simple, model. Therefore, the sometimes considered as the best alternative Gompertz (human mortality) law [27] is adapted to the present context, in Appendix B,—tuning the parameters into their size growth rather than decay value. *Only* the number of temples evolution case is reported within such a Gompertz approach; see Appendix B. Finally, in Appendix C, it is emphasized that social forces can be introduced at least in two different ways in an ABM, based on Vehulst and/or Gompertz analytic evolutions.

As hinted, this paper is based on a compilation of already published papers on econophysics aspects of the Belgium ACC [3, 4, 16], but new figures are here included.

4.2 The Data Set

4.2.1 Number of Temples Data

Since there is no data on adept adhesions, one indirect way of observing the evolution of the size of the ACC has been mapped into the counting of the number of temples as a function of time. Such data is meaningfully available for Belgium till 1935. Note that temples have been mainly constructed in Belgium, though others exist in France and Brazil. Though it might also be of interest to consider the data for the whole sect on a world wide basis, only the 27 temples constructed in Belgium during the main growth phase of the cult are here below considered, for coherence. The number of temples constructed as a function of time has been extracted both from the archives of the ACC and from [18] and [20] compilation and discussions.

Most of the times, the exact day of the *inauguration* or *consecration* is known,—sometimes (twice) only the month is known. To be more precise about the exact day of the event, for the latter cases, would request much time consuming, searching for this information through news media, without being certain of the success. When in doubt, the dates in an Appendix of a 1934 book by Debouxhtay [18] are used, because I consider them as the most reliable ones. In order to count the number of days and months between successive rise of temples, when the day is unknown, the date has been assumed to be the 15th of the month. The number of days between two events has been calculated, taking into account bissextile years if necessary; the number of months has also been calculated but rounded up to the nearest integer according to the sum between the number of days till the end of the first month and from the beginning of the last month corresponding to the two marginal events so considered. The x-axis for the figure discussed will thus be the cumulative number of months since the rise of the first temple in Jemeppe-sur-Meuse, on Aug. 15, 1910; the data extends up to the 27-th temple raised[3] in La Louvière, on Dec. 03, 1933. After a huge consecration gap, lasting more than 20 years, four other temples have been raised in more recent years, but three of these have been closed already. These temples are not included in the analysis.

For giving some perspective, let it be noticed that, in France, the first temple was consecrated in Paris (13ème) on Oct. 26, 1913, while the 15th was in Valenciennes, on Aug. 07, 1932. The most recent one, the 31st and 32nd, were inaugurated in Caen and Toulouse in 1991 and 1993 respectively, but, even after some local research, the days and months are unknown. Thus, the evolution of the number of temples in France is not studied here because of such uncertainties.

4.2.2 *Financial Data*

The financial data set has been extracted from the Belgian daily official journal, i.e. *Moniteur Belge*, when it was available in the archives of the Antoinist Cult Library in Jemeppe-sur-Meuse. A few issues are missing, i.e. *ca.* 1960–1965, without any known reason, but those do not appear *a posteriori*, from the subsequent data analysis, to impair the discussion and conclusion. The examined time range starts in 1920, i.e. since when it was mandatory to report it. However, due to the introduction of the EUR in 2000, in order to keep the Belgian Franc (BEF) as the usual unit, only the data till 2000 is discussed,—again without much apparent loss of content for the present discussion and conclusion. Due to an evolution in the Belgian legal rules for reporting income and expenses data over the last century, many detailed items can be found in the oldest reports. In order to have enough meaningful comparison over several decades, such detailed data on the years, i.e. grossly before 1940, has been concatenated such that only the yearly total *incomes* and total *expenses* are finally used, are displayed in Fig. 4.1, and discussed below. Notice that these so called *income* values do not take into account the left-over from the previous year(s). That is why, in Fig. 4.1, sometimes, it seems that there are more expenses than income in a given year.

Moreover, the financial data,—equivalent to the Belgium community data, here studied, but corresponding to the France Antoinist cult activity, is not available, and thus cannot be studied below.

[3]For 135 880.70 BEF of that year.

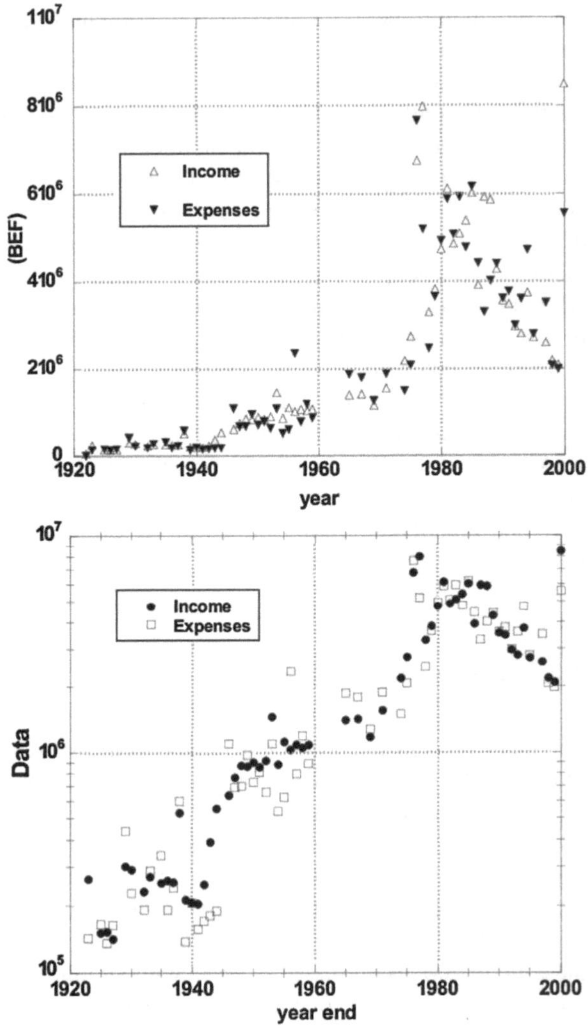

Fig. 4.1 (*top*) Yearly income and expenses of the Belgium Antoinist Cult Community as reported in the *Moniteur Belge*; (*bottom*) logarithmic scale display of yearly income and expenses suggesting three growth regimes before some marked decay

4.2.3 Agent Based Model Numerical Analysis Methodology. The Verhulst Logistic Function

For the modelization of an agent based community growth as a function of time t, let us take the (Verhulst) so called logistic function as the first approximation, i.e., a sigmoid curve,

$$z = z_\infty \frac{e^{r(t-t_m)}}{1+e^{r(t-t_m)}} = \frac{z_\infty}{1+e^{-r(t-t_m)}}, \tag{4.1}$$

where z_∞ is the upper limit of z as time t tends to infinity, t_m is the position of the inflection point, at mid z amplitude, such that $z_m = z_\infty/2$, and r is the supposedly

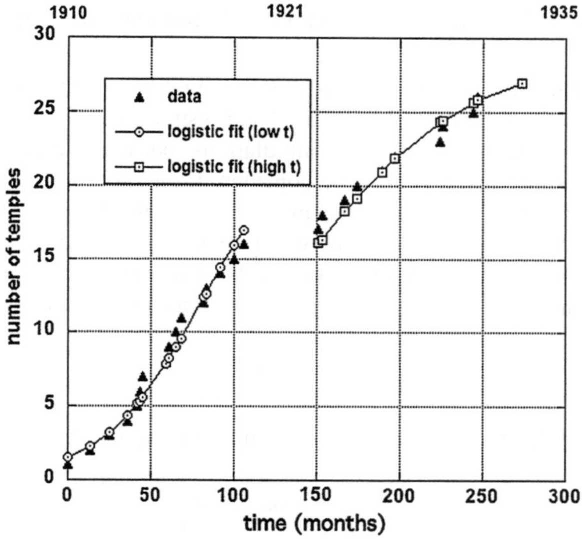

Fig. 4.2 Logistic fits, at low and high t, of the number of temples of the Antoinist Cult Community in Belgium as a function of the number of months, since the consecration of the first temple on Aug. 15, 1910

constant growth rate. This way of expressing the logistic curve has the advantage that the initial measure z_0, at $t = t_0$, is a rapidly fixed value: $z_0 = z_\infty \frac{e^{r(t-t_m)}}{1+e^{r(t_0-t_m)}} = \frac{2.z_m}{1+e^{-r(t_0-t_m)}}$, for estimating one of the three parameters in Eq. (4.1).

Above the inflexion point, one can use an asymptotic expansion i.e.

$$z \simeq z_\infty \left(1 - e^{-r(t-t_m)}\right). \tag{4.2}$$

Such a latter exponential growth behavior, Eq. (4.2), is sometimes referred to as the von Bertalanffy curve [53] curve,—of mass accumulation, in biology. It is also through an appropriate change of variable, nothing else than Malthus exponential growth rate, $y = y_0 e^{b\tau}$, i.e. if $r \equiv -b$, $\tau \equiv t - t_m$ and $z \equiv z_\infty(1 - y/y_0)$.

Note also that

$$\frac{z/z_\infty}{1 - z/z_\infty} = e^{r(t-t_m)}. \tag{4.3}$$

4.3 Number of Temples Evolution

The number of raised temples by the Belgium ACC is displayed in Fig. 4.2, as a function of the evolved month since the rise of the first temple on Aug. 15, 1910. For the sake of completeness, let it be mentioned that Dericquebourg [20] has given a sketch of this number of temple evolution, in Belgium and in France, in an appendix to his book, though without any qualitative nor quantitative discussion.

4.3.1 Numerical Analysis. Verhulst Logistic Law

The best fit behavior to a logistic law has been searched through a log-log plot method, based on Eq. (4.3). The upper value, z_∞, was imposed to be an integer. It has occurred after many simulations, but it could occur readily to many, that *two* distinct regimes must be considered: one at "*low t*", i.e. during the initial growth of the ACC, and another in *high t*, at "later times". The regimes are readily separated by a 4 year gap, between 1919 and 1923, during which there was *no* temple construction. The low t logistic fit of the number (c) of temples as a function of the number (m) of months (cumulated since the rise of the first temple) corresponds to

$$c(m) = \frac{24}{1 + e^{-0.03395*(m-80)}} \tag{4.4}$$

while the fit in the upper m regime corresponds to

$$c(m) = \frac{29}{1 + e^{-0.0195*(m-140)}}. \tag{4.5}$$

Both data and fits, in each regime, are combined and shown in Fig. 4.2.

It is remarkable that the initial growth rate for such data is about 0.03395, i.e. largely more than 3 temples every ten years, and reduces to 0.0195 in the latest years, i.e. about 2 temples per year. Nevertheless, although the initial logistic law should have led to expecting $\sim$ 24 temples at saturation, the latter one would predict 29 temples at most. It is interesting to recall here that Verhulst modification of Malthus (unlimited growth) equation was based on considering a "limiting carrying capacity" of the "country", for the considered population. *Mutatis mutandis*, such 24 and 29 z_∞ values reflect such an effect.

Notice that a *unique* logistic curve fit, over the whole time interval, would give a value of the growth rate $\sim$ 0.02355, but not fulfilling the Jarque-Bera (JB) test [30],—even when finite sample effects are taken into account [28].

One might debate whether the original logistic map is the most appropriate law to be considered. One might suggest a skewed logistic with extra parameters, as considered, e.g. in [41, 42, 57] studying various population growth cases. This has not been considered for this report, because the parameters entering such skewed curves are hardly meaningfully interpreted, in the present investigation context. It seems preferable in view of the data analysis and the framework of this investigation to further discuss the findings, in Sect. 4.3.2, as due to the influence of social forces [38] on agents, i.e. adepts.

Note also that the Gompertz (double exponential) growth law [27] is studied in Appendix B, for the above data.

4.3.2 Quantitative Measure of Social Forces

At first sight, the presently investigated growth regimes do not seem to overlap much. Moreover the rates of growth seem somewhat different in the successive regimes, indicating sequential rather than overlapping (and competitive) processes, in contrast to social and technological cases [24, 33], as well as botanical [1, 43] and other biological,—in which successive competing molecular reactions are of course involved. Therefore, one might rather consider, beside social endogenous contagion, an effect due to, in the words of Montroll, "social forces" [38]. It is worth to recall that Montroll argued that social evolutionary processes occur due to competition between new ideas and old ones. Moreover, deviations from the classical logistic map are often associated with intermittent events. In many cases, a few years after on such event, it can be abstracted as an instantaneous function impulse. Montroll argued that the most simple generalization of Verhulst equation, in such a respect, goes when introducing a force impulse, $F(t) = \alpha\delta(t-\tau)$, in the r.h.s. of the kinetic Verhulst equation so that the dynamical equation for some forced evolution process $X(\equiv z/z_\infty)$ reads

$$\frac{d\ln(X)}{dt} = k(1-X) + \alpha\delta(t-\tau). \tag{4.6}$$

In so doing, in the time regime after the withdrawal of the intermittent force, the evolutionary curve are parallel lines, on a semi-log plot, see Eq. (4.3): the unaccelerated one, above or below the latter depending whether the process is accelerated or deterred at time τ. The impulse parameter α is easily obtained as explained in [38] and for the present case in [4]. One finds $\alpha_1 = 0.103$, at $\tau_1 \sim 1914$, and $\alpha_2 = -0.405$, at $\tau_2 \sim 1922$, from Fig. 4.3. These are very reasonable orders of magnitude. It should have been obvious that the decelerating force should be higher in magnitude that the accelerating one. The fact that the forces are usually not instantaneous ones, and do not suddenly accelerate or decelerate the process, are approximations which are reexamined, in Appendix C,—where some emphasis is further made on different points of view: *mathematical-like*, at first sight, but fundamentally relevant for discussing the causes of evolution of many populations, along ABM ideas.

4.3.3 Agent Based Model Interpretation

The ACC present hierarchy interpretation and mine go along the lines of the historical points reported in the Introduction here above. The first acceleration, ca. 1914 can be historically connected to the first world war. The workers and their families needed some intra-community social support, and interestingly being satisfied by the healing of their soul and sometimes body, gave quite an amount of money to build structures, temples, replacing the mere "lecture rooms" where the adepts first gathered for cult activities.

After the war, income and housing taxes were implemented. However, social organizations attempted to be legally screened from such taxes. In Belgium, since its independence from The Netherlands, in 1830, the catholic priests, and officials of a few other cults nowadays, are paid as government employees, on a specially

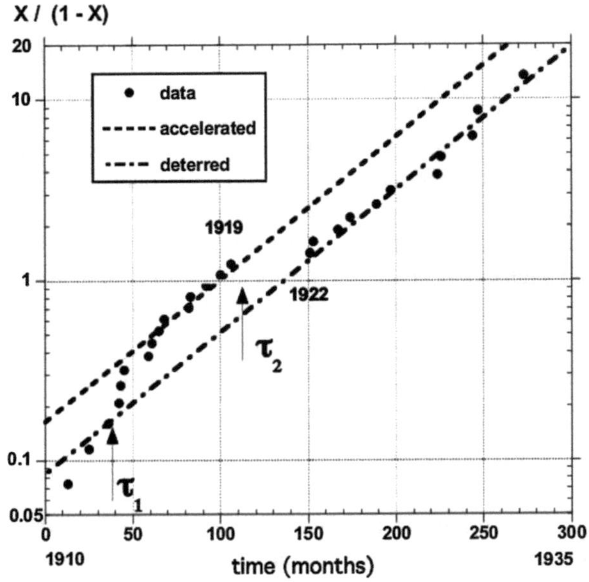

Fig. 4.3 Logistic variation ($X/(1-X)$) of the number of ACC temples (X) in Belgium as a function of the number of months (cumulated from the raise of the first temple, in 1910), indicating a "social force effect" at time τ_1, accelerating the process over a time span, and a decelerating force impulse at time τ_2

adapted scale. However, the introduction, and legal recognition, of a new active socio-religious group into the religious affairs of the country was not well appreciated by the catholic church leaders and adepts. Whence it took a while before the Belgian government, manipulated by the bishops and catholic members of the parliament, accepted to consider a new law establishing a role for social organizations, associated to some religious movement. During more than four years, as convincingly outlined by Dericquebourg [20], the intended law[4] on Organisations of Public Utility suffered many parliamentary delays, starting from 1919, *in fine* much decelerating the temple construction process. One should admit that it was quite natural from a tax point of view to wait for the rising of new temples. When the law was finally voted and applicable, accumulated money could be used for temple construction, over a few months (20 or so), after 1923. The growth could resume, as seen in Fig. 4.2. For completeness, note that the 1929 October crash occurs during the 230-th month; observe the small gap around such a time.

Later on, after 1938, the number of temples did not much increase, indirectly indicating the unnecessary need for such constructions because of the stability or even decay in the number of adepts. The recent closing of three temples seems to confirm such a number evolution.

Following such considerations, the relationship between psychological and sociological needs, at times of great economic difficulty, can be observed in such a cult adepts. It is remarkable that poor social conditions led to high cash gifts,—sufficiently as to build temples during a war time. The somewhat incredible rising level of gifts by adepts for the construction of temples at difficult times is an in-

[4]Nowadays, the law is applied much outside socio-religious groups, e.g. to museums.

teresting observation of the intracommunity *autocatalytic process*.[5] Yet, it could be argued that the ACC hierarchy was intending to build more temples in order to increase the acceptance of adepts, as "clients", as done in more financially prone churches nowadays. However, it should be re-emphasized that proselytism is far from any such goal in the ACC. Let it be repeated that the cult "desservants" are not paid. The (lack of) growth in the temple number is thus entirely due to the intra-community factor state, rather than a leadership manipulation.

4.4 Income and Expenses Evolution

Recall that yearly expenses and income are available over about 80 years during the 20-th century. Note that this so called "income" value does not take into account the left-over from the previous year(s). Further study has led to interesting analytic description and explanations about the evolutions of these financial matters [3].

4.4.1 Numerical Analysis

The raw data, Fig. 4.1 appears as if points are pretty scattered. However, after much fit searching, it appears that in *both* income and expenses cases, three growth and one decay regimes can be found, see Fig. 4.4, approximately over barely overlapping time spans. Because of the y-axis scale, crushing the first two regimes, it is fair to describe the figures in words. The presence of a maximum in 1929 can be observed before a smooth and short decay till the beginning of the first world war in 1940. The next growth regimes ends with a small decay at the end of the *golden sixties*. Finally, the last bump is seen to occur at the beginning of the eighties. A marked decay follows thereafter.

The fit parameter values are given in Table 4.1. In general, it appears that the income growth rate is slightly larger than the expenses growth rate; the more so except in the 3-rd regime, where the *absolute* difference is much larger. However, let it be stressed that the growth rate difference, in *relative* value, is close to 27 %,—in *each* case. Such correlations between expenses and income can be seen in Fig. 4.5. The relationship law is approximately linear, though sometimes, a few large income years can occur.

It can be observed that a one year time delay occurs in such cases, whence leading to conclude about sound financial management by the ACC hierarchy.

---

[5]The velocity of growth, or concentration growth, of an enzyme is depending on the concentration of substrate [35]. The rate of an enzyme-catalyzed reaction is proportional to the concentration of the enzyme-substrate complex.

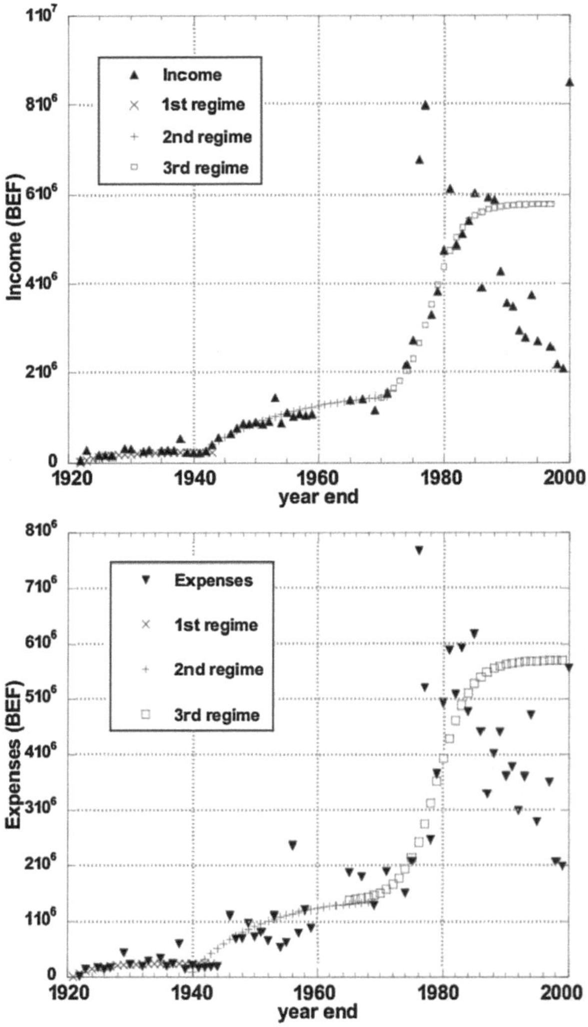

Fig. 4.4 Successive logistic fits of the Belgian Antoinism Cult Community income (*top*) and expenses (*bottom*) on mentioned year, taken from official *Moniteur Belge* journal, indicating three time dependent regimes; parameter values are found in Table 4.1

Table 4.1 Comparison of the growth rate r in yearly income and expenses of the Belgian Antoinist Cult Community in different time regimes; the growth rates correspond to the respective regimes in Fig. 4.4; t_m and z_∞ are fit parameters for the logistic function, Eq. (4.1), see text

Income				Expenses			
Time regimes	z_∞ 10^{-6}	r	t_m	Time regimes	z_∞ 10^{-6}	r	t_m
1922–1940	0.24	0.29	1921	1922–1946	0.22	0.22	1922
1940–1968	1.20	0.10	1941	1946–1968	1.45	0.08	1946
1968–1980	4.50	0.63	1979	1968–1980	4.50	0.50	1978

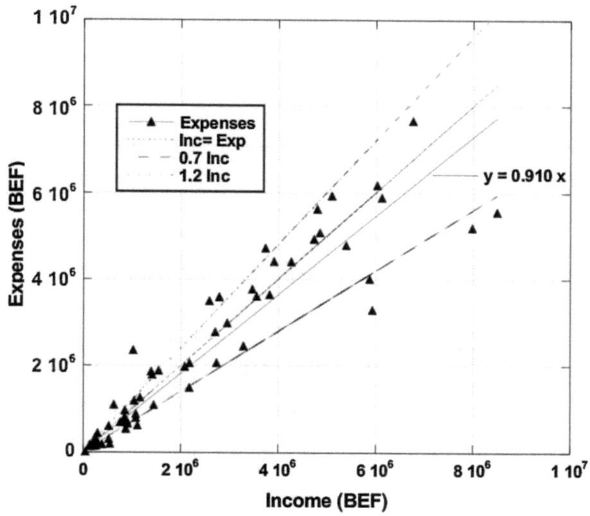

Fig. 4.5 Apparent *quasi*-linear correlation between the Belgium ACC expenses and income, demonstrated by: the best linear fit line going through the origin, the 45° slope line, and two linear envelopes

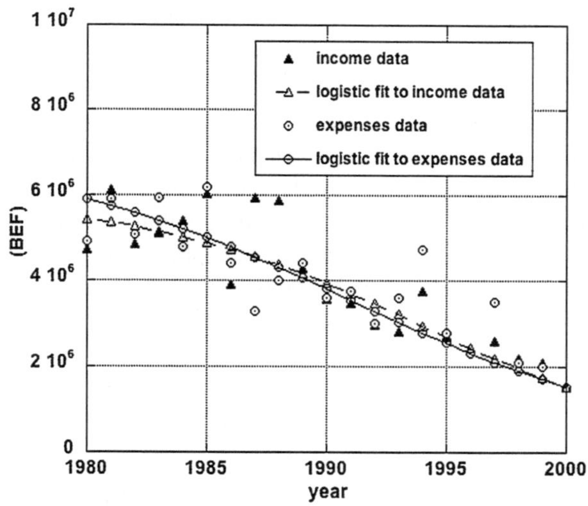

Fig. 4.6 Income and expenses data of the Belgium Antoinist Cult Community, from 1980 till 2000, likely indicating a slow logistic decay after *ca.* 1980

4.4.2 Most Recent Decay Regime

A sharp peak in the income and the expenses data is observed near 1980.[6] The decay has been searched whether it obeys a simple law.[7]

---

[6]One should remember that the available budget for expenses in a given year is the sum of the expected income for the year and the left-over from the previous year.

[7]Due to the legal change in reporting data after 2000, after the introduction of the EUR, only data previous to 2000 is considered.

Table 4.2 Parameter values for decay laws appropriate to the yearly income and expenses of the Belgium ACC

Income				Expenses			
	a	b	R^2		a	b	R^2
linear	6.93	0.0237	0.847	linear	6.87	0.0195	0.729
power	2.17×10^{24}	7.13	0.843	power	1.68×10^{20}	5.88	0.727
exp.	6.945	278.5	0.843	exp.	6.87	337.8	0.727

Three analytic decay laws have been tested to sketch[8] some relevant relation between the natural log (ln) of the income and expenses, after 1975 till (excluding) 2000: (i) a linear, (ii) a power, and (iii) an exponential law. The following forms, with only two parameters, were used

$$\ln(y) = a - b * (year - 1975), \tag{4.7}$$

$$\ln(y) = a * (year)^{-b}, \tag{4.8}$$

$$\ln(y) = a * e^{-(year-1975)/b}. \tag{4.9}$$

The best parameter values, with the global R^2, are given in Table 4.2. It appears that the three laws lead to rather similar variations, with $R^2 \sim 0.84$ or ~ 0.73 for the income and expenses data respectively. Note that a decaying law like $\ln(y) = a * (year - 1975)^{-b}$ does not give anything meaningful.

To remain within a logistic map philosophy, an *ad hoc* fit has been searched for on the income (y_i) and expenses (y_e) data for the years after 1980 till (excluding) 2000,—turning back the time axis in Verhulst equation. The following results are found

$$y_i \simeq \frac{5.9 \times 10^{-6}}{1 + e^{0.175*(year-1994)}}, \tag{4.10}$$

$$y_e \simeq \frac{7.1 \times 10^{-6}}{1 + e^{0.145*(year-1991)}}. \tag{4.11}$$

Yet such expressions and parameter values should be considered as merely indicative ones. They should be taken with some caution, since the position of the maximum, thus the first decay year of this decay regime, is rather ill defined in the data, because of wide fluctuations near this maximum.

Nevertheless, a definitive conclusion can be reached, after comparing such values with the r value found, see Table 4.1, for the 1968–1980 regime: the growth rate (~ 0.6) is approximately 4 times larger than the decay rate (~ 0.15). Note that this

[8] Extreme value data points have been excluded from the analysis indeed for keeping R^2 meaningful; compare Figs. 4.1 and 4.6.

asymmetry between growth and decay is similarly found in business cycles, see e.g. [6]. Thus one may suggest here to wonder whether some ABM simulation leading to such asymmetric features can be imagined. An open question!

One might also wonder whether some revival of the community, as in the phoenix effects discovered in [51] can appear in the future due to the present political and sociological constraints, and the new forces similar to those known at the beginning of the 20-th century.

4.5 Conclusions

The complexity of qualitatively studying agent based groups, like religious communities, through their social and historical aspects is known, but the quantitative, in particular financial, aspects are also challenging. Data is usually sparse and not necessarily reliable [16, 31, 37],—except when legal constraints are imposed. It was of interest to find a case with a growth regime *and* a decay regime, in order to have some insight on the causes of such behaviors for further deep modeling. Facing such a challenge, one goal has been to find a society evolving on a rather short time span, such that the data be reliable as much as possible.

It was of common knowledge that a religious community, the Antoinists, here called ACC, having appeared in the 19-th century, in Belgium, but not so flourishing nowadays, could present a bump feature in the number of adepts. A difficulty stems in the voluntary lack of such an information in the cult system. However, the number of temples, since the rising of the first one in 1910 are reliable data, and publication of financial data have been made mandatory after 1919. Their study has led to interesting features.

Since the logistic growth function has proven useful in modeling a wide variety of phenomena of growing systems, it has been used as the analytic equivalent of an ABM. However, complex social systems rarely follow a single S-shaped trajectory. They often present a simple extremum or (irregular) oscillations, whence implying the need to go beyond simple Verhulst-like models [3, 5]. Here, it has readily appeared that two regimes must be investigated for the temple number evolution: (I) one between 1010 and 1919, for 16 temples; (II) another between 1923 and 1935, for 11 temples. Three (asymmetric) regimes are found for the financial evolution.

The "model" indicates that such communities are markedly influence by external considerations ("external fields"), beside their intrinsic "religious" goals. Practically, in the present case, as illustrated, the crash of 1929 induces a drop in income, but the second world war increases the community strength. The golden sixties "reduce" the income: the adepts wealth is being increase, but they reduce their offering, becoming in some sense more egoistical. Therefore, one can deduce that there are two different causes for the drop in income: either a lack of money of the adepts, or in contrast, paradoxically, "too much" wealth. Similarly, the increase in cult income, at its legal beginning, may result from the thanking for healing the suffering,—but also occur due to the income explosion till 1985. The variation in expenses are immediately related with such income considerations. Therefore such an ABM, apparently leading to a "universal"-like interpretation, contains ingredients, with non-universal "amplitudes", but is expected to be applicable to other societies,

From a practical point of view, it has to be emphasized that the ACC was appealing because of the suffering of people, working under very hard conditions in the Liège, BE, area. The catholic social system was lacking convincing impact. Local people were searching within proto-science appeal for mind and soul healing through connections with spiritualistic phenomena. Thus, within a pure altruism, Père Antoine started to preach and to give psychological remedies, i.e., "first principles", for accepting one's life, sometimes "demonstrating" his "body healing powers". The initial seed of the ACC grew within a rather weak competitive framework, due to Père Antoine's charisma, simplicity, and affection. A follow-up by local people resulted, not far from recalling what happened a long time ago, in what are now more established religions. However, his charismatic leadership was most likely lost, at Père Antoine's death. Moreover, there was not much serious attack, nor martyrs appearing, which are two aspects for an increase in community strength and expansion [44]. No cult of saints was established, on one hand, and on the other hand, social and economic conditions largely improved, reducing the need for intra-community self-support. Therefore some size leveling off had to be expected; as well seen in the evolution of the number of temples.

A marked decay, in adepts, is known to occur since the end of the 20-th century, inciting one to conclude to a doomed situation, according to the theory of Stark [46]—in contrast to, e;g., the Jehovah's Witnesses [47]. Indeed, ideologies (whether religious or secular) seem to lack coherence and potency unless they are developed and promulgated by vigorous formal organizations and social movements. One crucial aspect of the ACC concerns its survival under much improved economic, social, and health conditions of workers to whom the Père Antoine "philosophy" appealed.

Finally, one might nevertheless wonder about the future of the community, i.e. whether some revival of the community, through phoenix-like effects [51] can appear in the future within present political, economic, and sociological conditions, upon the constraints of new forces on workers having hard job conditions, though not immediately similar to those known at the beginning of the 20-th century.

Thus, *as should most likely be really expected in fact*, it is concluded that social phenomena are very complex processes offering much challenge for quantitative mathematical modeling. Nevertheless the above findings lead to simple empirical laws. Even though the ACC is surely a marginal religious case, the merging of qualitative and quantitative considerations as done here above might appeal to finding other cases of interest among sects having reliable data and do suggest generalizations.

Acknowledgements Comments on preliminary versions by J. Hayward, A. Pękalski, and F. Schweitzer have surely improved the present paper. Infinite thanks go to Frère W. Dessers and Soeur S. Taxquet, presently President and Secretary of the Administration Board of the Antoinist Cult in Belgium, respectively, for their kindness, patience, availability, when I asked for data and historical points, and for trusting me,—when allowing me to remove archives from the library for scanning outside the office.

Appendix A: Historical Perspective

Louis Antoine was born in 1846 near Liège, Belgium, the youngest son of a 8 children family, baptised as catholics; his father was a coal miner. He followed the usual catholic education of the time, read much, became a coal miner and later on a metallurgist worker. After killing inadvertently a friendly soldier during his military service, he was punished, but used that accident as a basis for thinking about "good and bad"; he volunteered to work abroad for the Cockerill factory; got married, and had a son. At 42, he got more sedentary, continued to think much about religion, mankind role and life values, He read Kardec's Livre des Esprits (1857) and was enthused. He wishes to become medium becomes the founder of a spiritism group "Les Vignerons du Seigneur", but his son dies at 20 in 1893. He begins to discover that he can be "healing", writes a small book about healing, gathers friends and starts becoming a cult leader, and is known as a healer. But he never asks for any money for his miracles. Nevertheless he gets into trouble with justice, and medical doctors. Somewhat to avoid such problems, he reminds the patients that faith is the key; he transfers his good health into others by faith. Still he gets problems with the law: sometimes, 1200 people gather per day at his home for some so called "operation". Later on, considering that to heal a body is not enough, he turns toward more moral value rebirth. His predicator role increases, in the Jemeppe "temple", and in 1906 starts publishing notes taken by some scribe, outlining his doctrine. The concept of disease is denied, just as is that of death, and there is belief in the reincarnation: it is intelligence which creates suffering; only "faith in" removes it, and not the intervention of health professionals. He becomes the "Father", (le Père) and establishes some clothing rules (a long black dress for "desservants"). At Easter 1910, he is considered as a "prophet", and on Aug. 15, 1910 sanctifying the first temple. He will expand and clarify the doctrine (*Evil does not exist*), will raise more temples till his death in 1912. His wife, Catherine, who could not read, maintained and pursued the cult activities amongst schisms and heresies, till her death in Nov. 1940.

To put several references in evidence, let a few be quoted: [18–20, 52] where much can be found on the doctrine, rules (dressing), symbols (the tree), prophecies, testimonies, social roles and values, spiritual and philosophical (reincarnation) contexts, also with some historical and sociological perspectives; see also www.culteantoiniste.com.

In essence, let it be emphasized that there is no search for increasing the financial wealth of the hierarchy. There is some 8-year rotation in duties. It is known that a check, used as an offering, is not cashed in a bank because it has a personal connotation. Only anonymous offerings are accepted; there is no proselytism and one does not ask for money from followers. The "desservants" are not paid. There is neither exclusivity on religious adherence nor does one provide any prescription on social and political issues. The main goal is to worship and to heal, in line somewhat with Mary Baker Eddy's ideas [23].

In my opinion, the Antoinism, or the Antoinist Cult, cannot be considered as a sect,—though it was so in France ; see Journal Officiel, Commission d'enquêtes parlementaires sur les sectes en France, Rapport 2468, Dec. 1995 [32]. Is it a religion,

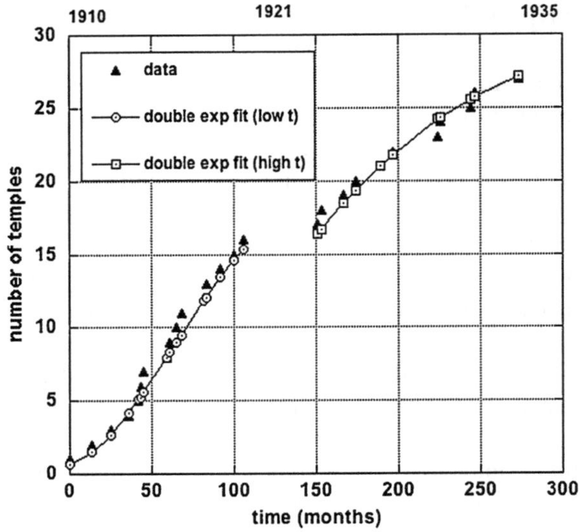

Fig. 4.7 Gompertz double exponential law fit of the number of consecrated ACC temples in Belgium as a function of the number of months (cumulated), in low and high t regimes since the consecration of the first temple in 1913

a church? Maybe, but there is no strong structural hierarchy and clerical body. It is surely, from a catholic Christian point of view, a heresy. Is it a cult? in the sense of Nelson [40], most likely yes; it should therefore disappear if there is no one to pick up the ideas and turned them toward some money getting religious scam.

Appendix B: Gompertz Double Exponential Law

Similarly to the analysis reported in the main text, a Gompertz double exponential law fit can be searched for the number of temples as a function of the number of months (cumulated again).

The best fit behavior to such a Gompertz double exponential law has been searched through a log-log plot method, imposing the amplitude to be an integer. It has occurred after many simulations that two distinct regimes must be considered, exactly as in the analysis along the Verhulst approach: one at *low* time, i.e. during the initial growth of the ACC, and another at later (*high*) time, with a 4 year gap, between 1919 and 1923. One obtains respectively :

$$c(m) = 23e^{-e^{-(m-62)/48.5}}, \tag{4.12}$$

$$c(m) = 31e^{-e^{-(m-116)/77.5}} \tag{4.13}$$

as shown on Fig. 4.7. Note that the upper (*absolute*) values of the possible number of temples to be expected slightly differ in the Verhulst and Gompertz approaches,— though in an opposite *relative* value for the low and high t regimes. The rates found in Eqs. (4)–(5) and here above, in Eqs. (12)–(13), are comparable: 0.034 $\sim$

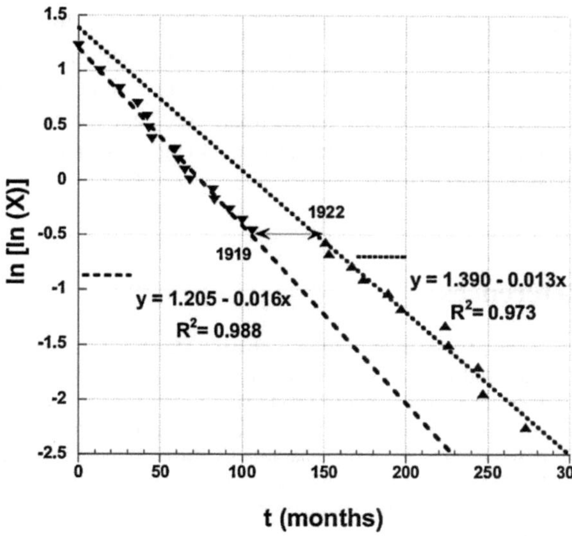

Fig. 4.8 Gompertz plot: Log(Log) variation of the (relative to some maximum possible) number of ACC temples (X) raised in Belgium as a function of the number of months, starting from the raise of the first temple indicating a "social force effect" influencing a variation in growth rates

(1/48.5 $\simeq$) 0.021, and 0.0195 $\sim$ (1/77.5 $\simeq$) 0.0129 respectively, depending on the regime.

Appendix C: On Social Forces

It is hereby emphasized that social forces can be introduced at least in two different ways in an ABM, based on Vehulst and/or Gompertz analytic evolutions. Indeed, a different adaptation of the ideas in [38] on the evolution of competing entities, economic or sociologic ones, occurs if, instead of Eq. (4.6), one writes

$$\frac{d\ln(X)}{dt} = k(1-X) + (\alpha/\theta)(1-X).\big[H(t-\tau) - H\big(t-(\tau+\theta)\big)\big] \qquad (4.14)$$

where $H(t)$ is the Heaviside function. In other words, one is (mathematically) letting Montroll's α to be X-dependent over the time interval $[\tau; \tau+\theta]$. However, the emphasis differs much from [38]: i.e., rather than modifying the (Malhtus) X term, one adapts the (Verhulst) $(1-X)$ term, to (economic or social) constraints.

Mathematically, on the same footing as Eq. (4.14), one can write

$$\frac{dX}{dt} = kX(1-X) + (\alpha/\theta)X(1-X).\big[H(t-\tau) - H\big(t-(\tau+\theta)\big)\big]. \qquad (4.15)$$

Readily, the rate before the pulse is k but is $k + \alpha/\theta$ after the "pulse" application. This "second rate" depends on the pulse strength and some time duration θ.

Within the Gompertz framework, starting from the differential equation

$$\frac{dy}{dt} = ry\log\left[\frac{k}{y}\right], \qquad (4.16)$$

it is "sufficient" to replace $(1 - X)$ by $\sim -\ln(k/y)$; compare Eqs. (4.6) and Eq. (4.16). A double log (X) plot as a function of time is shown in Fig. 4.8, i.e. an appropriate replot of Fig. 4.3,—the best fit equations being written in the figure. The fit is very precise. It is emphasized that the fit lines are *not* parallel anymore. From these, one can deduce $\theta_1 = 34$ (months) and $\theta_2 = 135$ (months), i.e. ~ 3 and 11 years respectively.

References

1. Aachi Mezghani M, Sahli A, Labidi F, Meddeb K, Jebari A, El-hadj B (2008) Analysis of primary and secondary growth and modelling growth dynamics of olive shoots (Olea europaea l). J Hortic Sci Biotechnol 83:411–418
2. Ausloos M (2010) On religion and language evolutions seen through mathematical and agent based models. In: Rangacharyulu C, Haven E (eds) Proceedings of the first interdisciplinary CHESS interactions conference. World Scientific, Singapore, pp 157–182
3. Ausloos M (2012) Econophysics of a religious cult: the Antoinists in Belgium [1920–2000]. Physica A 391:3190–3197
4. Ausloos M (2012) Another analytic view about quantifying social forces. Adv Complex Syst 15:1250088
5. Ausloos M (2013) Gompertz and Verhulst frameworks for growth AND decay description. arXiv:1109.1269
6. Ausloos M, Miskiewicz J, Sanglier M (2004) The durations of recession and prosperity: does their distribution follow a power or an exponential law? Physica A 339:548–558
7. Ausloos M, Petroni P (2007) Statistical dynamics of religions and adherents. Europhys Lett 77:38002
8. Ausloos M, Petroni P (2009) Statistical dynamics of religions. Physica A 388:4438–4444
9. Ausloos M, Petroni P (2010) On world religion adherence distribution evolution. In: Takayasu M, Watanabe T, Takayasu H (eds) Econophysics approaches to large-scale business data and financial crisis. Springer, Berlin, pp 289–312
10. Bainbridge WS (1978) Satan's power: a deviant psychotherapy cult. University of California Press, Berkeley
11. Beckford JA (1975) The trumpet of prophecy: a sociological study of Jehovah's witnesses. Blackwell, Oxford
12. Bird F (1979) The pursuit of innocence: new religious movements and moral accountability. Sociol Anal 40:335–346
13. Blackmore JT (1995) Conclusions. In: Ludwig Boltzmann: his later life and philosophy 1900–1906. Springer, Berlin, pp 206–207
14. Carbone A, Kaniadakis G, Scarfone AM (2007) Tails and ties: topical issue on physics in society. Eur Phys J B 57:121–125
15. Chakrabarti BK, Chakraborti A, Chatterjee A (eds) (2006) Econophysics and sociophysics: trends and perspectives. Wiley-VCH, Weinheim
16. Clippe P, Ausloos M (2012) Benford's law and Theil transform of financial data. Physica A 391:6556–6567
17. Congrève R (1891) The catechism of positive religion. Kegan Paul Trench Trubner and Co, London; reissued by Cambridge University Press, 2009
18. Debouxhtay P (1934) Antoine le guérisseur et l'Antoinisme. Gothier, Liège
19. Deregnaucourt F, Desart M (1909) Père Antoine révélation par Le père Antoine. Suivi de "Le couronnement de l'Oeuvre révélée". Presses du Culte Antoiniste, Jemeppe-sur-Meuse
20. Dericquebourg R (1993) Les Antoinistes. Brepols, Maredsous
21. Downton J Jr (1979) Sacred journeys: the conversion of young Americans to divine light mission. Columbia University Press, New York

22. Durkheim E (1912) Les formes élémentaires de la vie religieuse. Les Presses universitaires de France, Paris. (tr.) (1968) The elementary forms of religious life
23. Eddy MB (1875) Science and health with key to the scriptures. Christian Science, Boston
24. Fokas N (2007) Growth functions, social diffusion, and social change. Rev Sociol 13:5–30
25. Galam S (2008) Sociophysics: a review of Galam models. Int J Mod Phys C 19:409–440
26. Glock CY, Bellah RN (eds) (1976) The new religious consciousness. University of California Press, Berkeley
27. Gompertz R (1825) On the nature of the function expressive of the law of human mortality, and on a new mode of determining the value of life contingencies. Philos Trans R Soc Lond 115:513–583
28. Herzel H, Schmitt AO, Ebeling W (1994) Finite sample effects in sequence analysis. Chaos Solitons Fractals 4:97–113
29. Iannaccone L (1998) Introduction to economics of religion. J Econ Lit 6:1465–1496
30. Jarque CM, Bera AK (1987) A test for normality of observations and regression residuals. Int Stat Rev 55:163–172
31. Johnson TM, Barrett DB (2004) Quantifying alternate futures of religion and religions. Futures 36:947–960
32. Journal Officiel (1995) Commissions d'enquêtes parlementaires sur les sectes en France. Rapport 2468
33. Marchetti C (1997) Millenarian cycles in the dynamics of the Catholic church. Technol Forecast Soc Change 55:281–299
34. Marx K, Engels F ([1844] 1978) Contribution to the critique of Hegels philosophy of right. In: Tucker RC (ed) The Marx-Engels reader. Norton, New York, pp 53–65
35. Michaelis L, Menten ML (1913) Die Kinetik der Invertinwirkung. Biochem Z 49:333–369. (tr.) Johnson KA, Goody RS (2011) Biochemistry 50:8264–8269
36. Mimkes J (2006) A thermodynamic formulation of social science in econophysics and sociophysics. In: Chakrabarti BK, Chakraborti A, Chatterjee A (eds) Econophysics and sociophysics: trends and perspectives. Wiley-VCH, Weinheim, pp 279–310
37. Mir TA (2012) The law of the leading digits and the world religions. Physica A 391:792–798
38. Montroll EW (1978) Social dynamics and the quantifying of social forces. Proc Natl Acad Sci USA 75:4633–4637
39. Montroll EW, Badger WW (1974) Introduction to quantitative aspects of social phenomena. Gordon and Breach, New York
40. Nelson GK (1968) The concept of cult. Sociol Rev 16:351–362
41. Pearl R, Winsor CP, Barclay WF (1928) The form of the growth curve of the canteloup (Cucumis melo) under field conditions. Proc Natl Acad Sci USA 14:895–901
42. Pearl R, Winsor AA, Miner JR (1928) The growth of seedlings of the Canteloup Cucumis melo in the absence of exogenous food and light. Proc Natl Acad Sci USA 14:1–4
43. Reed HS (1920) The dynamics of a fluctuating growth rate. Proc Natl Acad Sci USA 6:397–410
44. Roach AP (2005) The devil's world: heresy and society 1100–1300. Pearson, Harlow
45. Shupe AD Jr, Bromley DG (1979) The Moonies and the anti-cultists: movement and countermovement in conflict. Sociol Anal 40:325–334
46. Stark R (1996) Why religious movements succeed or fail: a revised general model. J Contemp Relig 11:133–146
47. Stark R, Iannaccone LR (1997) Why the Jehovah's witnesses grow so rapidly: a theoretical application. J Contemp Relig 12:133–157
48. Stauffer D (2004) Introduction to statistical physics outside physics. Physica A 336:1–5
49. Tipton SM (1982) Getting saved from the sixties: moral meaning in conversion and cultural change. University of California Press, Berkeley
50. Verhulst PF (1845) Recherches mathématiques sur la loi d'accroissement de la population. Nouv Mém Acad R Sci B-Lett Brux 18:1–38
51. Vitanov NK, Dimitrova ZI, Ausloos M (2010) Verhulst-Lotka-Volterra (VLV) model of ideological struggle. Physica A 389:4970–4980

52. Vivier R (1936) Délivrez-nous du mal. Grasset, Paris
53. von Bertalanffy L (1938) A quantitative theory of organic growth. Inquiries on growth laws II. Hum Biol 10:181–213
54. Wallis R (1977) The road to total freedom: a sociological analysis of scientology. Columbia University Press, New York
55. Westley F (1978) The cult of man: Durkheim's predictions and new religious movements. Sociol Anal 39:135–145
56. Zablocki B (1971) The joyful community. Penguin, Baltimore
57. Zwanzig R (1973) Generalized Verhulst laws for population growth; hereditary effect; damped exponential decay kernel. Proc Natl Acad Sci USA 70:3048–3051

Chapter 5
Characterizing Financial Crisis by Means of the Three States Random Field Ising Model

Mitsuaki Murota and Jun-ichi Inoue

Abstract We propose a formula of time-series prediction by means of three states random field Ising model (RFIM). At the economic crisis due to disasters or international disputes, the stock price suddenly drops. The macroscopic phenomena should be explained from the corresponding microscopic view point because there are existing a huge number of active traders behind the crushes. Hence, here we attempt to model the artificial financial market in which each trader i can choose his/her decision among 'buying', 'selling' or 'staying (taking a wait-and-see attitude)', each of which corresponds to a realization of the three state Ising spin, namely, $S_i = +1, -1$ and $S_i = 0$, respectively. The decision making of traders is given by the Gibbs-Boltzmann distribution with the energy function. The energy function contains three distinct terms, namely, the ferromagnetic two-body interaction term (endogenous information), random field term as external information (exogenous news), and chemical potential term which controls the number of traders who are watching the market calmly at the instance. We specify the details of the model system from the past financial market data to determine the conjugate hyper-parameters and draw each parameter flow as a function of time-step. Especially we will examine to what extent one can characterize the crisis by means of a brand-new order parameter—'turnover'—which is defined as the number of active traders who post their decisions $S_i = \pm 1$, instead of $S_i = 0$.

5.1 Introduction

Individual human behaviour including human mental state is an attractive topic for both scientists and engineers. However, it is still extremely difficult for us to tackle the problem by making use of scientifically reliable investigation. This is because

M. Murota · J. Inoue (✉)
Graduate School of Information Science and Technology, Hokkaido University, N14-W-9, Kita-ku, Sapporo 060-0814, Japan
e-mail: jinoue@cb4.so-net.ne.jp

M. Murota
e-mail: murota@complex.ist.hokudai.ac.jp

F. Abergel et al. (eds.), *Econophysics of Agent-Based models*,
New Economic Windows, DOI 10.1007/978-3-319-00023-7_5,

there exists quite large person-to-person fluctuation in the observation of individual behaviour.

On the other hand, in our human 'collective' behaviour instead of individual, we sometimes observe several universal facts which seem to be suitable materials for computer scientist to figure out the phenomena through sophisticated approaches such as agent-based simulations. In fact, collective behaviour of interacting agents such as flying birds, moving insects or swimming fishes shows highly non-trivial properties. As well-known especially in the research field of engineering, as a simplest and effective algorithm in computer simulations for flocks of intelligent agents, say, animals such as starlings, the so-called BOIDS founded by Reynolds [1] has been widely used not only in the field of computer graphics but also in various other research fields including ethology, physics, control theory, economics, and so on [2]. The BOIDS simulates the collective behaviour of animal flocks by taking into account only a few simple rules for each interacting 'intelligent' agent.

In the literature of behavioral economics [3], a concept of the so-called information cascade is well-known as a result of such human collective behaviour. This concept means that at the financial crisis, traders tend to behave according to the 'mood' (atmosphere) in society (financial market) and they incline to take rather 'irrational' strategies in some sense.

Apparently, one of the key measurements to understand the information cascade is 'correlation' between ingredients in the societies (systems). For instance, in particular for financial markets, cross-correlations between stocks, traders are quite important to figure out the human collective phenomena. As the correlation could be found in various scale-lengths, from macroscopic stock price level to microscopic trader's level, the information cascade also might be observed 'hierarchically' in such various scales from prices of several stocks to ways (strategies) of trader's decision making.

Turning now to the situation of Japan, after the earthquake on 11th March 2011, Japanese NIKKEI stock market quickly responded to the crisis and quite a lot of traders sold their stocks of companies whose branches or plants are located in that disaster stricken area. As the result, the Nikkei stock average suddenly drops after the crisis [4, 5].

It might be quite important for us to make an attempt to bring out more 'microscopic' useful information, which is never obtained from the averaged macroscopic quantities such as stock average, about the market. As a candidate of such 'microscopic information', we can use the (linear) correlation coefficient based on the two-body interactions between stocks [6, 7]. To make out the mechanism of financial crisis, it might be helpful for us to visualize such correlations in stocks and compare the dynamical behaviour of the correlation before and after crisis.

In order to show and explain the cascade, we visualized the correlation of each stock in two-dimension [4, 5]. We specified each location of N stocks from a given set of the $N(N-1)/2$ distances by making use of the so-called multi-dimensional scaling (MDS) [8] (see Fig. 5.1).

On the other hand, the macroscopic phenomena should be explained from the corresponding microscopic view point because there are existing a huge number of

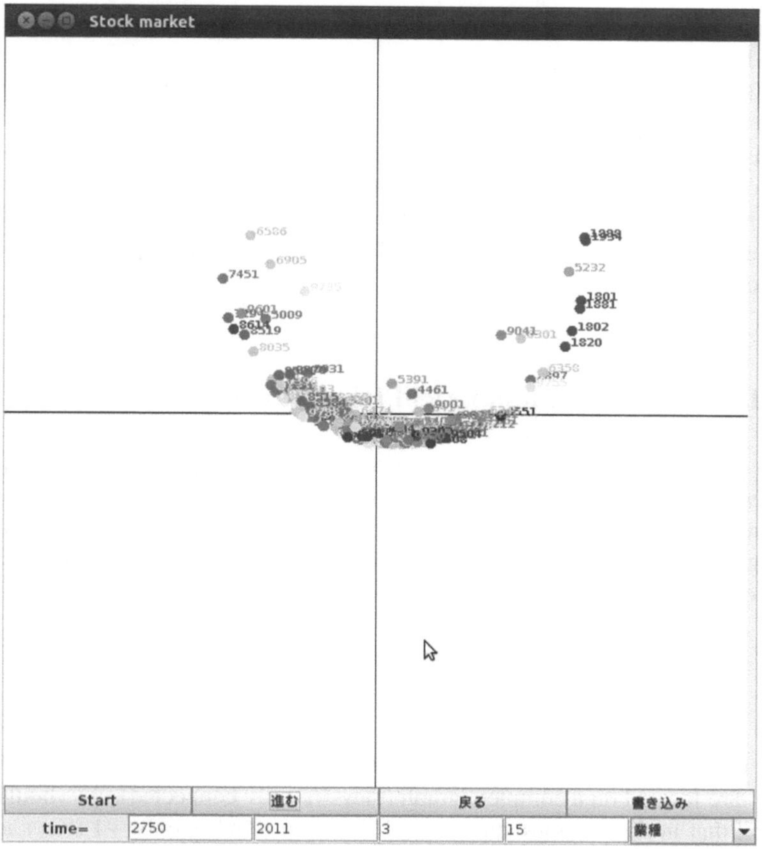

Fig. 5.1 Two-dimensional plot by means of the MDS. We picked up 200-stocks including the so-called TOPIX Core30 and the Nikkei stock average as empirical data set (the data set was taken from Yahoo! finance [9]). The figure is just after the earthquake (15th March 2011) (see [4, 5] for the details). The curious shape of cluster appears after the crisis

active traders behind the crushes. In Ref. [5], we proposed a theoretical framework to predict several time-series simultaneously by using cross-correlations in financial markets. The justification of this assumption was numerically checked for the empirical Japanese stock data, for instance, those around 11 March 2011, and for foreign currency exchange rates around Greek crisis in spring 2010.

However, in the previous study [5], inspired by the study of Kaizouji [10], we utilized Ising model and assumed that each trader does not stay at all for trading. Apparently, it is not realistic situation for trader's decision making. Hence, here we attempt to model the artificial financial market in which each trader i can choose his/her decision among 'buying', 'selling' or 'staying (taking a wait-and-see attitude)', each of which corresponds to a realization of the three states Ising spin, namely, $S_i = +1, -1$ and $S_i = 0$, respectively. Especially we will examine to what extent one can characterize the crisis by means of an order parameter—'turnover'—

which is defined as the number of 'active traders' who post their decision $S_i = \pm 1$, instead of $S_i = 0$.

This paper is organized as follows. In Sect. 5.2, we introduce the three states RFIM and explain the thermodynamic properties including the critical phenomena such as phase transitions. In Sect. 5.3, we construct a prediction formula based on the model introduced in Sect. 5.2. We introduce 'turnover' as an order parameter to characterize the crisis. In Sect. 5.4, we carry out computer simulations with the assistance of empirical data set to check the usefulness of our approach. Section 5.5 is concluding remark.

5.2 There States Random Field Ising Model

In this paper, we extend the prediction model based on Ising model given by [5, 10] by means of three states random field Ising model. Before we construct the prediction model for financial time-series, we consider the thermodynamics of the following Hamiltonian (energy function) that describes decision makings of N traders (each of the traders is specified by a label $i = 1, \ldots, N$).

$$\mathcal{H}(S) = -\frac{J}{N}\sum_{i,j=1}^{N} S_i S_j - h\sum_{i=1}^{N}\sigma(t)S_i - \mu\sum_{i=1}^{N}|S_i| \tag{5.1}$$

where each spin S_i $(i = 1, \ldots, N)$ can take ± 1 and 0, and here we assume that all traders are located on a complete graph (they are fully connected). We should keep in mind that in the previous studies [5, 10], a spin S_i takes only $+1$ (buy) and -1 (sell). However, in our model system, besides ± 1, S_i can take 0 which means that the trader i takes a wait-and-see attitude (stays). Namely,

$$S_i = \begin{cases} +1 & \text{(buying)}, \\ 0 & \text{(staying)}, \\ -1 & \text{(selling)}. \end{cases} \tag{5.2}$$

The first term in the right hand side of (5.1) causes the collective behavior of the traders because the Hamiltonian (5.1) decreases when all traders tend to take the same decision. In this sense, the first term is regarded as endogenous information for the traders. On the other hand, the second term denotes the exogenous information which is a kind of market information available for all traders. Here one can choose the following market trend during the past τ-steps as $\sigma(t)$.

$$\sigma(t) = \frac{q(t) - q(t-\tau)}{\tau} \tag{5.3}$$

where $q(t)$ denotes a real price at time t. The third term appearing in the right hand side of equation (5.1) controls the number of traders who are staying at the moment t. From the view point of spin systems, $\sigma(t)$ is regarded as a 'random field' on each spin because the $\sigma(t)$ might obey a stochastic process. Therefore, the spin system described by (5.1) should be refereed to as *random field Ising model (RFIM)*. Obviously, the parameter μ is regarded as 'chemical potential' in the literature of physics. For $\mu \gg 1$, most of the traders take 'buying' or 'selling' instead of 'staying' from the view point of minimization of the Hamiltonian (5.1). In the limit of $\mu \to \infty$, the fraction of traders who take $S_i = 0$ vanishes, namely, the system is identical to the conventional Ising model [10] in this limit. As we will see later, the set of parameters (what we call 'hyper-parameters') (J, h, μ) should be estimated (learned) from the past time-series.

In this paper, we shall focus on the modification by means of the above three states RFIM. We investigate to what extent the prediction performance is improved. Moreover, we attempt to quantify the number of traders who are staying at the crushes in order to characterizing the financial crisis.

5.2.1 Equations of State

To make a link between the prediction model and statistical physics of the three states RFIM, we should investigate the equilibrium state described by the Hamiltonian (5.1) at unit temperature. According to statistical mechanics, each microscopic state $\boldsymbol{S} = (S_1, \ldots, S_N)$ of the Hamiltonian (5.1) obeys the distribution $\exp[-\mathcal{H}(\boldsymbol{S})]/Z$, where the normalization constant $Z = \sum_{\boldsymbol{S}=0,\pm1} \exp[-\mathcal{H}(\boldsymbol{S})]$ is refereed to as *partition function* and it is given by

$$Z = \sum_{\boldsymbol{S}=0,\pm1} \exp\left[\left(\sqrt{\frac{J}{2N}}\sum_{i=1}^{N} S_i\right)^2 + h\sum_{i=1}^{N} \sigma(t) S_i + \mu \sum_{i=1}^{N} |S_i|\right] \tag{5.4}$$

where we defined $\sum_{\boldsymbol{S}=0,\pm1}(\cdots) = \sum_{S_1=0,\pm1} \cdots \sum_{S_N=0,\pm1}(\cdots)$. Here we should keep in mind that arbitrary two traders are connected each other. By using a trivial equality concerning the Gaussian integral

$$e^{\alpha^2} = \int_{-\infty}^{\infty} \frac{dx}{\sqrt{2\pi}} \exp\left(-\frac{x^2}{2} + \sqrt{2}\alpha x\right), \tag{5.5}$$

the system is reduced to a single spin 'S' problem in the limit of $N \to \infty$ as

$$\begin{aligned} Z &= \int_{-\infty}^{\infty} da \int_{-\infty}^{\infty} \frac{d\tilde{a}}{2\pi/N} \int_{-\infty}^{\infty} \frac{dm}{\sqrt{2\pi/JN}} e^{-\frac{NJ}{2}m^2 + N\mu a - N\tilde{a}a} \\ &\quad \times \left\{ \sum_{S=0,\pm1} e^{[Jm + h\sigma(t)]S + \tilde{a}|S|} \right\}^N \\ &\simeq \exp\bigl[N\Phi(m, a, \tilde{a})\bigr] \end{aligned} \tag{5.6}$$

where we used the saddle point method to evaluate the integrals with respect to $a, \tilde{a}$ and m. $\Phi(m, a, \tilde{a})$ appearing in the final form (5.6) is regarded as a free energy density and it is given by

$$\Phi(m,a,\tilde{a}) = -\frac{J}{2}m^2 + \mu a - a\tilde{a} + \log\bigl\{1 + \mathrm{e}^{\tilde{a}} 2\cosh\bigl[Jm + h\sigma(t)\bigr]\bigr\}. \quad (5.7)$$

Then, the saddle point equation $\partial\Phi/\partial m = 0$ leads to

$$m = \frac{1}{N}\sum_{i=1}^{N} S_i = \frac{\mathrm{e}^{\tilde{a}} 2\sinh[Jm + h\sigma(t)]}{1 + \mathrm{e}^{\tilde{a}} 2\cosh[Jm + h\sigma(t)]}. \quad (5.8)$$

Apparently, the above m stands for 'magnetization' in the literature of statistical physics, however, as we will see later, it corresponds to the 'return' in the context of time-series prediction for the price. This is because the number of buyers is larger than that of the sellers if the m is positive, and as the result, the price increases definitely. Another saddle point equation $\partial\Phi/\partial\tilde{a} = 0$ gives

$$a = \frac{1}{N}\sum_{i=1}^{N} |S_i| = \frac{\mathrm{e}^{\tilde{a}} 2\cosh[Jm + h\sigma(t)]}{1 + \mathrm{e}^{\tilde{a}} 2\cosh[Jm + h\sigma(t)]}. \quad (5.9)$$

It should be noticed that from $\partial\Phi/\partial a = 0$, we have $\tilde{a} = \mu$. Hence, by substituting the $\tilde{a} = \mu$ into (5.8) and (5.9), we immediately have the following equations of state

$$m = \frac{2\mathrm{e}^{\mu}\sinh[Jm + h\sigma(t)]}{1 + 2\mathrm{e}^{\mu}\cosh[Jm + h\sigma(t)]}, \quad (5.10)$$

$$a = \frac{2\mathrm{e}^{\mu}\cosh[Jm + h\sigma(t)]}{1 + 2\mathrm{e}^{\mu}\cosh[Jm + h\sigma(t)]}. \quad (5.11)$$

We should bear in mind that a is a 'slave variable' and it is completely determined by m. However, a itself has an important meaning to characterize the market because the a is regarded as the number of traders who are actually trading (instead of staying). In this sense, the a could be 'turnover' in the context of financial markets. In other words, the turnover a is a measurement to quantify the activity of the market, and a large a means high activity of the market. Strictly speaking, the a could not be regarded as 'turnover' because in our modeling, we assumed that each trader posts unit volume to the market. However, by introducing v_i as volume for each trader i and replacing the spin variables in (5.1) as $S_i \to v_i S_i, v_i \in \mathbb{R}^+$, $a = (1/N)\sum_{i=1}^{N} |v_i S_i|$ is regarded as turnover in its original meaning.

Obviously, from Eqs. (5.10) and (5.11), the equation of state for the conventional Ising model [5, 10] is recovered in the limit of $\mu \to \infty$ as

$$m = \tanh\bigl[Jm + h\sigma(t)\bigr], \quad a = 1. \quad (5.12)$$

In following, we analyze the above Eqs. (5.10), (5.11) to investigate the equilibrium properties of our model system.

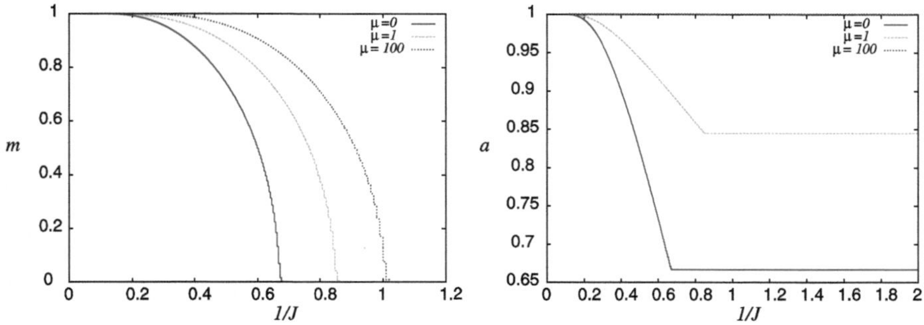

Fig. 5.2 $(1/J)$-dependence of magnetization m and turnover a for the case of $h = 0$. At the critical point $(1/J)_c \equiv 2e^{\mu}/(1+2e^{\mu})$, the second order phase transition takes place

5.2.2 Equilibrium States and Phase Transitions

We first consider the case of $h = 0$ in (5.1) or (5.10) and (5.11). For this case, we can solve the equations of states (5.10), (5.11) numerically. We show the $(1/J)$-dependence of magnetization m in Fig. 5.2 (left). From this panel, we find that the magnetization m monotonically decreases as $1/J$ increases for arbitrary finite μ and it drops to zero at the critical point $(1/J)_c$. The critical point is dependent on the value of μ. In order to investigate the μ-dependence of the critical point $(1/J)_c$, we expand the right hand side of (5.10) up to the first order of m. Then, we have

$$(1/J)_c = \frac{2e^{\mu}}{1+2e^{\mu}}. \tag{5.13}$$

It should be noted that $(1/J)_c = 1$ for the conventional Ising model [5, 10] is recovered in the limit of $\mu \to \infty$.

We next plot the $(1/J)$-dependence of the turnover a in the right panel of Fig. 5.2. From this panel, we are confirmed that above the critical point, the turnover a takes a constant value:

$$a = \frac{e^{\mu}}{1+e^{\mu}} \tag{5.14}$$

Here we should notice again that $a = 1$ is recovered in the limit of $\mu \to \infty$, which means that there is no trader who is staying at the moment.

We next evaluate the behavior of magnetization m as a function of h keeping the value of J as $J = 1.2$ (we set $\sigma(t) = 1$ for simplicity). Then, one observes from Fig. 5.3 (left) that the system undergoes a first order phase transition which is specified by a transition between bi-stable states in the free energy, namely, the states $m > 0$ and $m < 0$. The critical values m_* at the critical point is determined by

$$1 - m_*^2 = \frac{1}{4e^{2\mu}-1}\left\{\frac{1-J(1-m_*^2)}{2-2(1-m_*^2)}\right\} + \frac{1}{J}. \tag{5.15}$$

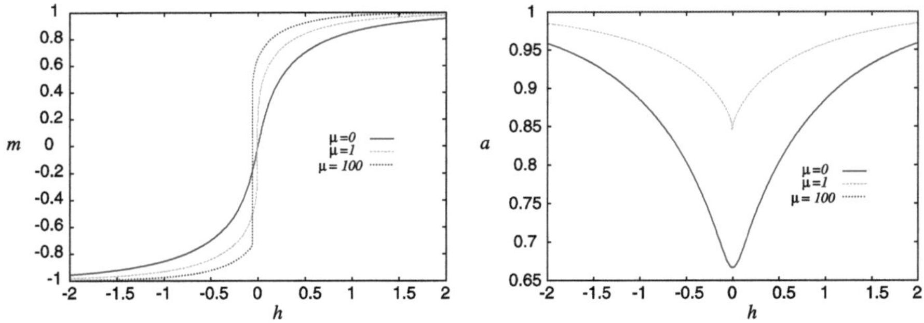

Fig. 5.3 Magnetization a and turnover a as a function of h. We plot each of them by keeping the J as $J = 1.2$

We should notice that one recovers $m_* = \pm\sqrt{(J-1)/J}$, which is the result for the conventional Ising model [10], in the limit of $\mu \to \infty$. For the case of $\mu < \infty$, the critical values m_* is given by

$$m_* = \begin{cases} \pm\sqrt{\frac{J-1}{J}} & (\mu \ge \frac{1}{2}\log(\frac{J^2+2}{8})), \\ \pm\frac{\sqrt{J^2-2(4e^{2\mu}-1)}}{J} & (\mu < \frac{1}{2}\log(\frac{J^2+2}{8})). \end{cases} \tag{5.16}$$

Then, the critical point h_c is obtained as a solution of the following equation.

$$\frac{1}{J} = \frac{4e^{2\mu} + 2e^{\mu}\cosh(Jm_* + h_c)}{\{1 + 2e^{\mu}\cosh(Jm_* + h_c)\}^2}. \tag{5.17}$$

In the next section, taking into account the above equilibrium properties and phase transitions, we shall construct the prediction model based on the Hamiltonian (5.1) and evaluate the statistical performance by means of computer simulations with the assistance of empirical data analysis.

5.3 The Prediction Model

In this section, we construct our prediction model. Let us define $p(t)$ as the price at time t. Then, the return, which is defined as the difference between prices at successive two time steps t and $t+1$, is given by

$$p(t+1) - p(t) = \Delta(t). \tag{5.18}$$

To construct the return $\Delta(t)$ from the microscopic view point, we assume that each trader $(i = 1, \ldots, N)$ buys or sells unit-volume, or stays at each time step t. Then, let us call the group of buyers as $\mathcal{A}_+(t)$, whereas the group of sellers is referred to as $\mathcal{A}_-(t)$. As we are dealing with three distinct states including 'staying', we define the group of traders who are staying by $\mathcal{A}_0(t)$. Thus, the total volumes of buying, selling and staying are explicitly given by

$$\phi_+(t) \equiv \sum_{i \in \mathcal{A}_+(t)} 1, \qquad \phi_-(t) \equiv \sum_{i \in \mathcal{A}_-(t)} 1, \qquad \phi_0(t) \equiv \sum_{i \in \mathcal{A}_0(t)} 1, \tag{5.19}$$

respectively. Apparently, the total number of traders should be conserved, namely, the condition $\mathcal{A}_+(t) + \mathcal{A}_-(t) + \mathcal{A}_0(t) = N$ ($\equiv$ Total # of traders) holds.

Then, the return $\Delta(t)$ is naturally defined by means of (5.19) as

$$\Delta(t) = \lambda\big(\phi_+(t) - \phi_-(t)\big) \tag{5.20}$$

where λ is a positive constant. Namely, when the volume of buyers is greater than that of sellers, $\phi_+(t) > \phi_-(t)$, the return becomes positive $\Delta(t) > 0$. As the result, the price should be increased at the next time step as $p(t+1) = p(t) + \Delta(t)$.

5.3.1 The Ising Spin Representation

The making decision of each trader ($i = 1, \dots, N$) is now obtained simply by an Ising spin (5.2). The return is also simplified as

$$\Delta(t) = \lambda\big(\phi_+(t) - \phi_-(t)\big) = \lambda \sum_{i=1}^{N} S_i^{(t)} \equiv m_t \tag{5.21}$$

where we set $\lambda = N^{-1}$ to make the return:

$$m_t = \frac{1}{N} \sum_{i=1}^{N} S_i^{(t)} \tag{5.22}$$

satisfying $|m_t| \leq 1$. Thus, m_t corresponds to the so-called 'magnetization' in statistical physics, and the update rule of the price is written in terms of the magnetization m_t as

$$p_{t+1} = p_t + m_t \tag{5.23}$$

as we mentioned before.

5.3.2 The Boltzmann-Gibbs Distribution

It should be noticed that the state vectors of the traders: $\boldsymbol{S} = (S_1, \dots, S_N)$ are determined so as to minimize the Hamiltonian (5.1) from the argument in the previous section. For most of the cases, the solution should be unique. However, in realistic financial markets, the decisions by traders should be much more 'diverse'. Thus, here we consider statistical ensemble of traders $\boldsymbol{S}$ and define the distribution of the ensemble by $P(\boldsymbol{S})$. Then, we shall look for the suitable distribution which maximizes the so-called Shannon's entropy

$$H = -\sum_{\boldsymbol{S}=0,\pm 1} P(\boldsymbol{S}) \log P(\boldsymbol{S}) \tag{5.24}$$

under two distinct constraints:

$$\sum_{\boldsymbol{S}=0,\pm 1} P(\boldsymbol{S}) = 1, \qquad \sum_{\boldsymbol{S}=0,\pm 1} P(\boldsymbol{S})\mathcal{H}(\boldsymbol{S}) = \mathcal{H} \tag{5.25}$$

and we choose the $P(\boldsymbol{S})$ which minimizes the following functional $f\{P(\boldsymbol{S})\}$:

$$\begin{aligned} f\{P(\boldsymbol{S})\} = &-\sum_{\boldsymbol{S}=0,\pm 1} P(\boldsymbol{S}) \log P(\boldsymbol{S}) - \lambda_1 \Big(\sum_{\boldsymbol{S}=0,\pm 1} P(\boldsymbol{S}) - 1 \Big) \\ &- \lambda_2 \Big(\sum_{\boldsymbol{S}=0,\pm 1} P(\boldsymbol{S})\mathcal{H}(\boldsymbol{S}) - \mathcal{H} \Big) \end{aligned} \tag{5.26}$$

where λ_1, λ_2 are Lagrange's multipliers. After some easy algebra, we immediately obtain the solution

$$P(\boldsymbol{S}) = \frac{\exp[-\beta \mathcal{H}(\boldsymbol{S})]}{\sum_{\boldsymbol{S}=0,\pm 1} \exp[-\beta \mathcal{H}(\boldsymbol{S})]} \tag{5.27}$$

where β stands for the inverse-temperature. In following, we choose unit temperature $\beta = 1$.

Here we should assume that the magnetization as a return at time $t+1$ is given by the expectation of the quantity $(1/N)\sum_{i=1}^{N} S_i^{(t)}$ over the distribution (5.27), that is,

$$m_{t+1} = \sum_{\boldsymbol{S}^{(t)}=0,\pm 1} \left\{ \frac{1}{N} \sum_{i=1}^{N} S_i^{(t)} \right\} P(\boldsymbol{S}^{(t)}) = \frac{2\mathrm{e}^{\mu_t} \sinh[J_t m_t + h_t \sigma(t)]}{1 + 2\mathrm{e}^{\mu_t} \cosh[J_t m_t + h_t \sigma(t)]} \tag{5.28}$$

where we defined

$$\mathcal{H}(\boldsymbol{S}^{(t)}) = -\frac{J_t}{N} \sum_{i,j=1}^{N} S_i^{(t)} S_j^{(t)} - h_t \sum_{i=1}^{N} \sigma(t) S_i^{(t)} - \mu_t \sum_{i=1}^{N} |S_i^{(t)}| \tag{5.29}$$

in (5.27) and used (5.10) to evaluate the expectation.

Thus, we have the following prediction formula

$$p(t+1) = p(t) + m_t, \tag{5.30}$$

$$m_t = \frac{2e^{\mu_t}\sinh[J_t m_{t-1} + h_t\sigma(t)]}{1 + 2e^{\mu_t}\cosh[J_t m_{t-1} + h_t\sigma(t)]}, \tag{5.31}$$

$$J_{t+1} = J_t - \eta \frac{\partial \mathcal{E}(J_t, h_t, \mu_t)}{\partial J_t}, \tag{5.32}$$

$$h_{t+1} = h_t - \eta \frac{\partial \mathcal{E}(J_t, h_t, \mu_t)}{\partial h_t}, \tag{5.33}$$

$$\mu_{t+1} = \mu_t - \eta \frac{\partial \mathcal{E}(J_t, h_t, \mu_t)}{\partial \mu_t} \tag{5.34}$$

where we introduced the cost function $\mathcal{E}$ to determine the parameters (J_t, h_t, μ_t) by means of gradient descent learning as

$$\mathcal{E}(J_t, h_t, \mu_t) = \frac{1}{2}\sum_{l=1}^{t}\left[\overline{\Delta q(l)} - \frac{2e^{\mu_t}\sinh[J_t\overline{\Delta q(l-1)} + h_t\sigma(t)]}{1 + 2e^{\mu_t}\cosh[J_t\overline{\Delta q(l-1)} + h_t\sigma(t)]}\right]^2 \tag{5.35}$$

and η is a learning rate. To obtain the explicit form of the learning equations, we take the derivatives as

$$\frac{\partial \mathcal{E}}{\partial J_t} = -\sum_{l=1}^{t}\left[\overline{\Delta q(l)} - \frac{2e^{\mu_t}\sinh[J_t\overline{\Delta q(l-1)} + h_t\sigma(t)]}{1 + 2e^{\mu_t}\cosh[J_t\overline{\Delta q(l-1)} + h_t\sigma(t)]}\right] \times \frac{2e^{\mu_t}\overline{\Delta q(l-1)}\cosh[J_t\overline{\Delta q(l-1)} + h_t\sigma(t)] + 4e^{2\mu_t}\overline{\Delta q(l-1)}}{\{1 + 2e^{\mu_t}\cosh[J_t\overline{\Delta q(l-1)} + h_t\sigma(t)]\}^2},$$

$$\frac{\partial \mathcal{E}}{\partial h_t} = -\sum_{l=1}^{t}\left[\overline{\Delta q(l)} - \frac{2e^{\mu_t}\sinh[J_t\overline{\Delta q(l-1)} + h_t\sigma(t)]}{1 + 2e^{\mu_t}\cosh[J_t\overline{\Delta q(l-1)} + h_t\sigma(t)]}\right] \times \frac{2e^{\mu_t}\sigma(t)\cosh[J_t\overline{\Delta q(l-1)} + h_t\sigma(t)] + 4e^{2\mu_t}\sigma(t)}{\{1 + 2e^{\mu_t}\cosh[J_t\overline{\Delta q(l-1)} + h_t\sigma(t)]\}^2}, \tag{5.36}$$

$$\frac{\partial \mathcal{E}}{\partial \sigma_t} = -\sum_{l=1}^{t}\left[\overline{\Delta q(l)} - \frac{2e^{\mu_t}\sinh[J_t\overline{\Delta q(l-1)} + h_t\sigma(t)]}{1 + 2e^{\mu_t}\cosh[J_t\overline{\Delta q(l-1)} + h_t\sigma(t)]}\right] \times \frac{2e^{\mu_t}\sinh[J_t\overline{\Delta q(l-1)} + h_t\sigma(t)]}{\{1 + 2e^{\mu_t}\cosh[J_t\overline{\Delta q(l-1)} + h_t\sigma(t)]\}^2}. \tag{5.37}$$

In the above expressions, $\overline{\Delta q(t)}$ is evaluated for the true price $q(t)$ by

$$\overline{\Delta q(t)} \equiv \frac{1}{M}\sum_{i=t-M+1}^{t}\big[q(i+1) - q(i)\big]. \tag{5.38}$$

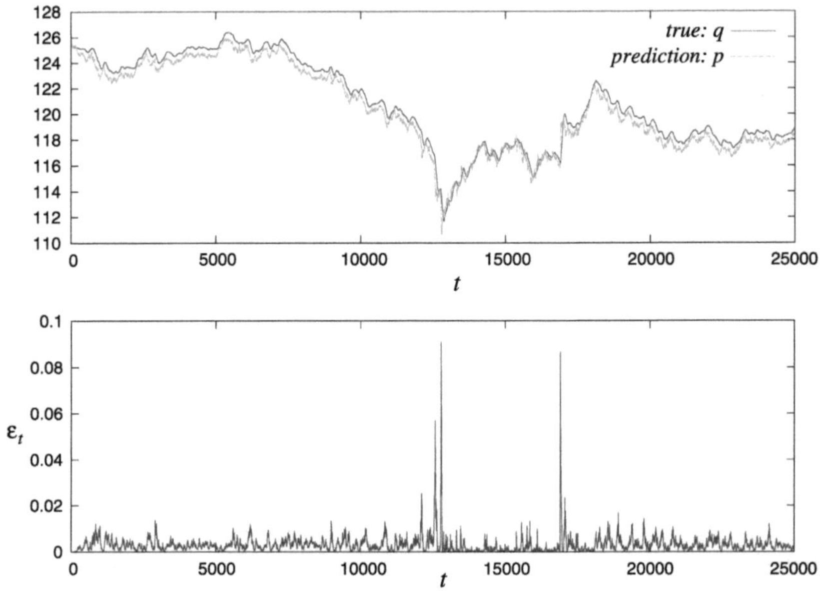

Fig. 5.4 The EUR/JPY exchange rate $q(t)$ (high frequency tick-by-tick data) from 25th April 2010 to 13th May 2010, which was used in the reference [5] and the prediction $p(t)$. The lower panel shows the mean-square error $\varepsilon_t \equiv \{q(t) - p(t)\}^2 / \max q(t)$

By substituting m_t and the set of parameters (J_t, h_t, μ_t) into (5.11), we obtain the turnover a at each time step as

$$a_t = \frac{2e^{\mu_t} \cosh[J_t m_t + h_t \sigma(t)]}{1 + 2e^{\mu_t} \cosh[J_t m_t + h_t \sigma(t)]}. \tag{5.39}$$

We should remember that we defined the exogenous information $\sigma(t)$ by the trend $\sigma(t) = [q(t) - q(t - \tau)]/\tau$ and here we choose $M = \tau$ to evaluate the trend and $\overline{\Delta q(t)}$.

5.4 Computer Simulations

In Fig. 5.4, we show the true time-series $q(t)$ which contains a crush and the prediction $p(t)$ with mean-square error $\varepsilon_t \equiv \{q(t) - p(t)\}^2 / \max q(t)$. The empirical true time-series $q(t)$ is chosen from EUR/JPY exchange rate (high frequency tick-by-tick data) from 25th April 2010 to 13th May 2010 (it is the same data set as in Ref. [5]). We set $\tau = M = 100$ [ticks], and chose $J_0 = 0.1, h_0 = 0.6, \mu_0 = 0.1$ as the initial values of parameters. From these panels, we confirm that the mean-square error takes small value within at most several percent although the error increases around the crush. Thus, we might conclude that our three states RFIM works well on the prediction of financial data having a crush.

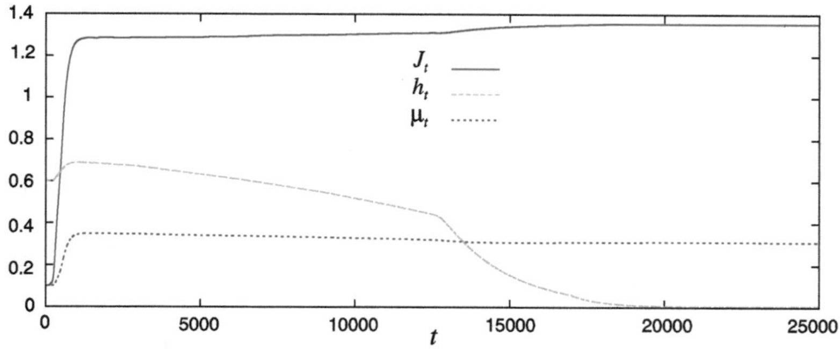

Fig. 5.5 Time-evolution of parameters (J, h, μ). The flow converges to the critical point

We next consider the flow of parameters (J_t, h_t, μ_t) which evolve across the crush. The result is shown in Fig. 5.5. From this figure, we clearly find that the strength of exogenous information h_t drops to zero after the crush. Chemical potential μ_t and strength of endogenous information J_t converge to 0.31 and 1.35, respectively. In our previous study [5], as the critical point was $(1/J)_c = 1$ for $\mu \to \infty$, the strength of endogenous information J converged to the critical value 1. However, in the three states RFIM with $h = 0$ and $\mu = 0.31 < \infty$, the critical point is sifted to $(1/J)_c = 1/1.35 \simeq 0.74$ (see Fig. 5.2 (left)). Therefore, in this simulation, these two parameters converge to the corresponding critical point $(J_c, \mu_c) = (1.35, 0.31)$. From the result, we conclude that the system described by the Hamiltonian (5.1) automatically moves to the critical point after the crush.

We next utilize the USD/JPY exchange rate from 12th August 2012 to 24th August 2012 as true time-series. It should be noted that the duration of this data is 1 minutes, hence, the data is not tick-by-tick data. In the simulation for this data set, we set $\tau = M = 10$ [min]. We show the simulated turnover a with the corresponding true value in Fig. 5.6. From these panels, we find that the empirical data for the turnover increases instantaneously around the crush, whereas the simulated turnover a does not show such striking feature although it possess a relatively large peak just before the crush. In order to convince ourselves that the simulated turnover a can characterize the crush, we should carry out much more extensive simulations for various empirical data. It should be addressed as our future study.

5.4.1 Comparison with the Conventional Ising Model

Finally we compare our result with that of the conventional Ising model [10]. Here we used the USD/JPY exchange rate from 1st March 2012 to 31st July 2012, whose minimum duration is 30 minutes. We choose the width of the time-window $\tau = M = 10$ [min]. The initial values of parameters are set to $J_0 = 0.1, h_0 = 0.6$ for the conventional Ising model, whereas are chosen as $J_0 = 0.1, h_0 = 0.6, \mu_0 = 0.1$

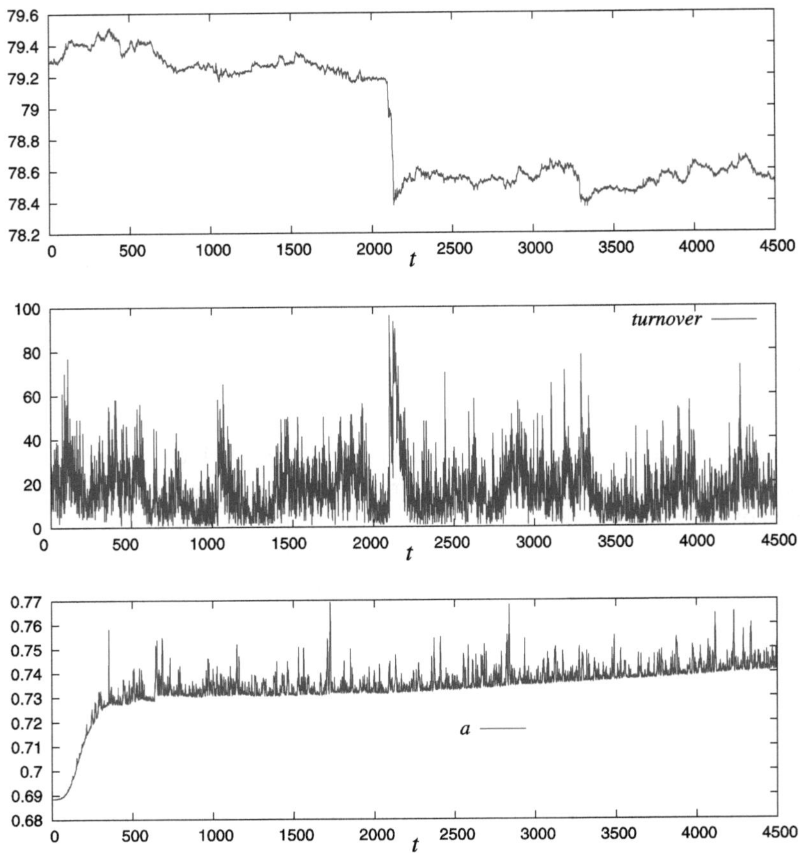

Fig. 5.6 USD/JPY exchange rate from 12th August 2012 to 24th August 2012 as true time-series (the *upper panel*) and the true turnover (the *middle panel*). The *lower panel* shows the simulated turnover a evaluated by our prediction model

for the three states RFIM. The results are shown in Fig. 5.7. From this figure as a limited result, we find that the performance of the prediction by the three states RFIM is superior to that of the conventional Ising model.

5.5 Concluding Remark

In this paper, we extended the formulation of time-series prediction using Ising model given by Kaizouji [10] or Ibuki *et al.* [5] by means of three states RFIM. We found that the crisis could be ‘partially” characterized by the simulated turnover. We also confirmed that the three states $S_i = 0, \pm 1$ in each trader's decision making apparently improves the statistical performance in the prediction.

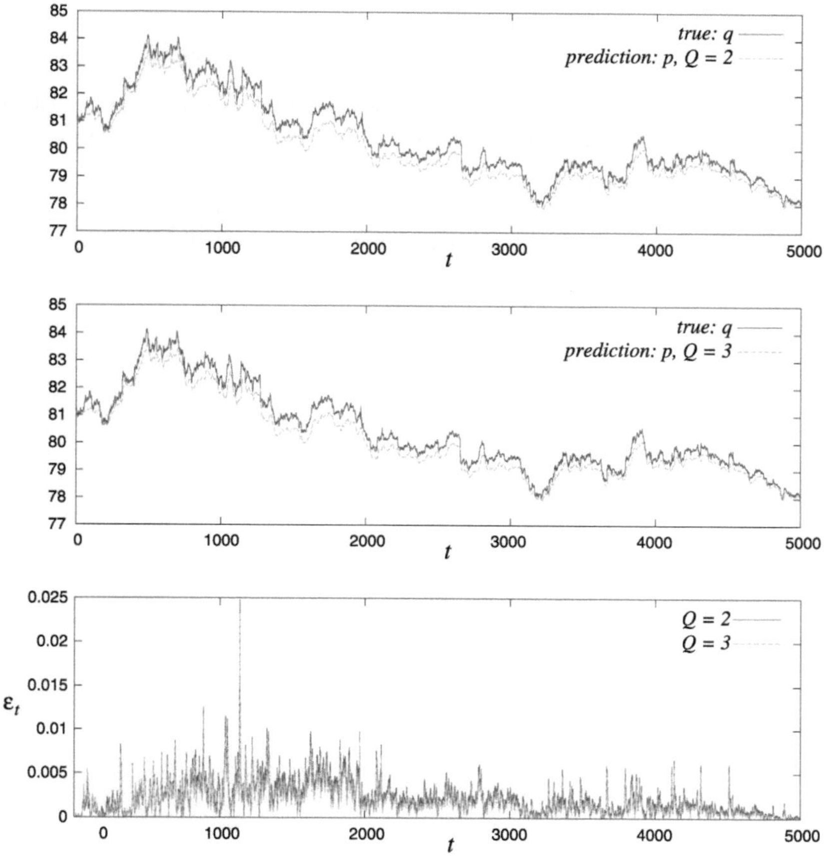

Fig. 5.7 Comparison with the conventional Ising model [5, 10]. The *upper panel* shows the result of the conventional Ising model, whereas in the *middle panel*, the result of three states RFIM is exhibited. The *lower panel* shows the corresponding mean-square errors

Acknowledgements This work was financially supported by Grant-in-Aid for Scientific Research (C) of Japan Society for the Promotion of Science, No. 22500195. The authors acknowledge Takero Ibuki, Shunsuke Higano and Sei Suzuki for fruitful discussion and useful comments. We thank organizers of *Econophysics-Kolkata VII*, especially, Frederic Abergel, Anirban Chakraborti, Asim K. Ghosh, Bikas K. Chakrabarti and Hideaki Aoyama.

References

1. Reynolds CW (1987) Flocks, herds, and schools: a distributed behavioral model. Comput Graph 21:25
2. Makiguchi M, Inoue J (2010) Numerical study on the emergence of anisotropy in artificial flocks: a BOIDS modelling and simulations of empirical findings. In: Proceedings of the operational research society simulation workshop 2010 (SW10), pp 96–102 (CD-ROM the preprint version, arXiv:1004.3837)

3. Kahbeman D, Tversky A (1979) Econometrica 47(2):263
4. Ibuki T, Suzuki S, Inoue J (2013) New economic windows. In: Proceedings of econophysics-Kolkata VI. Springer, Milan, pp 239–259
5. Ibuki T, Higano S, Suzuki S, Inoue J (2012) Hierarchical information cascade: visualization and prediction of human collective behaviour at financial crisis by using stock-correlation. ASE Hum J 1(2):74–87
6. Mantegna R (1999) Eur Phys J B 11:193
7. Onnela J-P, Chakraborty A, Kaski K, Kertesz J, Kanto A (2003) Phys Rev E 68:056110
8. Borg I, Groenen P (2005) Modern multidimensional scaling: theory and applications. Springer, New York
9. http://finance.yahoo.co.jp/
10. Kaizoji T (2000) Physica A 287:493

Chapter 6
Themes and Applications of Kinetic Exchange Models: Redux

Asim Ghosh, Anindya S. Chakrabarti, Anjan Kumar Chandra, and Anirban Chakraborti

Abstract In this article, we briefly discuss the general formalism of kinetic exchange models and their various applications in economics and sociology. Inspired from the kinetic theory of gases in statistical physics, the kinetic exchange model for closed economic systems were first proposed by simply considering the agents as gas molecules, and wealth of agents as kinetic energy exchanged amongst the gas molecules. The formalism had been successfully applied to modeling of wealth distributions in 2000s. This has further spurred new research in recent times in various areas of soft sciences—firm dynamics, opinion formation in the society, etc.

6.1 Introduction

The essential theme of the kinetic exchange models is the exchange of energy due to collisions amongst a collection of inanimate particles. Here, we will present that story and its economic and social counterparts to provoke some more collisions between economists and physicists that *may* lead to exchange of ideas (exchanging kinetic energy between these two arrogant groups might not be good idea to begin with!). On a more serious note, the kinetic exchange models have been one of the

A. Ghosh · A.K. Chandra
Theoretical Condensed Matter Physics Division, Saha Institute of Nuclear Physics, 1/AF Bidhannagar, Kolkata 700 064, India

A. Ghosh
e-mail: asimghosh066@gmail.com

A.K. Chandra
e-mail: anjanphys@gmail.com

A.S. Chakrabarti
Economics Department, Boston University, 270 Bay State Road, Boston, MA, USA
e-mail: anindya@bu.edu

A. Chakraborti (✉)
Laboratoire de Mathématiques Appliquées aux Systèmes, École Centrale Paris, 92290 Châtenay-Malabry, France
e-mail: anirban.chakraborti@ecp.fr

F. Abergel et al. (eds.), *Econophysics of Agent-Based models*, New Economic Windows, DOI 10.1007/978-3-319-00023-7_6,

most widely used formalisms in the growing interdisciplinary field of econophysics [1–4] and sociophysics [5–7]. The concept of kinetic exchange was taken from statistical physics which was proposed towards the end of the nineteenth century. First, Bernoulli gave a picture of kinetic theory of gases in his paper 'Hydrodynamica.' After that Maxwell and Boltzmann derived energy distribution function for kinetic theory of gas. The other pioneers in this field were Max Planck, Rudolf Clausius, Josiah Willard Gibbs. The kinetic exchange model is one of the simplest models in statistical mechanics, which attempts in deriving the average macroscopic behaviors from the microscopic properties of particles.

The kinetic exchange models[1] had been primarily used to explain income/wealth distributions [8, 9]. More specifically, the target was to write down a minimal set of stochastic equations that gave rise to a distribution mimicking the actual income/wealth distributions. Even though this target has been satisfactorily achieved, as discussed below, the framework suffers from a major drawback in terms of *ad hoc*-ism. The explanations (or the terminology) that has been used to describe the exchange processes is not exactly what one would call 'economic'. The researchers working on this topic essentially took a solvable model from statistical mechanics and made analogies of certain quantities. For example, energy was interpreted as wealth, particles were substituted by agents, etc. and needless to say, such abstraction and *ad hoc* approach attracted its fair share of criticism [10]. However, the same abstraction may also prove to be one of the strongest features of this whole literature. Since the terms are not tied to some specific economic quantity, there is little reason to confine them in the area of income/wealth distributions only. This motivation led to applications of the same basic framework to explain different economic and social phenomena.

What we will discuss in this paper is roughly as follows. We start by describing the simple observation that a simple random scattering-like interaction amongst the agents gives a wealth distribution similar to the 'Boltzmann-Gibbs' type. However in our real society, each of the agents have a "saving propensity." We discuss that when saving is introduced in the model, depending upon distribution of the saving propensity amongst agents, different wealth distributions can be generated. Further, we will review how a kinetic exchange model may give a "phase transition" by introducing a "threshold," where the associated phase transition is of the "active-absorbing" type. Then we will discuss the applications of the same type of formalism in firm dynamics and later in the opinion dynamics in the society. The basic aim of this article is to enthuse the readers in the use of the simple yet powerful formalism of kinetic exchange models in related areas. By no means this is an exhaustive or technical review. We would like to refer the readers to the original books and articles for further details and references.

---

[1] We shall often use in this article, the full form or the shorter acronym, KEM, interchangeably.

6.2 Kinetic Wealth Exchange Models (KWEM)

Since, the KEM was first applied to explain the origin of inequality that is seen in the income/wealth distributions, it is a natural starting point for us to indicate the regularities of those distributions. The first well known observation was by Pareto who showed that distribution of wealth for the richer section of the economy follows a power law [11]. He observed that roughly 20 % people who were in the tail, owned about 80 % of the total wealth of the economy. After that preliminary but extremely significant study, many others have conducted research to know the exact distribution of wealth or income in an economy, but it remains as one of the elusive problems in economics. For the other end of the income/wealth spectrum it has been observed that the people of low income or wealth in our society follow a log-normal or gamma-like distribution [3] though there is some ambiguity over the fit of the theoretical distributions to the real data. However, one surprising fact is that the general features of the distributions do not change from country to country i.e., they are robust to the exact specification of the economy/country.

To understand the precise origin and nature of these robust features in the income or wealth distribution, concepts of kinetic theory of gas molecules have been used with success. One can easily map the problem with kinetic exchanges, by considering an economic agent to be like a gas molecule and wealth of that agent as similar to the kinetic energy of the molecule. The different exchange dynamics that we discuss below, when combined effectively, produce different features of the empirical wealth distribution.

6.2.1 Basics of Kinetic Wealth Exchange Models

6.2.1.1 Boltzmann-Gibbs Distribution in Economic Systems

Independent of the modeling efforts of social scientists,[2] a distinct approach was taken by physicists Drăgulescu et al. [9, 16] who considered a toy model where the agents simply reshuffled a part of their wealth in a closed economy. The benefit of having a toy model is that everybody knows it is a toy model and it can be changed very easily. The most important departure from the standard economic theory is that they got rid of all microeconomic decision-making processes. That came at the cost of losing all perspective of why should any trade occur at all. However, the benefit exceeds the cost. One can then think of the economic system as comprising of only the agents and their characterizing quantity, wealth instead of keeping track of preferences, beliefs, market mechanisms, etc. (which are the usual burden of most, if not all, neo-classical models in economics).

[2]It was discovered later that economists, like Bennati, and sociologists, like Angle, were independently using similar tools and models since the 1980s; one can refer to Refs. [13–15] for details.

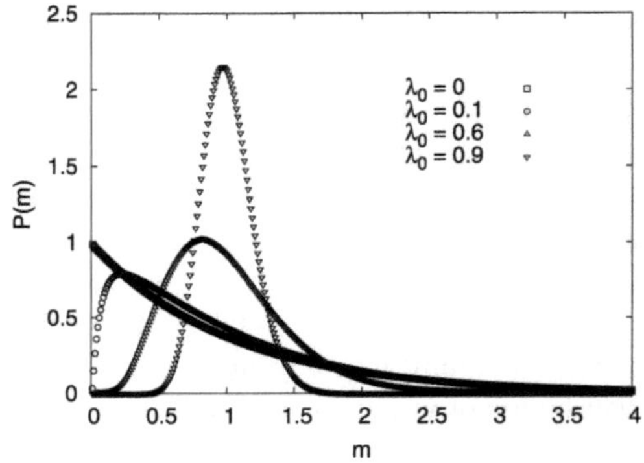

Fig. 6.1 Steady state wealth distribution $P(m)$ vs. m for CC model for $\lambda_0 = 0, 0.1, 0.6, 0.9$. Monte Carlo simulations were done by taking $N = 100$ agents and average wealth $M/N = 1$. Taken from [8]

In the toy model, all the agents traded with each other through pair-wise interaction similar to random energy scattering of gas molecules, thus losing or gaining a certain amount of wealth. Suppose two agents i and j having wealth m_i and m_j respectively, this trade dynamics can be written as

$$m_i' = r(m_i + m_j), \tag{6.1a}$$

$$m_j' = (1 - r)(m_i + m_j) \tag{6.1b}$$

where r is any random number drawn from 0 to 1, and after trading the wealth of the two agents i and j are m_i' and m_j', respectively. From the above equation, it is clear that the total wealth of the two agents before and after trading remains constant i.e., $m_i + m_j = m_i' + m_j'$.

The resultant wealth distribution of this model can be derived analytically.[3] One can also do simple Monte Carlo simulations to find the resulting steady state distribution. The N agents are each given initially 1 unit of wealth (so, total wealth $M = N$ thus fixing the average wealth in the economy),[4] and they trade with each other according to the dynamics given by Eq. (6.1b). It is observed that in steady state, the distribution of wealth is similar to 'Boltzmann-Gibbs' type distribution for kinetic theory of gases, i.e, $P(m) \sim \exp(-m/T)$, where $P(m)$ is the probability of an agent having wealth between m and $m + dm$, and T is average wealth of the model (here $T = M/N = 1$). The Monte Carlo simulation results for this model is shown in Fig. 6.1 ($\lambda_0 = 0$ case).

[3]Standard tools of statistical mechanics like Boltzmann transport equation, Pauli's master equation, maximization of entropy principle, etc. can be used to derive the steady state distribution of the 'Boltzmann-Gibbs' type (see, e.g., Refs. [3, 17]).

[4]It should be noted that the initial wealth distribution does not affect the steady state distribution, as long as the average wealth of the system remains the same. Essentially the system is *ergodic*.

6.2.1.2 CC Model

An important modification of the previous toy model was done by Chakraborti et al. [18] by introducing savings amongst all the agents. In that model, one considers a close economic system where total wealth and total number of agents are conserved. All the agents exchange their wealth through a trading. This much is identical to the basic model. The distinct feature is that before trading both of the participants save a fraction of wealth (this feature kind of mimics the reality that we do not put all of our wealth on the mercy of the market mechanism every now and then!). So the trading equation between two agents i and j can be written as

$$\begin{aligned} m_i' &= \lambda_0 m_i + r(1-\lambda_0)(m_i+m_j), \\ m_j' &= \lambda_0 m_j + (1-r)(1-\lambda_0)(m_i+m_j) \end{aligned} \tag{6.2}$$

where $0 \le \lambda_0 < 1$ is the 'saving propensity' (fraction of wealth that is being saved) of the agents. For simplicity, debt was not permitted in this model. By running simulations, it was observed that the steady state distribution is completely different from Boltzmann-Gibbs like, for any positive λ_0 value. The shape of the distribution looks like Gamma-like distributions [17, 20–22], and the most probable value depends upon the value of λ_0. Results corresponding to different values of the saving propensities are shown in Fig. 6.1. The analytical closed form of the steady state distribution remains an open problem.

6.2.1.3 CCM Model

In the last case, the CC model, all agents had a fixed saving propensity λ_0, i.e., the savings propensity does amongst the (homogeneous) agents. But for modeling purpose and also for the sake of reality, one can go one step forward and assume that the agents are heterogeneous. So a natural generalization is to consider saving propensities to be different for different agents. Precisely this modification was done by Chatterjee et al. [23]. They made the same assumptions in their model as the previous ones, but the only difference was that each agent i had a characteristic saving propensity λ_i, which could take value $0 \le \lambda_i < 1$ drawn randomly from a uniform random distribution. Each λ_i (for $i = 1, \dots, N$) is fixed over time, and is thus a quenched variable. The trading equation of this model can be written as

$$\begin{aligned} m_i' &= \lambda_i m_i + r\big((1-\lambda_i)m_i + (1-\lambda_j)m_j\big), \\ m_j' &= \lambda_j m_j + (1-r)\big((1-\lambda_i)m_i + (1-\lambda_j)m_j\big) \end{aligned} \tag{6.3}$$

where λ_i and λ_j are the saving propensities for agent i and agent j, respectively. By doing this apparently simple modification, an interesting phenomenon emerged—the steady state wealth distribution gave rise to a power law tail with exponent 2

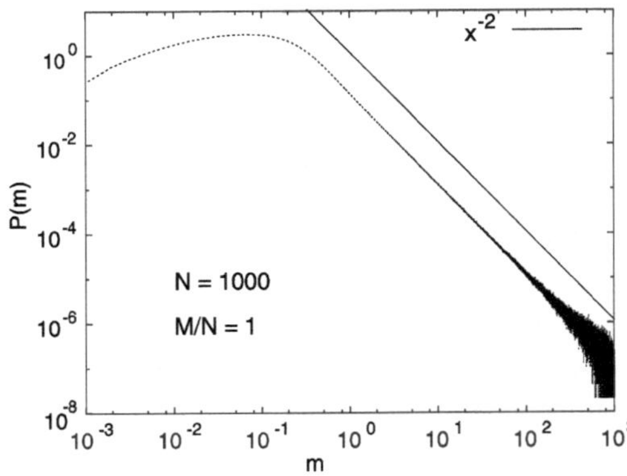

Fig. 6.2 Steady state wealth distribution $P(m)$ vs. m for CCM model ($0 \leq \lambda_i < 1$). *Dashed line* represents the results of Monte Carlo simulations, for $N = 10^3$ and $M/N = 1$. The power law is fitted with x^{-2} (*solid line*). Adapted from [8]

(see Fig. 6.2).[5] The steady state wealth distribution statistics for a single realization of quenched set (fixed propensities λ_i's) is observed to be significantly different in nature with respect to the statistics averaged over a large number of *independent* quenched configurations (variable sets). The peculiarities of the statistics from any one realization is independent of the sample size, as observed in Refs. [14, 24]. This feature of the model suggests that the observed power law tail is essentially a *convolution* of the single member distributions [14, 24]. Thus the power law tail can be explained by a set of overlapping Gamma-distributions arising from the agents with very high propensities ($\lambda_i \to 1$) [14, 19]. Another interesting feature is that the wealth that an agent accumulates is correlated to the saving propensity, first observed numerically in Eq. (14) of Ref. [14]. These observations allow the steady state distribution to be easily derived analytically [19, 37].[6]

As we have pointed out before, the empirical income or wealth distributions do display both the exponential part and the power-law decay. These two models, CC and CCM, and simple other variants [3, 26], are then able to capture (or at least reproduce) the basic features of the whole income/wealth distribution.

6.2.1.4 An Extension of CCM Model

Here we discuss another extension of the kinetic exchange model, studied in [27, 28]. The model can be described as follows: using the same framework as above, the only difference is that a trade takes place between two agents investing the *same* amount of wealth. Therefore in every transaction, the agents take an "effective" saving propensity λ which changes over time. Suppose, any two agents i

[5]Detailed numerical studies [24] showed that while the first two, the toy model studied in Sect. 6.2.1.1 and the CC model in Sect. 6.2.1.2, are *ergodic* and *self-averaging*, the third one (CCM) is not, which makes it very difficult to be studied numerically. This is an advisory note to students and beginners who want to study this numerically.

[6]For another attempt using master equation, see Ref. [25].

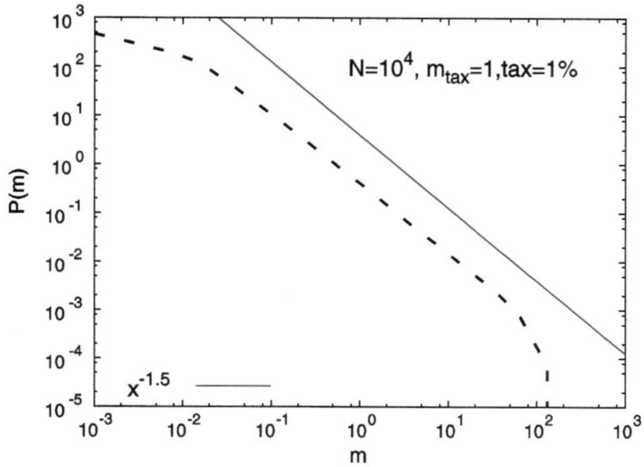

Fig. 6.3 Steady state distribution wealth for the extension model described in Sect. 6.2.1.4. The Monte Carlo simulation was done by taking $N = 10^4$, $M/N = 1$, and 1 % wealth tax was collected from the agents having wealth ($m_{tax} = 1$) greater than the average wealth of the model, after every 10 Monte Carlo time steps. The numerical simulations are plotted using a *dashed line*. The power law is fitted with $x^{-1.5}$ (*solid line*). Taken from [27]

and j have wealth m_i and m_j, respectively, and they go for a trading. If the saving propensity for agent i is assumed to be $\lambda_i = m_i/(m_i + m_j)$ and for agent j, to be $\lambda_j = m_j/(m_i + m_j)$, then it is clear from trading equation Eq. (6.3), that the total wealth is conserved before and after trade, and both agents invest the wealth: $m_i m_j/(m_i + m_j)$. For this model, it is observed that in steady state, the wealth condenses to a single agent, a feature very similar to the results obtained by Chakraborti [29]. The condensation can be avoided, if taxation is introduced into the system. Suppose, the tax is applied for the agents who have wealth greater than the average wealth, and this tax is collected periodically after a constant time interval. For this model it is found that the distribution of wealth again has a power law tail.

In the Monte Carlo simulations, $N = 10^4$ agents were considered and everybody was initially given $M/N = 1$ unit of wealth. All agents traded among themselves according to rules described above. Also, 1 % of total wealth was taken as tax, after every 10 Monte Carlo time steps (one Monte Carlo time step is defined as equivalent to N numbers of random trades among the agents) from the agents who have wealth ($m_{tax} = 1$) greater than the average value. The collected wealth is then *re-distributed uniformly* over all the agents. By doing this, it was observed (see Fig. 6.3) that the wealth distribution follows a power law tail with exponent 1.5. This is another way of recovering the power law using the kinetic exchange framework.

6.2.2 Phase Transitions in Kinetic Exchange Model

Here, we will describe another variant [30] of the above kinetic exchange models by introducing a threshold value, inspired by the concept of 'poverty line' in eco-

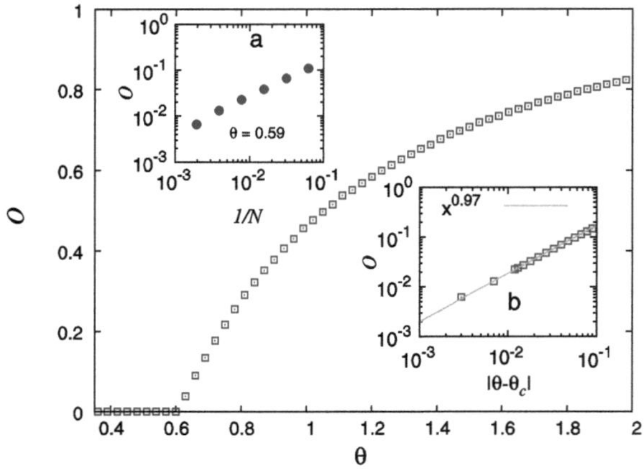

Fig. 6.4 The threshold values versus order parameters plot for $N = 10^4$. (*Right inset*): Near critical point the order parameter fits with scaling form $O = (\theta - \theta_c)^\beta$ with $\beta = 0.97$, $\theta_c = 0.6075$. (*Left inset*): It is shown that below critical point $\theta = 0.59$ the order parameter goes to zero in the thermodynamic limit. Taken from [30]

nomics. The model can be described as follows: Here agents exchange their wealth as described in Eq. (6.1a). But the only difference is that a threshold value of wealth θ is defined, and a trade between two agents occurs, if *at least one* of the two agents has wealth less than θ. Since there is a value of threshold, if all agents accumulate wealth greater than θ, then in such a situation the dynamics stops. The maximum limit of the threshold value θ below which the dynamics is stopped within some finite time, defines as critical value θ_c. The order parameter O is defined as the average total number of agents having wealth less than θ i.e., $O = \int_0^\theta P(m)dm$, where $P(m)$ is the probability distribution function of wealth. To make the system *ergodic*, a perturbation is applied into the system whenever the dynamics is stopped, and a particle having energy above θ is selected at random and its energy fully transfered to any other particle. For characterization, the model was studied for mean field (MF), one dimensional (1D) and two dimensional (2D) square lattices.

The mean field results are discussed here. Suppose after a time step τ, called the "relaxation time", the dynamics reaches a steady state. After the system reaches steady state, the order parameters are measured for different values of θ, and plotted as shown in Fig. 6.4. From the figure, it is observed that after the point $\theta = 0.6075$ (critical point) the order parameter increases. The order parameter near the critical point obeys a scaling form as $O \sim (\theta - \theta_c)^\beta$, where β is order parameter exponent, and the value $\beta = 0.97$ fits well with the scaling form. Also at critical point, the time variation of order parameter fits with the scaling form $O(t) \sim t^\delta$ with $\delta = 0.93$ (see Fig. 6.5). To confirm the existence of the transition, the relaxation times τ are measured for different values of θ. It is observed that there exists a clear time scale divergence behavior with scaling form $\tau \sim |\theta - \theta_c|^{-z}$, with $z = 0.83$ (see Fig. 6.6). All these observations and behaviors suggest that there exists a "phase transition" at $\theta = \theta_c$. To determine the exact universality class, the model was studied for 1D and 2D square lattices too, and the obtained scaling exponents suggested that the universality class is close to the Manna universality [30–32].

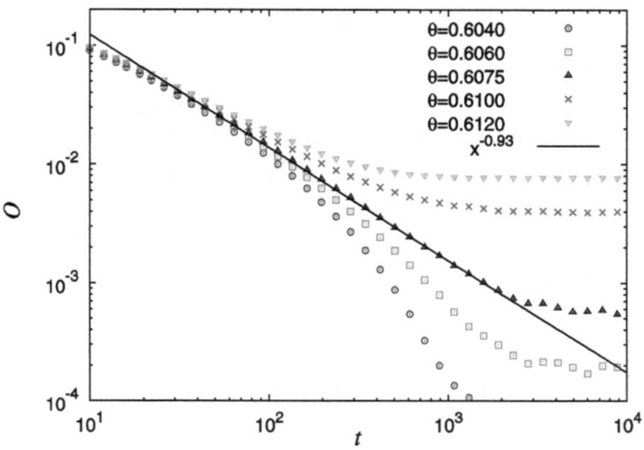

Fig. 6.5 Time variation of order parameters for different values of θ are shown. Near critical point, the order parameter fits the scaling form $O = t^{-\delta}$ with $\delta = 0.93$. Taken from [30]

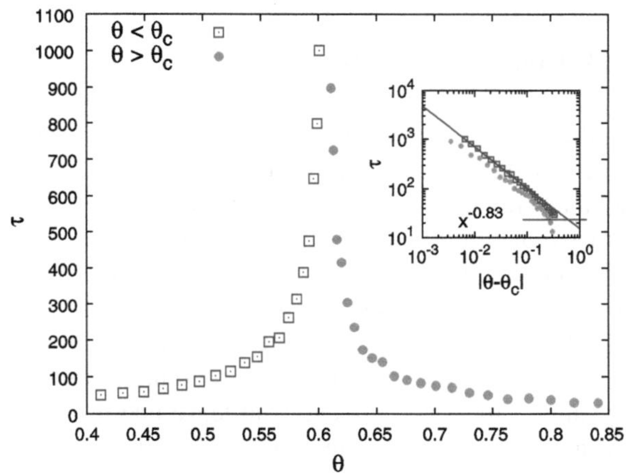

Fig. 6.6 The relaxation times are plotted for different θ values. It is observed that at critical point, the relaxation time diverges clearly. (*Inset*) At critical point, the relaxation time τ diverges as $\tau = (\theta - \theta_c)^{-z}$ with $z = 0.83$. Taken from [30]

6.2.3 A Brief Summary of the KWEM

In this section, we have discussed different kinetic exchange models of wealth for closed economic systems. First we have considered the random reshuffling exchange dynamics and observed that the wealth distribution for such a closed economic system obeys Boltzmann-Gibbs distribution. Next we have discussed that the shape of distribution changes from Boltzmann-Gibbs to Gamma-like if a (homogeneous) saving propensity is introduced in the model [17, 20–22]. We have then discussed a model (CCM) whereby assigning uniformly random saving propensities (i.e., introducing a particular type of heterogeneity in terms of savings propensity) to all the agents, a wealth distribution with a power law tail having exponent 2 is generated. Later an extension of the CCM model has been proposed where it is assumed that both agents invest the same amount wealth for a trading. For this model, if one considers taxation for the agents who have wealth greater than the average wealth in the model, then interestingly the wealth distribution again gives a power law tail, albeit with a different exponent (namely 1.5). These models reproduce features qualitatively and (one might argue) quantitatively similar to those of the empirical wealth distributions in the economy [3, 9]. In short, we have a formalism that reproduces both the basic features of income/wealth distributions (observed since the pioneering studies by Pareto [11] and Gibrat [12]): a low income region that resembles a Gamma-like distribution and the tail region that follows a power law.

Furthermore, we have discussed a kinetic exchange model where the agents who have wealth lower than a threshold value are able to initiate trade with any other agent. In this model, we have observed that there is a certain threshold value of wealth, θ_c, below which there is no agent below that particular threshold value. In other words, there is a critical value of $\theta = \theta_c$ above which this 'absorption' never happens. From critical finite size scaling, it was observed that scaling exponent for this transition is close to Manna universality class.

6.3 Firm Dynamics

In this section, we will present some recent progress made on explaining the firm size distributions in a similar framework. Of course, we cannot literally consider a binary random collision model to be a *replica* of all economic phenomena. But in case of firm dynamics, the essential mechanism that describes a redistribution of a certain quantity (wealth in the standard KEMs) still makes considerable sense. Think of firms as agents (instead of buyers and sellers) and workers in place of wealth. In the wealth exchange models, the buyers and sellers exchange wealth intentionally while carrying out economic transactions, whereas in case of firm-dynamics, there are workers who leave one firm and join others (be it intentional or unintentional). So again we have a redistribution of a certain quantity, workers w (instead of wealth). Assuming no migration, birth or death of workers, the economy remains conserved. The dynamics is technically identical to that of usual KEM except that one has to consider a n-particles collision process instead of binary collision. In case of wealth exchange a binary collision makes sense since usually the number of participants in one particular transaction is two. However, in case of firms, if one worker leaves one firm there is no particular reason to think that (s)he goes to another pre-specified firm. So we need to generalize the standard model to incorporate the possibility of having an n-ary interaction. Apart from this economic reason, there is another important advantage of such a generalization. The usual binary collision model with constant savings propensity λ cannot be solved analytically. At least it has not been solved so far. However, one can easily show that if there is a system-wide redistribution, the system can easily be solved analytically. Reference [38] proposed such a model and gave the solutions.

Before describing the model, we have to explain what we can expect from such an unconventional application of KEM. There are a couple of specific targets. The first one is obvious. Since, the KEM generates a power law very easily and since firm size distribution shows a power law, that is reason enough to apply the model. There is another empirical regularity not as well known. Firm growth process in the developing economies is known to produce divergence in their growth path giving rise to bimodality in the size distribution [40]. A surprising fact is that such bimodality has been observed in wealth distribution as well [41]. Hence, if we have a model which can accommodate binary trading as well as the whole system-side trading, then the same framework can be used to explain the non-standard features of wealth distribution as well as firm size distribution. It appears that KEM is well-suited to explain these features as well (see Ref. [39]).

6.3.1 Regularities in Firm-Size Distribution

The remarkable robustness of the long tail of the size distribution of the firms, is known from the early works by Gibrat [12]. Intuitively, a few very large firms can operate side by side with a large number of small firms. However, it was Axtell [33] who presented clear evidence that the distribution can be characterized very well by a power law. The same paper also remarked on the stability of the power law feature which has survived changes in the political, regulatory and social regimes (e.g., demographic changes in the workforce due to the influx of women in the labor force). Numerous innovations and technological changes in the production process had taken place within the same time period which were unable to affect it. Lastly, the changes in the market structures, policy changes, firm mergers, acquisitions, death and birth of firms, thousand of tweaks in the corporate laws did not affect this feature. This indicates that the statistical features of the firm growth process may well be independent of microeconomic decision-making processes like, why people choose to leave their jobs (or why and how firms decide to lay off for that matter) etc. Hence, any microeconomic foundation for the firm dynamics is not needed (at least at a first approximation). However, the rate at which the firms gain and lose workers is of interest to us as that determines the size of the firm. This rate is called the *turnover rate* in the economics literature. Another way to look at the same thing is that it measures how long the employees stay in their respective jobs.

Apart from the power law, Ref. [40] presents evidence that the developing economies are characterized by a bimodal distribution of firms. There is a bunch of very small firms creating a peak and there is another bunch of very big firms creating another peak with little mass of firms in between. This particular feature is known as the 'missing middle'. A very interesting feature is that as an economy develops usually this 'missing middle' disappears. Hence, this is somehow related to the economic condition of the country under consideration.

A Little Digression on Theory We intend to show that the turnover rates play a crucial role in the firm size distribution. But since turnover rate dictates not only inflow but also outflow, we need another parameter to describe outflow only. Hence, we describe the model in terms of the 'retention rate' which has a role identical to that of savings propensity that we used in KWEM, discussed in the previous section. The retention rate refers to the fraction of workers that stays with the firm at the end of the period. Clearly the turnover rate and the retention rate are related. As is evident, one aspect of job separation and worker hiring is that the process follows the rule of local conservation. If one worker goes from one firm to another then the total workforce remains unchanged but the workers' distribution across the firms change. Since the workers at any given year (or quarter) move around in a very large number of firms, we model this process as a repeated interaction between a large number of agents (firms) which exchanges a finite mass of (number of) workers between themselves. Clearly, the idea of the kinetic exchange model is suitable for this purpose.

Now we can discuss the economic interpretation of the terms used in the model. First, the economy consists of a large number of firms populated by workers. By firms we mean each and every production units capable of producing any kind of goods and services. Therefore there is no formal unemployment in the model. We adopt this idea in order to simplify the model so that we do not have to keep track of the mass of workers who are moving in and out of the employed workers' pool. Note that this does not affect the tail of the distribution (Zipf's law) since the tail is formed only by large firms. Secondly, we have made another simplifying assumption which is that the workers are treated as a continuous variable. While it is certainly true that there is an integer constraint on workers' head-counts, we have an advantage in treating the workers as a continuous flow in and out of the firms as it is easier (mathematically) in this case to derive the size distributions as we do not have to worry about integer constraints. Thirdly, we follow the definition that the firm's size is just the mass of workers working in the firm. There are other measures (like stock valuation, amount of goods or services produced etc.). But the number (mass) of workers has the most unambiguous definition. Hence we stick to it.

6.3.2 A Model with Constant Retention Rate

Reference [38] described the model in the following way. There is exactly one exogenous variable (the retention rate λ) and one endogenous variable (the firm sizes) in the model. We assume that time is discrete. The economy consists of an array of N firms with perfectly elastic demand for labor i.e. any firm can absorb any mass of workers that come to it. At the very beginning of the process, all firms have exactly one unit of workers (more formally, the measure of workers is one for each firm). Technically this means the total (and hence average) mass of w is constant over time. The fraction of workers that decides to stay back in their firm (which we call the *retention rate*), is denoted by λ which may in principle, vary between the firms. For the time being, we treat them as given and constant across the firms. This treatment is pioneered by Ref. [18] in the context of modeling income/wealth distributions as we have discusses above in great details. Let us denote the firm size of the i-th firm (we measure a firm's size by its workforce) by w_i ($i \leq N$ where N is the set of firms). Also, suppose that the number of firms from which the workers are leaving and moving into, is n. At each time point $(1 - \lambda)$ fraction of the workforce of those

n firms wants to leave (or the firms wanted them to leave, whatever appeals to the reader!). As mentioned above, we do not explain why they choose to do so. Hence there would be a total pool of workers that wants to change their workplace. Next, this pool of workers is randomly divided into those n firms. Hence, the dynamics is given by the following set of equations,

$$
\begin{aligned}
w_1(t+1) &= \lambda w_1(t) + \varepsilon_{1(t+1)}(1-\lambda)\sum_j^n w_j(t), \\
&\ldots \\
w_i(t+1) &= \lambda w_i(t) + \varepsilon_{i(t+1)}(1-\lambda)\sum_j^n w_j(t), \\
&\ldots \\
w_n(t+1) &= \lambda w_n(t) + \varepsilon_{n(t+1)}(1-\lambda)\sum_j^n w_j(t)
\end{aligned}
\tag{6.4}
$$

such that $\sum_j^n \varepsilon_{j(t)} = 1$ for all t. As is evident from above, this is a straight generalization of the usual kinetic exchange models (with $n = 2$) that has primarily been used to study the income/wealth distribution models (see Ref. [8] and the above sections). A little note on the notations: we use t within the first bracket when referring to the endogenous variables like the size of the firm ($w(t)$) and we use the same in subscript when referring to the exogenous random variables (e.g., ε_t).

Construction of the Division Factor ε We impose some meaningful restrictions on ε (as described by Ref. [38]).

1. $\varepsilon_i \geq 0\ \forall i$ and the sum of all ε_is has to be equal to one. Otherwise, the economy would not be conserved.
2. The distributions of all ε_i are identical which implies that the expection $E(\varepsilon_i) = 1/n$ for all i.
3. If $n = 2$, $\varepsilon_i \sim$ uniform [0, 1]. We impose this constraint so that at the lower limit of n, we get back the usual CC-CCM models (see Ref. [8]).

Formally, the problem then boils down to that of sampling uniformly from the unit simplex (see Refs. [35, 36]). We follow the standard algorithm and below we show the corresponding distribution of ε.

1. Create a vector of independent random variables drawn from uniform distribution over [0, 1], $\xi_1, \xi_2, \ldots, \xi_n$.
2. Take logarithm of all the elements of the vector.
3. Divide each element by the sum of all the elements. Call the i-th result ε_i for all i.

One can derive the probability density function of the ε_i which is the following:

$$f(\varepsilon_i) = (n-1)(1-\varepsilon_i)^{n-2}, \tag{6.5}$$

that is, ε has a beta pdf with parameters 1 and $n-1$. Clearly, when $n=2$ the distribution of ε is uniform [0, 1] as expected.

6.3.2.1 Solution of the Model

We follow Ref. [38] to describe the solution. First, we note that the solution to the usual kinetic exchange model with binary interaction is not known yet (see Refs. [8, 15]). However, one can derive an exact result for the case where the number of interacting firms is in the order of the system size N i.e., if one considers the case where $2 \ll n \le N$.

Note that if n is of the order of N, $\sum_j^n w_j$ is well approximated by n (recall that $E(w_j) = 1$ for all j). To make sure, note that $\sum_j^N w_j = N$ by specification of the model. To derive an exact result (instead of approximation), we shall assume that all firms interact at every step, i.e., $n = N$. Evidently the system of equation becomes

$$\begin{aligned} w_1(t+1) &= \lambda w_1(t) + \varepsilon_{1(t+1)}(1-\lambda)N, \\ &\dots \\ w_i(t+1) &= \lambda w_i(t) + \varepsilon_{i(t+1)}(1-\lambda)N, \\ &\dots \\ w_N(t+1) &= \lambda w_N(t) + \varepsilon_{N(t+1)}(1-\lambda)N \end{aligned} \tag{6.6}$$

with each ε_i is beta distributed as has been shown in Eq. (6.5) (see Construction of ε in Sect. 6.3.2). Note that in this form, we get rid of the effects of $w_j(t)$ in the evolution equation of $w_i(t)$ for all $j \neq i$ thus enabling us to uncouple the system of equations describing the coupled system (note that technically this system is still coupled). One more simplification is possible.

Let $\mu = N(1-\lambda)\varepsilon$ ignoring the subscripts. For a given N, it is easy to verify that the probability distribution of μ for large N is

$$\lim_{N\to\infty} f(\mu) \simeq \psi e^{-\psi\mu} \quad \text{where } \psi = \frac{1}{1-\lambda}. \tag{6.7}$$

Therefore, the system effectively reduces to

$$\begin{aligned} w_1(t+1) &= \lambda w_1(t) + \mu_{1(t+1)}, \\ &\dots \\ w_i(t+1) &= \lambda w_i(t) + \mu_{2(t+1)}, \\ &\dots \\ w_N(t+1) &= \lambda w_N(t) + \mu_{N(t+1)}, \end{aligned} \tag{6.8}$$

which is a system of autoregressive type equations with the distribution of errors (μ) given by Eq. (6.7). This is clearly solvable now.

6.3.2.2 Steady State Distributions

Let us now describe the steady state behavior of the system. First, we can consider the moments and show how they differ from the usual binary exchange mechanism. One writes the k-th moment of the distribution (without subscript) as

$$E\left((w-1)^k\right) = E\left(\sum_{l=0}^{k} \binom{k}{l} (-w)^l\right). \tag{6.9}$$

One simplifying assumption we make here is that w_i and w_j are independent variables (technically, they are not since the sum of all w_i's has to be a constant by structure of the model, N in this case; but for large number of interacting firms, this is a good approximation). It is easy to verify that with all firms interacting ($n = N$), the variance is given by

$$V(w) = \frac{(1-\lambda)}{(1+\lambda)}$$

whereas in the case of binary interaction [34]

$$V(w) = \frac{(1-\lambda)}{(1+2\lambda)}.$$

Note that for $\lambda = 0$, variance is unity in both cases which indicates that the distribution is the same (exponential) in both cases (not proven here; please see Ref. [38] for a derivation). To solve the system, let us write it as

$$w(t+1) = \lambda w(t) + \mu_{t+1}$$

which can, in turn, be rewritten with the lag operator L as $(1-\lambda L)w(t) = \mu_t$ and hence,

$$w(t) = \mu_t + \lambda\mu_{t-1} + \lambda^2\mu_{t-2} + \lambda^3\mu_{t-3} + \cdots .$$

Note that we already know the distribution of μ from Eq. (6.7),

$$f(\mu) \simeq \frac{1}{1-\lambda} e^{-\frac{1}{1-\lambda}\mu}.$$

Therefore by transforming the variable we can write

$$w = \tilde{\mu}_0 + \tilde{\mu}_1 + \tilde{\mu}_2 + \tilde{\mu}_3 + \cdots$$

where $\tilde{\mu}_j = \lambda^j \mu_{t-j}$ is distributed as

$$f(\tilde{\mu}_j) = \frac{1}{\lambda^j(1-\lambda)} e^{-\frac{\tilde{\mu}_j}{\lambda^j(1-\lambda)}}.$$

One can neglect the terms with high powers (more than say $\bar{k}$) of λ. Then firm-size w is the nothing but the sum of $\bar{k}$ exponentially distributed random variables with different parameters. Note that the Laplace transformation $L(s)$ of μ_j is $\phi_j/(\phi_j+s)$ where $\phi_j = 1/(\lambda^j(1-\lambda))$. Since the μ_j's are *i.i.d.* by definition (since the division factor ε was *i.i.d.*), pdf of w would be the convolution of the pdfs of the $\bar{k}$ random variables. By property of Laplace transformation, one can verify that the distribution of w would be (by taking limit on $\bar{k}$)

$$f(w) = \lim_{\bar{k}\to\infty} \sum_{i=1}^{\bar{k}} \phi_i \exp(-\phi_i w) \prod_{j=1, j\neq i}^{\bar{k}} \left(\frac{\phi_j}{\phi_j - \phi_i}\right) \tag{6.10}$$

where ϕ_i defined as $\phi_i = 1/(\lambda^i(1-\lambda))$.

6.3.3 Distributed Retention Rates

So far we have considered only a fixed retention rate λ. In this section we consider distributed λ (i.e., the retention rates differ across firms but they are fixed over time) following Ref. [8]. Specifically, we assume that the retention rates are uniformly distributed over the interval [0, 1] across the firms. The new system of equation is

$$w_1(t+1) = \lambda_1 w_1(t) + \varepsilon_{1(t+1)} \sum_j^n (1-\lambda_j) w_j(t),$$

$$\ldots$$

$$w_i(t+1) = \lambda_i w_i(t) + \varepsilon_{i(t+1)} \sum_j^n (1-\lambda_j) w_j(t), \tag{6.11}$$

$$\ldots$$

$$w_n(t+1) = \lambda_n w_n(t) + \varepsilon_{n(t+1)} \sum_j^n (1-\lambda_j) w_j(t).$$

To solve Eq. (6.11) in the steady state, note that $(1-\lambda_i)E(w_i) = C$, a constant, solves the problem. Hence, by following Ref. [37] one can easily show that the resultant distribution of the above model is a power law. Essentially, the argument is if λ is distributed uniformly across the firms, then the average mass of workers is the inverse of a uniform distribution which is known to be the Zipf's law. We have already encountered this argument in the CCM [8] model discussed at the beginning of this chapter.

6.3.4 A Model with Time-Varying Retention Rate: Emergence of Bimodality

We have already discussed how one can model firm dynamics with the tools provided by KEM. Now we are in a position to tackle heterogeneity in the retention rate both over time and across agents. More precisely, we will describe the retention rate λ as a function of current size and thus induce a non-trivial time-dependence on the retention rate as size of any firm fluctuates over time. Reference [39] defines the dynamics by the following set of equations,

$$w_1(t+1) = \lambda\big(w_1(t)\big)w_1(t) + \varepsilon_1(t+1)\sum_j^N\big(1-\lambda\big(w_j(t)\big)\big)w_j(t),$$
$$\dots$$
$$w_i(t+1) = \lambda\big(w_i(t)\big)w_i(t) + \varepsilon_i(t+1)\sum_j^N\big(1-\lambda\big(w_j(t)\big)\big)w_j(t),$$
$$\dots$$
$$w_n(t+1) = \lambda\big(w_n(t)\big)w_n(t) + \varepsilon_n(t+1)\sum_j^N\big(1-\lambda\big(w_j(t)\big)\big)w_j(t)$$

such that $\sum_j^n \varepsilon_j(t) = 1$ for all t. As is evident from above, this is a further generalization of the model discussed above. Here, the retention rate λ_i not only characterizes the agents but also explicitly becomes a function of time, $\lambda_i = \lambda_i(t)$ due to dependence on the current level of employment $w_i(t)$. Following Ref. [39], let us assume the following functional form of λ,

$$\lambda(w) = c_1\big(1-\exp(-c_2 w)\big) \quad c_1 \text{ and } c_2 \text{ being constants.} \tag{6.12}$$

Note that the retention rate increases as the current work-force increases. This equation basically captures the more realistic scenario that as a firm increases its workforce, the more workers it retains; or in the context of wealth distribution, a richer person saves more (see also Ref. [42] for interesting discussions). The exact solution is not known for this system. SO we perform Monte Carlo simulation which shows emergence of bimodality for certain parameter configurations (see Fig. 6.7).

It should be emphasized that this whole exercise distinguishes the KEM approach to the problem of 'missing middle' from other approaches that put the importance either on size-dependent or size-independent dynamics. We take the position that the firm-level dynamics is size-dependent or independent, depending on the level

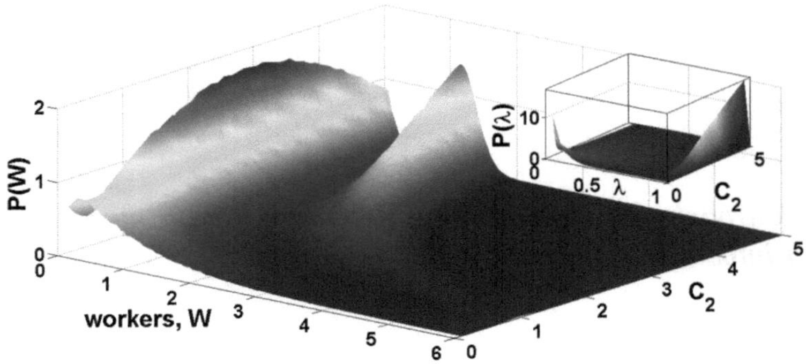

Fig. 6.7 Emergence of bimodality with Eq. (6.12) is shown with the variation of the parameter c_2 with $c_1 = 0.95$. The corresponding distribution of the retention rate λ is shown in the *inset*. Taken from [39]

of development of the economy as a whole. More specifically, we conjecture that the firms in the developed economies have fixed-heterogeneity whereas for poorer economies, the firms have size-dependent heterogeneity in the retention rates. In support of that conjecture, we note that the selection effects are important mostly for the micro firms (mostly unregistered, very small firms in the developing economies) and such effects are less prominent for larger firms (mostly found in the developed economies). The selection effects can produce heterogeneity depending on many factors, e.g., access to credit markets, pool of entrepreneurs, mobility of inputs, etc. Note that such facilities are mostly absent in the poorer economies. Hence, the firms in the poorer economies can have size-dependent dynamics. For example, a large firm can have access to credit market whereas a small firm may not have any access whatsoever; but in the developed economies all firms have access to credit markets, i.e., the access to credit market itself can act as a barrier to a small firm for expansion in size. This justifies the conjecture of scale-dependent heterogeneity for firms in poorer economies. However, we abstract from all such microeconomic details and posit that the heterogeneity is reflected solely in the retention rate which determines the firm's size in the this context. This simplification enables us to economize on the number of variables we study. We see that ex-post heterogeneity (agents are ex-ante identical, but because of the dependence of the retention rate on w, they are ex-post heterogeneous) induces bimodality. However, as the economy develops the heterogeneity becomes ex-ante as in the CCM model [8] giving rise to a power law distribution.

6.4 Opinion Dynamics

6.4.1 Kinetic Opinion Exchange Model

In this section, we deal with the emergence of consensus, which is an important issue in social science problems [5, 43–46]. The key question is of course, how a group of interacting individuals select between different options (vote, language, culture, opinions, etc.), leading to a state of 'consensus' in one such option, or a state of coexistence of many of them. Consensus, e.g., in opinion formation, is an "ordered phase" where the majority of the system is biased to a particular opinion. Though the influence of opinions in society has been an important field of study for a long time, the dynamics of opinion spreading has attracted the attention of physicists only recently, and there has been already several significant attempts to model such behavior in the light of rather well understood topics of physics like phase transitions and critical phenomena. These models have helped us to understand how *global* consensus emerges out of e.g., individual opinions [47–53]. In most of such formulations, opinions are usually modeled as variables, discrete or continuous, and are subject to changes due to many factors—binary interactions, global feedback, etc. or even external factors. Usually the interest in these studies lie in the distinct steady state properties, usually one phase characterized by individuals with widely different opinions and another phase with a finite fraction of individuals with similar opinions.

Here, we intend to focus our attention to a specific class of simple models proposed recently [54, 55], having apparent similarity with kinetic wealth exchange models (KWEMs) discussed earlier [8, 18, 23]. The tuning parameter in these models, analogous to the saving propensity in KEM, is 'conviction' (λ), which determines the extent to which an individual remains biased to its own opinion while interacting with somebody else, and the 'influence' parameter (μ), which is a measure of the influencing power or the ability of an individual to impose its opinion on some other individual. In the original model [54, 55], the two parameters were taken to be identical. The opinions of individuals are continuous variables in $[-1, 1]$ and change due to binary interactions. It was observed that if the conviction parameter was fixed above a threshold, the system reaches a state of consensus ("symmetry-broken" phase or "active" phase) and below this threshold value all individual opinions were equal to zero ("absorbing" phase). Later, a generalised version of this model was studied [56], where two parameters, λ and μ, were taken to be non-identical. It was found that in that case, the symmetric and symmetry-broken phases were separated by a phase boundary given by $\lambda = 1 - \mu/2$. Biswas et al. [57] studied some variants of the above discussed models and estimated the critical points by mean field theory (MFT), which were supported by numerical simulations. The critical exponents associated with the phase transition were also estimated. Later the discretized version of the LCCC model was exactly solved [58], which also showed an "active-absorbing phase transition" as was seen in the continuous version. Apart from the two- agent or binary interaction, the three-agent interaction were also taken into account. While the phase diagram of the two-agent interaction led to a continuous transition line, the three-agent interaction showed a discontinuous transition. Continuous opinion dynamics with both positive and negative mutual interactions were also studied [59].

In the model introduced earlier by Deffaunt et al., opinion exchange between two agents took place only when the difference between their opinions was less than or equal to a pre-assigned quantity δ [51]. This idea of bounded confidence was implemented in the LCCC model (controlled by the only parameter λ) [60]. Three distinct regions were identified in δ–λ phase diagram.

Percolation transitions of geometrical clusters (group of adjacent sites with an opinion value equal to or above a pre-assigned threshold value Ω) in the square lattice LCCC model, had also been studied by varying conviction and influencing parameters [61]. The transition point was different from that found for the transition of the order parameter. Although the transition point was also dependent on Ω, the critical exponents were independent of the threshold opinion value, conviction and influencing parameters. The exponents also suggested that percolation in LCCC model belongs to a separate universality class. We will now discuss in some details the above cases.

6.4.2 LCCC Model

In the original model, a discussion between two persons were viewed as a simple two-body *scattering* process in physics and at any time t, only two persons were allowed to discuss. In a society consisting of N persons each of the i-th person was assigned with an opinion value at a time t as $o_i(t) \in [-1, +1]$. Binary interactions took place as follows:

$$\begin{aligned} o_i(t+1) &= \lambda\big[o_i(t) + \varepsilon o_j(t)\big], \\ o_j(t+1) &= \lambda\big[o_j(t) + \varepsilon' o_i(t)\big], \end{aligned} \tag{6.13}$$

where ε and ε' are uncorrelated random numbers uniformly distributed between 0 and 1. λ was the conviction parameter which determines to which extent a person retains his own previous opinion and is independent of time. It was assumed here that the conviction parameter was also equal to the influence parameter which quantified how much an individual influenced another person. The agents were taken to be homogeneous in the sense of having the same or uniform conviction parameter.

Social interactions followed by Eq. (6.13) lead to consensus formation depending upon the value of the conviction parameter λ. The steady state value of the average opinion after a long time t, $O = |\sum_i o_i|/N$ represents the "ordering" in the system. Starting from a random disordered state (o_is are uniformly distributed with positive and negative values and $O \simeq 0$) after a certain time $t = \tau$ (relaxation time) the system either evolves to the "para" state (all individual agents having zero opinion) for $\lambda \leq 2/3$ or continuously changes to a symmetry broken state (all individuals having opinion of same sign) for $\lambda \geq 2/3$ (Fig. 6.8). The variance of O does not diverge but shows a cusp near $\lambda = 2/3$. The fraction of agents (p) having opinion $o_i = \pm 1$ was also measured in the steady state as a function of λ and the growth behaviour was found to be similar to O as discussed above.

The relaxation behaviours were studied for both O and p and a divergence of relaxation time $\tau \sim |\lambda - \lambda_c|^{-z}$ was observed for both. The relaxation behaviour for

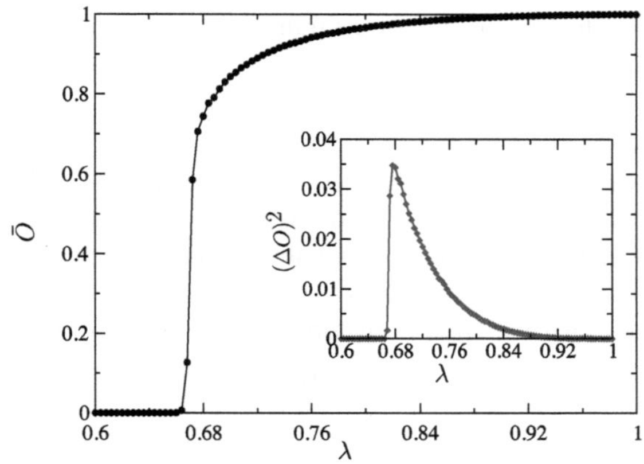

Fig. 6.8 Simulation results for average opinion as a function of λ. (*Inset*) Simulation results for the variance of O with λ. Taken from [54, 55]

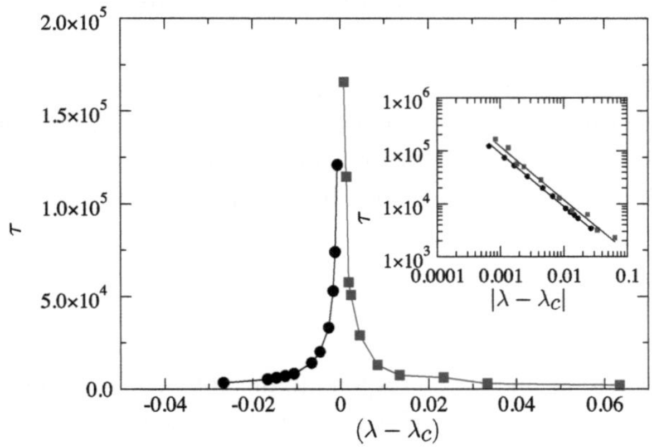

Fig. 6.9 Numerical results for relaxation time behaviors τ versus $\lambda - \lambda_c$, for multi-agent model with O. (*Insets*) Determination of exponent z from numerical fits of $\tau \sim |\lambda - \lambda_c|^{-z}$. Taken from [54, 55]

O has been shown in Fig. 6.9. The value of z corresponding to O and p are 1.0 ± 0.1 and 0.7 ± 0.1 respectively.

Apart from the exponents β and z (both for O and p), two more exponents were measured for 1D LCCC [57]. At the critical point ($\lambda_c = 2/3$), due to critical slowing down the system relaxes algebraically with time

$$O(t) \sim t^{-\delta} \tag{6.14}$$

The order parameter p also shows a similar form at the critical point. The value of δ for O is 1.00 ± 0.05 and p is 1.15 ± 0.01.

From finite size scaling theory, an order parameter X is expected to follow a scaling relation of the form

$$X(t) \approx t^{-\delta} \mathcal{F}\big(t^{1/\nu_{\|}} \Delta\big), \tag{6.15}$$

where $\Delta = \lambda - \lambda_c$ and $\mathcal{F}$ is a universal scaling function of a form such that for large argument, the time dependence drops out ($\mathcal{F}(x) \sim x^{\delta \nu_{\|}}$). Both O and p follow the scaling relation.

The basic nature of the transition produced by Eq. (6.13) was also obtained from a simple iterative map:

$$y(t+1) = \lambda(1+\varepsilon_t)y(t) \tag{6.16}$$

with the restriction that $y(t) \leq 1$, which is ensured by assuming that if $y(t) \geq 1$, $y(t)$ is set equal to 1. ε_t is a stochastic variable uniformly distributed between 0 and 1. In a mean-field approach, the above equation reduces effectively to a multiplier map like $y(t+1) = \lambda(1+\langle\varepsilon_t\rangle)y(t)$, where $\langle\varepsilon_t\rangle = 1/2$. For $\lambda \leq 2/3$, $y(t)$ converges to zero. An analytical derivation for the critical point was also given where it was found that $\lambda_c = \exp\{-(2\ln 2 - 1)\} \approx 0.6796$ [62].

6.4.3 The Generalised LCCC Model

In the generalised model [56], a second parameter representing the influencing power of an agent was treated distinctly from the conviction parameter, because in most realistic scenarios, a person with a strong retention power may not always have the same power to influence others. Thus, the interaction here is as follows:

$$\begin{aligned} o_i(t+1) &= \lambda_i o_i(t) + \varepsilon\mu_j o_j(t), \\ o_j(t+1) &= \lambda_j o_j(t) + \varepsilon'\mu_i o_i(t), \end{aligned} \tag{6.17}$$

where λ_i and μ_i correspond to the conviction and influencing parameter for the i-th agent. In the simpler version of the model studied, a homogeneous society with uniform λ and μ were assumed. Considering $\lambda = \mu$ gives the original LCCC model, as discussed earlier.

Again, the average opinion was studied both as functions of λ and μ. The average opinion showed spontaneous symmetry breaking in the λ–μ plane. In the steady state for non-zero solution of O the condition is

$$(1-\lambda)^2 = \langle\varepsilon\varepsilon'\rangle\,\mu^2, \tag{6.18}$$

which reduces to

$$\lambda = 1 - \mu/2. \tag{6.19}$$

The phase boundary obtained numerically satisfies Eq. (6.19) (Fig. 6.10). The relaxation behaviour was studied along two paths (A and B), (A) keeping the value of μ constant and changing the value of λ and (B) keeping the value of λ constant and changing the value of μ. The relaxation behaviour of O showed interesting feature along both the paths. The relaxation time along path A diverged algebraically along the phase boundary and more importantly the critical exponent changed with the tuning parameters (λ and μ). For $\mu_c = 0.4$, $z = 1.04 \pm 0.01$, for $\mu_c = 2/3$, $z = 1.10 \pm 0.03$ while for $\mu_c = 0.9$, $z = 1.21 \pm 0.01$ which indicated a

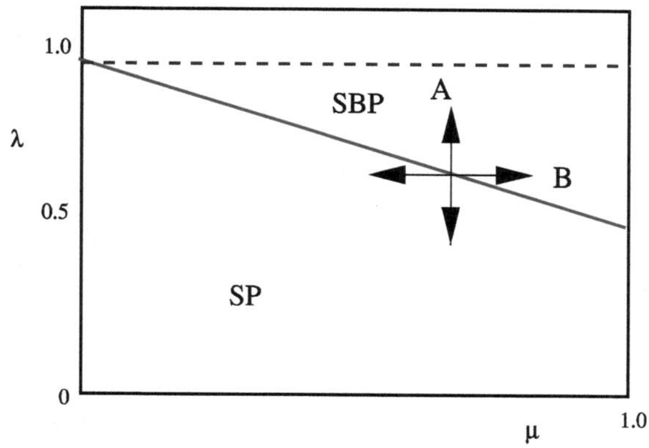

Fig. 6.10 The phase boundary obtained by numerical simulation coincides exactly with that given in Eq. (6.19). The acronyms SP and SBP denote the symmetric phase and the symmetry-broken phase, respectively. The paths A and B are possible trajectories along which the different studies can be made. Along the *dashed line* $\lambda = 1$, the opinions of all the agents are equal and take extreme values in two possible ways, either $o_i = 1$ or $o_i = -1$ for all i. Taken from [56]

non-universal behaviour. The order parameter O also showed power-law behaviour along the phase boundary,

$$O \propto (\lambda - \lambda_c)^{\beta} \tag{6.20}$$

where β also varied strongly with λ_c, μ_c. $\beta = 0.079 \pm 0.001$ at $\mu_c = 0.4$ and $\beta = 0.155 \pm 0.001$ at $\mu_c = 0.9$. The critical exponents corresponding to the condensate fraction p were z_p and β_p, which also showed non-universal behaviour along the phase boundary. When λ was kept constant and μ was varied near the phase boundary line, the magnitude of the time scales were about twice compared to those in path A along the path, although the values of the exponents were very close.

6.4.4 Variants of the LCCC Model

A simpler version of the LCCC model was studied [57], where an individual i upon meeting with another individual j retained his own opinion proportional to his conviction parameter and picked up a random fraction of j's opinion (model C hereafter). The interaction can be written as

$$\begin{aligned} o_i(t+1) &= \lambda o_i(t) + \varepsilon o_j(t), \\ o_j(t+1) &= \lambda o_j(t) + \varepsilon' o_i(t) \end{aligned} \tag{6.21}$$

where the symbols carry their usual meaning, as mentioned earlier.

Numerically, it was observed that below a critical value λ_c, $o_i = 0\ \forall i$ giving $O = 0$ while for $\lambda > \lambda_c$, $O > 0$ and went to 1 as $\lambda \to 1$, a *symmetry broken* phase

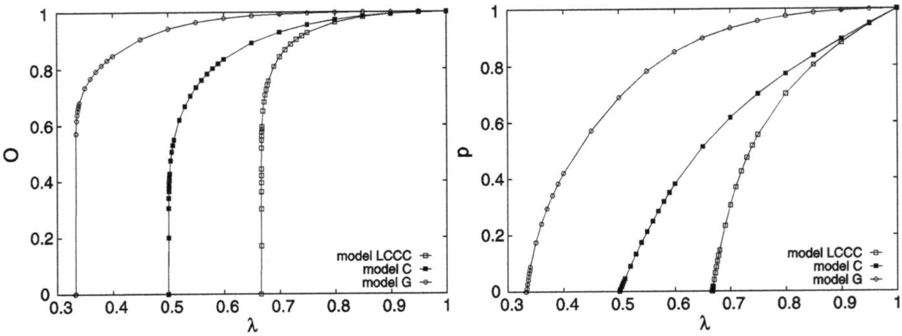

Fig. 6.11 The phase diagrams for the three models: LCCC, C and G, described in the text. *Left*: Behavior of order parameter O. *Right*: Behavior of condensation fraction p. Taken from [57]

with $\lambda_c \approx 1/2$. Mean field estimate gave for the stable value of O,

$$O\big(1 - \lambda - \langle\varepsilon\rangle\big) = 0, \tag{6.22}$$

and hence $\lambda_c = 1/2$.

The effect of global feedback on an agent's personal opinion during an interaction was also investigated [57]. An agent while taking part in a social interaction, apart from being influenced by the other person was also stochastically influenced by the "average opinion" of the entire society at that time (model G). Mathematically, this is represented as,

$$\begin{aligned} o_i(t+1) &= \lambda\big[o_i(t) + \varepsilon o_j(t)\big] + \varepsilon' O(t), \\ o_j(t+1) &= \lambda\big[o_j(t) + \eta o_i(t)\big] + \eta' O(t), \end{aligned} \tag{6.23}$$

where η and η' are random numbers drawn from uniform distribution [0, 1]. In this case, the *symmetry broken* phase $O \neq 0$ appeared for $\lambda > 1/3$, and for $\lambda \leq 1/3$ the system was in a *symmetric* phase, with $o_i = 0\ \forall i$ and all individual agents had the opinion 0. This was also explained by a mean-field approach as O reached a steady state value,

$$O = \lambda\big(1 + \langle\varepsilon\rangle\big)O + \big\langle\varepsilon'\big\rangle O \tag{6.24}$$

which gave $\lambda_c = 1/3$.

The comparative phase diagrams for the three models (1D LCCC, model C and G) according to behaviour of order parameter O and p has been shown in Fig. 6.11. Both for the above discussed models (1D LCCC, model C and G) the critical exponents were measured numerically (Table 6.1).

Table 6.1 Table comparing the different quantities for the 3 models (1D LCCC, model C and G)

Model	λ_c (Mean field)	Measured quantity	β	z	δ	$\nu_{\|\|}$
LCCC	2/3	O	0.10(1)	0.97(1)	1.00(5)	1.2(1)
		p	0.95(2)	1.1(1)	1.2(1)	1.1(1)
C	1/2	O	0.17(1)	1.58(1)	0.500(5)	0.10(1)
		p	0.98(2)	1.34(1)	0.521(5)	2.00(2)
G	1/3	O	0.081(1)	1.2(1)	0.585(1)	1.6(1)
		p	0.85(1)	1.75(1)	0.585(1)	2.0(1)

6.4.5 Discrete LCCC Model

An exact solution for the LCCC model was done for a discretized version (reproduced below in a similar form, from Ref. [58]). The dynamics evolves as

$$o_i(t+1) = \lambda o_i(t) + \mu\varepsilon o_j(t), \tag{6.25}$$

where μ represented the j-th agent's ability to influence others. Note that in the limit $\lambda = \mu$, one recovers the LCCC model. In the discrete version $\lambda = 1$ with probability ϕ and 0 with probability $1-\phi$. The parameter ε is either 1 or 0 with equal probability, and agents could have three possible opinion values ($o_i \in \{-1, 0, +1\} \forall i$). In the generalised case $\mu = 1$ with probability q and 0 with probability $1-q$.

If f_0, f_1 and f_{-1} be the fractions of agents having opinions 0, +1 and −1, then the evolution equation can be written as,

$$\begin{aligned} \frac{dO}{dt} &= f_{-1}^2(1-\phi) + f_{-1}f_1\left(1-\frac{\phi}{2}\right) + \frac{f_0 f_1 \phi}{2} + f_{-1}f_0(1-\phi) \\ &\quad - f_1^2(1-\phi) - f_1 f_{-1}\left(1-\frac{\phi}{2}\right) - \frac{f_0 f_{-1}\phi}{2} - f_1 f_0(1-\phi). \end{aligned} \tag{6.26}$$

In the steady state, the left hand side will be zero. This gives either $f_1 = f_{-1}$, (which implies disorder) or

$$f_0 = \frac{2(1-\phi)}{\phi}. \tag{6.27}$$

It was shown that in the ordered state $f_1 f_{-1} = 0$. This condition and the disordered state condition ($f_1 = f_{-1}$) should both be valid at the critical point. This is possible only when $f_1 = f_{-1} = 0$ at the critical point. This implied, at the critical point $f_0 = 1$. Furthermore, for the sake of continuity of f_1 and f_{-1}, $f_0 = 1$ for the entire disordered phase. This condition along with Eq. (6.27) gave $\phi_c = 2/3$.

Therefore, the order parameter should be (using $f_1 + f_{-1} + f_0 = 1$)

$$O = \pm(1 - f_0) \tag{6.28}$$

where the sign will depend on whether f_1 or f_{-1} is non-zero, in the ordered (symmetry-broken) phase. Using Eq. (6.27), the above expression yields

$$O = \pm \frac{3(\phi - \frac{2}{3})}{\phi}. \tag{6.29}$$

Therefore, Eq. (6.29) gives $\beta = 1$ (since $\phi_c = 2/3$).

Similar calculation in case of the discrete generalised LCCC model yields

$$O = \pm \frac{2(\phi - \phi_c) + (q - q_c)}{q\phi}. \tag{6.30}$$

which gives that the order parameter exponent is $\beta = 1$.

The three body opinion exchange was also solved exactly. Three agents were chosen randomly and an agent changes his opinion only when the other two agree among themselves. If they contradicted, then the first agent considered the group to be neutral and only retained a fraction of his opinion, depending upon his/her conviction parameter. This can be represented mathematically as

$$o_i(t+1) = \lambda o_i(t) + \lambda \varepsilon \theta_{jk}(t), \tag{6.31}$$

where, $\theta_{jk}(t) = o_j(t)$ if $o_j(t) = o_k(t)$, $\theta_{jk}(t) = 0$ otherwise. It was shown that in the ordered state,

$$f_0 = \frac{1}{2} - \frac{3\sqrt{\phi - 8/9}}{2\sqrt{\phi}}, \tag{6.32}$$

and the order parameter takes the form

$$O = \pm\left(\frac{1}{2} + \frac{3\sqrt{\phi - 8/9}}{2\sqrt{\phi}}\right). \tag{6.33}$$

This gives $O = 0$ for $\phi < 8/9$ and in the ordered phase minimum value of O can be $1/2$ which shows that the order-disorder transition is discontinuous.

6.4.6 LCCC Model with Bounded Confidence

In the models discussed so far, there were no restrictions imposed on the interactions between any two agents. A restricted LCCC model was studied, where two agents interact according to Eq. (6.13) only when $|o_i - o_j| \leq 2\delta$ [60], where δ is the parameter that represents the 'confidence' level and can vary from zero to 1. There are two extreme limits corresponding to this model: (a) $\delta = 1$ is identical to the original LCCC model, and (b) $\delta = 0$ is the case when two agents interact only when their opinions are exactly same. Three different states were defined to identify the status of the system. When $o_i = 0$ for all i it was called neutral state, $o_i \neq 0$ for all i, but

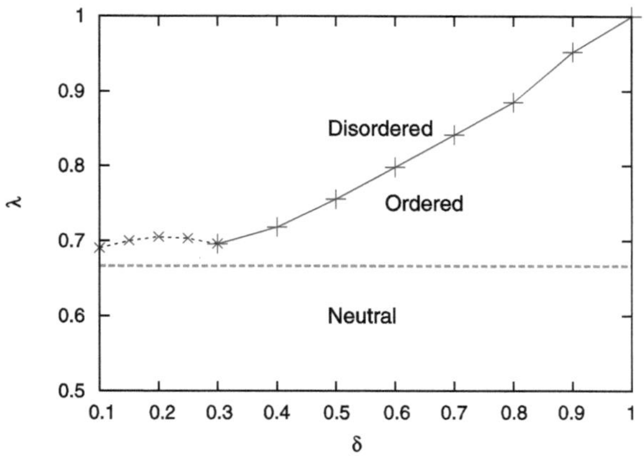

Fig. 6.12 The phase diagram in the δ–λ plane shows the existence of the neutral region (for $\lambda \le \lambda_{c1} \simeq 2/3$), the ordered region and the disordered region. The ordered and disordered regions are separated by a first order boundary (*continuous line in red*) for $\delta \ge 0.3$ obtained using a finite size scaling analysis. For $\delta < 0.3$, the phase boundary (*broken line in blue*) has been obtained approximately only from the behaviour of the order parameter (see text). Taken from [60]

$O \simeq 0$, it was called disordered state, and when $O \neq 0$ it was an ordered state. The three states were located in the δ–λ plane (Fig. 6.12).

It is quite obvious from the figure that the ordered state appears for $\lambda_{c1} \simeq 2/3$ (as in the original LCCC model) and is independent of the value of δ. For a fixed value of δ, the value of order parameter O increases with λ and decreases to zero as λ is increased further ($\ge \lambda_{c2}$). The decrease becomes steeper with both δ and the system size N. For $\lambda_{c1} \le \lambda \le \lambda_{c2}(\delta)$ an ordered region exists where opinions of one sign exist. At λ_{c2} a transition to a disordered state was observed and the transition point was dependent on the value δ. It was found that at least for $\delta > 0.3$, the order-disorder transition is first-order in nature. For $\delta < 0.3$, the ordered phase shrunk to a narrow region of the phase diagram.

6.4.7 Percolation in LCCC Model

The spreading of an opinion through a society is a very important issue. The cluster formation by groups of people acquiring similarity in opinion value is significant regarding this issue. The spreading of opinion among social agents may be compared with the percolation problem in physics. In order to have an insight of the spreading phenomena in LCCC model, percolation of geometrical clusters (comprised by a group of adjacent sites with an opinion value equal to or above a preassigned threshold value (Ω)) was studied on a square lattice where agents were located on the lattice sites [61]. The opinion exchange between pair of agents was same as that of LCCC (Eq. (6.13)). It was observed that the average value of the largest cluster size was controlled by the conviction/influencing parameter λ and for a fixed value

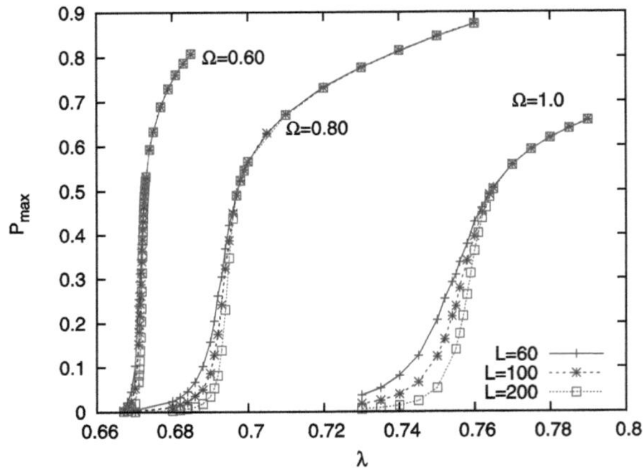

Fig. 6.13 Comparative plots for the largest cluster size with conviction parameter for three different system sizes and at three various values of the opinion threshold ($\Omega = 1.0$, 0.80 and 0.60). Taken from [61]

of Ω, at a critical value of $\lambda = \lambda_c^p$, the percolation transition occurs. One way to determine the percolation transition is to measure the relative size of the largest cluster which is designated by P_{max}. When the steady state is reached, P_{max} is calculated as S_L/L^2, where S_L is the size of the largest cluster and L is the linear size of the 2D system. The value of the critical point (λ_c^p) decreases with Ω (Fig. 6.13) and coincides with that for the transition point $\lambda_c = 2/3$ (as $\Omega \to 0.0$) at which the average opinion transition takes place (discussed in Sect. 6.4.2). Although the system does not show any finite size effect in case of the transition of the average opinion, the percolation transition shows prominent finite size effect for a given threshold opinion value Ω (Fig. 6.13).

The critical exponents were determined from the finite-size scaling relations [63, 64]. The order parameter follows the scaling form

$$P_{max} = L^{-\beta/\nu} \mathcal{F}\left[L^{1/\nu}\left(\lambda_c^p - \lambda\right)\right], \tag{6.34}$$

where $\mathcal{F}$ is a suitable scaling function. $P_{max} L^{\beta/\nu}$ were plotted against λ (at a fixed Ω) for different system sizes and then by tuning the value of β/ν, all the curves were made to cross at a single point which gives the critical conviction parameter (λ_c^p). A typical plot (for $\Omega = 0.80$ and $\lambda = \mu$) has been shown in Fig. 6.14. The finite-size scaling of the reduced fourth-order Binder cumulant of the order parameter defined as

$$U = 1 - \frac{\langle P_{max}^4 \rangle}{3\langle P_{max}^2 \rangle^2}, \tag{6.35}$$

was also studied where $\langle X \rangle$ means ensemble average of the parameter X. The Binder cumulant follows the scaling form

$$U = \mathcal{U}\left(\left(\lambda_c^p - \lambda\right) L^{1/\nu}\right), \tag{6.36}$$

where $\mathcal{U}$ is a suitable scaling function. The critical point corresponding to $\Omega = 0.80$ was $\lambda_c^p = 0.6955 \pm 0.0005$, which varies with the value of Ω. But the critical

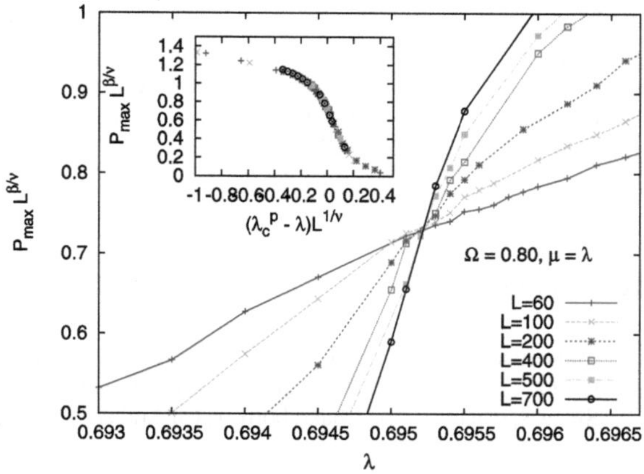

Fig. 6.14 $P_{max} L^{\beta/\nu}$ plotted against the conviction parameter λ where $\Omega = 0.80$ and $\mu = \lambda$. The curves for different system sizes ($L = 60, 100, 200, 400, 500$ and 700) cross at $\lambda_c^P = 0.6955 \pm 0.0005$ for $\beta/\nu = 0.130 \pm 0.005$. In the *inset* the data collapse for P_{max} with $\lambda_c^P - \lambda$ has been shown for $\Omega = 0.80$ giving $1/\nu = 0.80 \pm 0.01$ and $\beta/\nu = 0.130 \pm 0.005$. Taken from [61]

exponents $\beta/\nu = 0.130 \pm 0.005$ and $1/\nu = 0.80 \pm 0.01$ were independent of the value of Ω. They were also different from that obtained for the percolation transition in case of static Ising, dynamic Ising and standard percolation, indicating the LCCC dynamics to belong to a separate universality class.

The percolation transition was also studied in the case of generalised LCCC model (discussed in Sect. 6.4.3). Once again the critical exponents were found to be same as that obtained for the original LCCC model.

6.5 Final Remarks

In this article, we have tried to give a flavor of the many different kinetic exchange models, applied in various contexts such as in modeling of wealth distributions, or firm dynamics, or opinion formation in the society. There has been a flurry of activities in diverse domains, and several interesting observations and explanations have resulted, based on the common framework of simple exchanges of some quantity. It is interesting to see how the kinetic theory of gases which had played a substantial role in the initial development of the field of statistical mechanics, has inspired many more novel approaches in fields far away from the physics of gas molecules. There already exists a number of review articles, books, tutorials, etc. which have dealt with most of these topics. Keeping in mind the quote:

> "Dripping water hollows out stone, not through force but through persistence"—Ovid,

we have made another modest attempt! We would like to emphasize the effectiveness of the kinetic exchange models as serving as a skeleton for many diverse applications and implications. Hopefully, in the near future one will be able to put some more flesh on the skeleton to make it more human-like (or more reasonable, if you want)!

Acknowledgements We thank all our collaborators whose works we have reproduced here, and the critics who continue to motivate us to produce more. The authors would like to give special thanks to A. Chatterjee and S. Biswas for their useful comments and criticisms. AKC acknowledges the financial support from DST (India) under the SERC Fast Track Scheme for Young Scientists [Sanction no. SR/FTP/PS-090/2010(G)].

References

1. Mantegna RN, Stanley HE (1999) Introduction to econophysics: correlations and complexity in finance. Cambridge University Press, Cambridge
2. Sinha S, Chatterjee A, Chakraborti A, Chakrabarti BK (2011) Econophysics: an introduction. Wiley-VCH, Berlin
3. Chakrabarti BK, Chakraborti A, Chakravarty SR, Chatterjee A (2013) Econophysics of income and wealth distributions. Cambridge University Press, Cambridge
4. Aoyama H, Fujiwara Y, Ikeda Y, Iyetomi H, Souma W (2011) Econophysics and companies: statistical life and death in complex business networks. Cambridge University Press, New York
5. Chakrabarti BK, Chakraborti A, Chatterjee A (eds) (2006) Econophysics and sociophysics: trends and perspectives. Wiley-VCH, Berlin
6. Sen P, Chakrabarti BK (2013) Sociophysics: an introduction. Oxford University Press, Oxford
7. Galam S (2012) Sociophysics: a physicist's modeling of psycho-political phenomena. Springer, Berlin
8. Chatterjee A, Chakrabarti BK (2007) Eur Phys J B 60:135
9. Yakovenko VM, Barkley Rosser J Jr (2009) Rev Mod Phys 81:1703
10. Gallegati M, Keen S, Lux T, Ormerod P (2006) Physica A 370:1
11. Pareto V (1897) Cours d'economie politique. Rouge, Lausanne
12. Gibrat R (1931) Les inégalités économiques. Libraire du Recueil Sirey, Paris
13. Lux T (2005) In: Chatterjee A, Yarlagadda S, Chakrabarti BK (eds) Econophysics of wealth distributions. Springer, Milan, pp 51–60
14. Patriarca M, Chakraborti A, Kaski K, Germano G (2005) In: Chatterjee A, Yarlagadda S, Chakrabarti BK (eds) Econophysics of wealth distributions. Springer, Milan, pp 93–110
15. Patriarca M, Heinsalu E, Chakraborti A (2010) Eur Phys J B 73:145
16. Drăgulescu A, Yakovenko VM (2000) Eur Phys J B 17:723
17. Lallouache M, Jedidi A, Chakraborti A (2010) Sci Cult 76:478
18. Chakraborti A, Chakrabarti BK (2000) Eur Phys J B 17:167
19. Chakraborti A, Patriarca M (2009) Phys Rev Lett 103:228701
20. Patriarca M, Chakraborti A, Kaski K (2004) Physica A 340:334
21. Patriarca M, Chakraborti A, Kaski K (2004) Phys Rev E 70:016104
22. Chakraborti A, Patriarca M (2008) Pramana J Phys 71:233
23. Chatterjee A, Chakrabarti BK, Manna SS (2004) Physica A 335:155
24. Bhattacharya K, Mukherjee G, Manna SS (2005) In: Chatterjee A, Yarlagadda S, Chakrabarti BK (eds) Econophysics of wealth distributions. Springer, Milan, pp 111–119
25. Chatterjee A, Chakrabarti BK, Stinchcombe RB (2005) Phys Rev E 72:026126
26. Patriarca M, Chakraborti A (2013) Am J Phys 81(8):618. doi:10.1119/1.4807852
27. Ghosh A (unpublished)
28. Independent studies (unpublished) made by S Sinha and also A Chatterjee in 2005–2006, had similar observations; private communications
29. Chakraborti A (2002) Int J Mod Phys C 13:1315
30. Ghosh A, Basu U, Chakraborti A, Chakrabarti BK (2011) Phys Rev E 83:061130
31. Manna SS (1991) Physica A 179:249
32. Manna SS (1991) J Phys A 24:L363

33. Axtell R (2001) Science 293:1818
34. Chakrabarti AS, Chakrabarti BK (2009) Physica A 388:4151
35. Onn S, Weissman I (2011) Annu Oper Rev 189:331
36. Smith N, Tromble R (2004) tech report, John Hopkins University; available at http://www.cs.cmu.edu/~nasmith/papers/smith+tromble.tr04.pdf
37. Mohanty PK (2006) Phys Rev E 74:011117
38. Chakrabarti AS (2012) Physica A 391:6039
39. Chakrabarti AS (2013) Eur Phys J B 86:255. doi:10.1140/epjb/e2013-40114-4
40. Tybout J (2000) J Econ Lit 38:11
41. Ferrero JC (2005) In: Chatterjee A et al (eds) Econophysics of wealth distribution. Springer, Milan
42. Patriarca M, Chakraborti A, Germano G (2006) Physica A 369:723
43. Stauffer D (2009) In: Meyers RA (ed) Encyclopedia of complexity and systems science. Springer, New York
44. Stauffer D (2003) Comput Sci Eng 5:71
45. Castellano C, Fortunato S, Loreto V (2009) Rev Mod Phys 81:591
46. Galam S, Gefen Y, Shapir Y (1982) J Math Sociol 9:1
47. Liggett TM (1999) Interacting particle systems: contact, voter and exclusion processes. Springer, Berlin
48. Sznajd-Weron K, Sznajd J (2000) Int J Mod Phys C 11:1157
49. Galam S (2008) Int J Mod Phys C 19:409
50. Hegselman R, Krause U (2002) J Artif Soc Soc Simul 5:2
51. Deffuant G, Neau N, Amblard F, Weisbuch G (2000) Adv Complex Syst 3:87
52. Fortunato S (2005) Int J Mod Phys C 16:17
53. Toscani G (2006) Commun Math Sci 4:481
54. Lallouache M, Chakraborti A, Chakrabarti BK (2010) Sci Cult 76:485
55. Lallouache M, Chakrabarti AS, Chakraborti A, Chakrabarti BK (2010) Phys Rev E 82:056112
56. Sen P (2011) Phys Rev E 83:016108
57. Biswas S, Chandra AK, Chatterjee A, Chakrabarti BK (2011) J Phys Conf Ser 297:012004
58. Biswas S (2011) Phys Rev E 84:056106
59. Biswas S, Chatterjee A, Sen P (2012) Physica A 391:3257
60. Sen P (2012) Phys Rev E 84:016115
61. Chandra AK (2012) Phys Rev E 85:021149
62. Marsili M, private communication
63. Tsakiris N, Maragakis M, Kosmidis K, Argyrakis P (2010) Phys Rev E 82:041108
64. Biswas S, Kundu A, Chandra AK (2011) Phys Rev E 83:021109

Chapter 7
Kinetic Exchange Opinion Model: Solution in the Single Parameter Map Limit

Krishanu Roy Chowdhury, Asim Ghosh, Soumyajyoti Biswas, and Bikas K. Chakrabarti

Abstract We study a recently proposed kinetic exchange opinion model (Lallouache et. al., Phys. Rev E 82:056112, 2010) in the limit of a single parameter map. Although it does not include the essentially complex behavior of the multiagent version, it provides us with the insight regarding the choice of order parameter for the system as well as some of its other dynamical properties. We also study the generalized two-parameter version of the model, and provide the exact phase diagram. The universal behavior along this phase boundary in terms of the suitably defined order parameter is seen.

7.1 Introduction

Dynamics of opinion and subsequent emergence of consensus in a society are being extensively studied recently [1–9]. Due to the involvement of many individuals, this type of dynamics in a society can be treated as an example of a complex system, thus enabling the use of conventional tools of statistical mechanics to model it [10–14]. Of course, it is not possible to capture all the diversities of human interaction through any model of this kind. But often it is our interest to find out the global perspectives of a social system, like average opinion of all the individuals regarding an issue, where the intricacies of the interactions, in some sense, are averaged out. This

K.R. Chowdhury · A. Ghosh · S. Biswas (✉) · B.K. Chakrabarti
Saha Institute of Nuclear Physics, 1/AF, Bidhannagar, Kolkata 700064, India
e-mail: soumyajyoti.biswas@saha.ac.in

K.R. Chowdhury
e-mail: krishanu.1987@gmail.com

A. Ghosh
e-mail: asim.ghosh@saha.ac.in

B.K. Chakrabarti
e-mail: bikask.chakrabarti@saha.ac.in

Present address:
K.R. Chowdhury
Max Planck Institute for the Physics of Complex System, Dresden 01187, Germany

F. Abergel et al. (eds.), *Econophysics of Agent-Based models*,
New Economic Windows, DOI 10.1007/978-3-319-00023-7_7,

is similar to the approach of kinetic theory, where the individual atoms, although following a deterministic dynamics, are treated as randomly moving objects and the macroscopic behaviors of the whole system are rather accurately predicted.

Indeed, there have been several attempts to realize the human interactions in terms of kinetic exchange of opinions between individuals [10, 11, 14]. Of course, there is no conservations in terms of opinion. Otherwise, this is similar to momentum exchange between the molecules of an ideal gas. These models were often studied using a finite confidence level, i.e., agents having opinions close to one another interact. However, in a recently proposed model [15], unrestricted interactions between all the agents were considered. The single parameter in the model described the 'conviction' which is a measure of an agent's tendency to retain his opinion and also to convince others to take his opinion. It was found that beyond a certain value of this 'conviction parameter' the 'society', made up of N such agents, reaches a consensus, where majority shares similar opinion. As the opinion values could take any values between $[-1:+1]$, a consensus means a spontaneous breaking of a discrete symmetry.

There have been subsequent studies to generalize this model, where the 'conviction parameter' and 'ability to influence' were taken as independent parameters [16]. In that two-parameter version, similar phase transitions were observed. However, the critical behaviors in terms of the usual order parameter, the average opinion, were found to be non-universal. There have been other extensions in terms of a phase transition induced by negative interactions [17], an exact solution in a discrete limit [18], the effect of non-uniform conviction and update rules in these discrete variants [19], a generalized map version [20], a percolation transitions in a square lattices [21] and the effect of bounded confidence [22] in these models.

In the present study, we investigate the single parameter map version of the model, also proposed in Ref. [15]. Although the original model is difficult to tackle analytically, in this mean field limit, it can simply be conceived as a random walk. Using standard random walk statistics, several static and dynamical quantities have been calculated. We show that the fraction of extreme opinion behaves like the actual order parameter for the system, and the average opinion shows unusual behavior near critical point. The critical behavior of the order parameter and its relaxation behavior near and at the critical point have been obtained analytically which agree with numerical simulations.

7.2 Model and Its Map Version

Let the opinion of any individual (i) at any time (t) is represented by a real valued variable $O_i(t)$ ($-1 \leq O_i < 1$). The kinetic exchange model of opinion pictures the opinion exchange between two agents like a scattering process in an ideal gas. However, unlike ideal gas, there is no conservation of the total opinion. This is similar to the kinetic exchange models of wealth redistribution, where of course the conservation was also present [23–26]. The (discrete time) exchange equations of the model read

$$O_i(t+1) = \lambda O_i(t) + \lambda \varepsilon O_j(t), \tag{7.1}$$

and a similar equation for $O_j(t+1)$, where $O_i(t)$ is the opinion value of the ith agent at time t, λ is the 'conviction parameter' (considered to be equal for all agents for simplicity) and ε is an annealed random number drawn from a uniform and continuous distribution between $[0:1]$, which is the probability with which i and j interact (see [15]). Note that the choice of i and j are unrestricted, making the effective interaction range to be infinite. The opinion values allowed are bounded between the limits $-1 \le O_i(t) \le +1$. So, whenever the opinion values are predicted to be greater (less) than $+1$ (-1) following Eq. (7.1), it is kept at $+1$ (-1). This bound, along with Eq. (7.1), defines the dynamics of the model.

This model shows a symmetry breaking transition at a critical value of λ ($\lambda_c \approx 2/3$). The critical behaviors were studied using the average opinion $O_a = |\sum_{i=1}^{N} O_i(t \to \infty)|/N$ [27]. An alternative parameter was also defined in Ref. [15], which is the fraction of agents having extreme opinion values. This quantity also showed critical behaviors at the same transition point.

The model in its original form is rather difficult to tackle analytically (it can be solved in some special limits though [18]). However, as it is a fully connected model, a mean field approach would lead to the following evolution equation for the single parameter opinion value (cf. [15])

$$O(t+1) = \lambda(1+\varepsilon)O(t). \tag{7.2}$$

This is, in fact, a stochastic map with the bound $|O(t)| \le 1$. For all subsequent discussions, whenever an explicit time dependence of a quantity is not mentioned, it denotes the steady state value of that quantity and a subscript a denotes the average over the randomness (i.e., ensemble average). As we will see from the subsequent discussions, this map can be conceived as a random walk with a reflecting boundary. As in the case of the multiagent version, the distribution of ε does not play any role in the critical behavior. We have considered two distributions, one is continuous in the interval $[0:1]$ and the other is 0 and 1 with equal probability. Both of these give similar critical behavior.

We also briefly discuss the two-parameter model, where the 'conviction' of an agent and the ability to convince others were taken as two independent parameters [16]. In that context, the map would read

$$O(t+1) = (\lambda + \mu\varepsilon)O(t), \tag{7.3}$$

where μ is the parameter determining an agent's ability to influence others. As before, $|O(t)| \le 1$.

7.3 Results

7.3.1 Random Walk Picture

One can study the stochastic map in Eq. (7.2) by describing it in terms of random walks. Writing $X(t) = \log(O(t))$ (for all subsequent discussions we always take $O(t)$ to be positive), Eq. (7.2) can be written as

$$X(t+1) = X(t) + \eta, \tag{7.4}$$

where, $\eta(t) = \log[\lambda(1+\varepsilon)]$. As is clear from the above equation, it actually describes a random walk with a reflecting boundary at $X = 0$ to take the upper cut-off of $O(t)$ into account. Depending upon the value of λ, the walk can be biased to either ways and is unbiased just at the critical point. As one can average independently over these additive terms in Eq. (7.4), this gives an easy way to estimate the critical point [15]. An unbiased random walk would imply $\langle \eta \rangle = 0$ i.e.,

$$\int_0^1 \log\big[\lambda_c(1+\varepsilon)\big] d\varepsilon = 0 \tag{7.5}$$

giving $\lambda_c = e/4$, where we have considered a uniform distribution of ε in the limit $[0:1]$. This estimate matches very well with numerical results of this and earlier works [15]. In order to guess the λ dependence of O_a in the ordered region, we first estimate the "average return time" T (return time is the time between two successive reflections from $X = 0$) as a function of bias of the walk. For this uniform distribution of ε, the average position to which the walker goes following a reflection from the barrier is $(\lambda+1)/2$. The average amount of contribution in each step is given by $\int_o^1 \log[\lambda(1+\varepsilon)]d\varepsilon = \log(\lambda/\lambda_c)$. This, in fact, is a measure of the bias of the walk, which vanishes linearly with $(\lambda - \lambda_c)$ as $\lambda \to \lambda_c$. So, in this map picture, one would expect that on average by multiplying this λ/λ_c factor T times (i.e., adding $\log(\lambda/\lambda_c)$ T times in the random walk picture), $O(t)$ would reach 1 from $(\lambda+1)/2$. Therefore,

$$\frac{\lambda+1}{2}\left(\frac{\lambda}{\lambda_c}\right)^T = 1, \tag{7.6}$$

giving

$$T = -\frac{\log\lambda}{\log\lambda - \log\lambda_c} \approx -\frac{\log\lambda}{\lambda - \lambda_c} \tag{7.7}$$

for $\lambda \to \lambda_c$. Clearly, the average return time diverges near the critical point obeying a power law: $T \sim (\lambda - \lambda_c)^{-1}$. In Fig. 7.1 we have plotted this average return time as a function of λ. The power-law divergence agrees very well with the prediction.

The steady state average value of $X(t)$ i.e., X_a (and correspondingly O_a) is expected to be proportional to $\sqrt{T}$ in steps of $\log\lambda$:

$$X_a \sim \sqrt{T}\log\lambda = k\sqrt{T}\log\lambda, \tag{7.8}$$

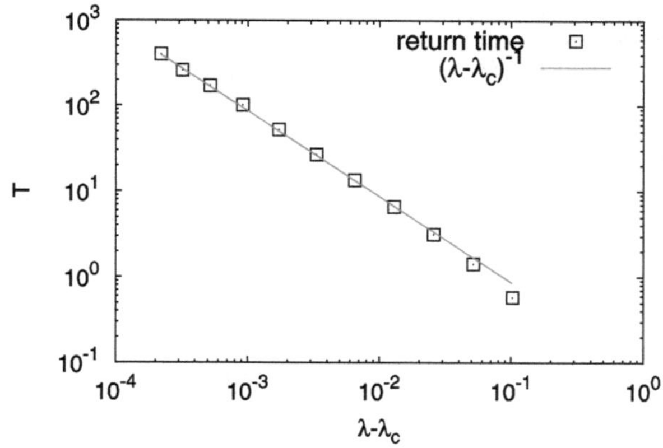

Fig. 7.1 The average return time T of $O(t)$ to 1 in the map described in Eq. (7.2) is plotted with $(\lambda - \lambda_c)$. It shows a divergence with exponent 1 as is predicted from Eq. (7.7)

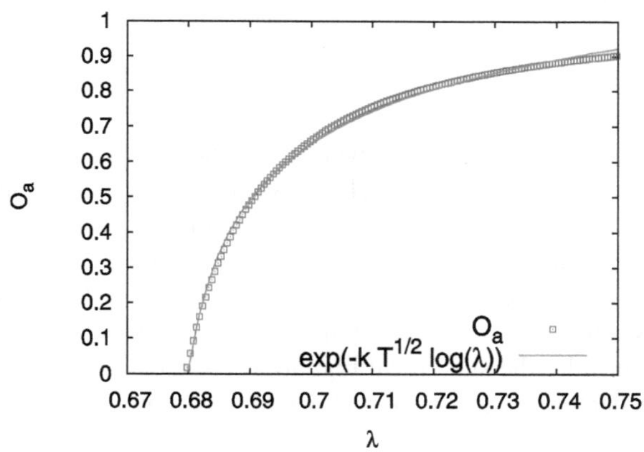

Fig. 7.2 O_a is plotted with λ. The data points are results of numerical simulations, which fits rather well with the *solid line* predicted from Eq. (7.9), with $k = 0.7$

where k is a constant. This gives

$$O_a = \exp\bigl[-k|\log\lambda|^{3/2}(\lambda - \lambda_c)^{-1/2}\bigr]. \tag{7.9}$$

The above functional form fits quite well (see Fig. 7.2) with the numerical simulation results near the critical point. It may be noted that the numerical results for the kinetic opinion exchange Eq. (7.2) also fits quite well with this expression (Eq. (7.9)). We note that O_a increases from zero at the critical point and eventually reaches 1 at $\lambda = 1$. But its behavior close to critical point cannot be fitted with a power-law growth usually observed for order-parameters. Such peculiarity in the critical behavior of O_a compels us to exclude it as an order parameter though it satisfies some other good qualities of an order parameter. Instead, we consider the average 'condensation fraction' ρ_a as the order parameter. In the multi-agent version, it was defined as the fraction of agents having extreme opinion values i.e., -1 or $+1$. In this case it is defined as the probability that $O(t) = 1$. We denote this

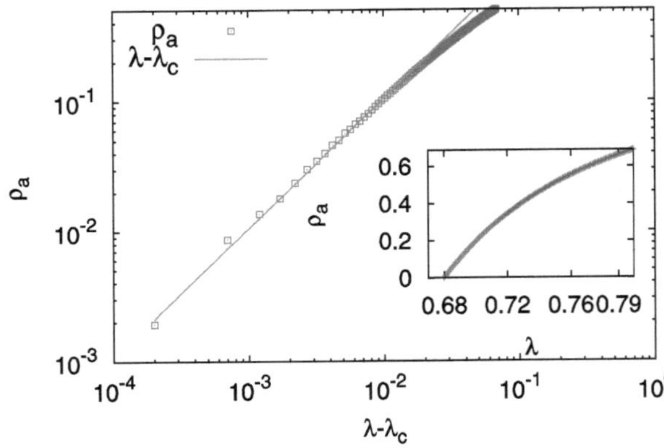

Fig. 7.3 The average condensation fraction (probability that $O = 1$) ρ_a is plotted with $(\lambda - \lambda_c)$. A linear fit in the log-log scale gives the growth exponent 1, as predicted from Eq. (7.10). *Inset* shows the variation of ρ with λ

quantity by $\rho(t)$. As is clear from the definition, one must have

$$\rho_a \sim \frac{1}{T}, \tag{7.10}$$

where T is the return time of the walker. As $T \sim (\lambda - \lambda_c)^{-1}$, $\rho_a \sim (\lambda - \lambda_c)^{\beta}$ with $\beta = 1$. This behavior is clearly seen in the numerical simulations (see Fig. 7.3).

Also, the relaxation time shows a divergence as the critical point is approached. We argue that there is a single relaxation time scale for both $O(t)$ and $\rho(t)$. So we calculate the divergence of relaxation time for $O(t)$ and numerically show that the results agree very well with the relaxation time divergence for both $O(t)$ and $\rho(t)$. Consider the subcritical regime where the random walker is biased away from the reflector (at the origin) and would have a probability distribution for the position of the walker as

$$p(X) = \frac{A}{\sqrt{t}} \exp\bigl[-B(X - vt)^2/t\bigr], \tag{7.11}$$

where $v \sim 1/T \sim (\lambda - \lambda_c)$ is the net bias and constants A, B do not depend on t. One can therefore obtain the probability distribution P of O using $p(X)dX = P(O)dO$,

$$P(O) = \frac{A}{\sqrt{t}} \frac{1}{O} \exp\bigl[-B(\log O - vt)^2/t\bigr]. \tag{7.12}$$

Hence

$$\begin{aligned} O_a(t) &= \int_0^1 O P(O) dO, \\ &= \frac{A}{\sqrt{t}} \int_0^1 \exp\bigl[-B(\log O - vt)^2/t\bigr] dO \\ &\sim \frac{A}{\sqrt{t}} \exp\bigl(-Bv^2 t\bigr), \end{aligned} \tag{7.13}$$

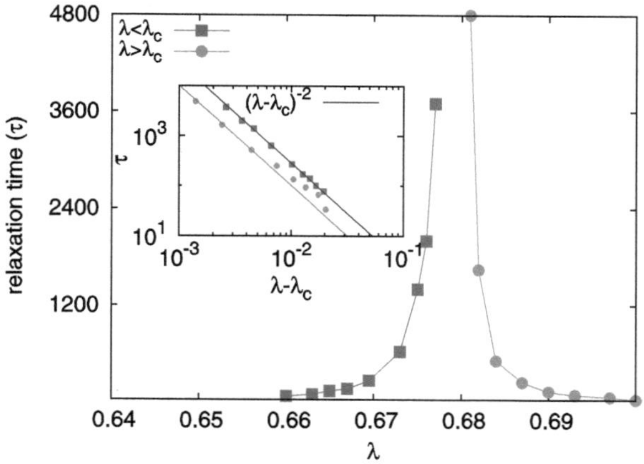

Fig. 7.4 The average relaxation time for $O_a(t)$ is plotted with λ. This shows a prominent divergence as the critical point is approached. In the *inset*, the relaxation time is plotted in the log-log scale against $(\lambda - \lambda_c)$. The exponent is 2 as is expected from Eq. (7.13)

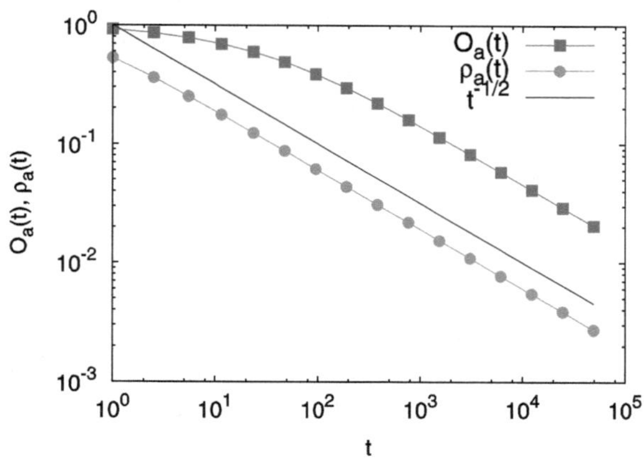

Fig. 7.5 The time dependence of both $O_a(t)$ and $\rho(t)$ are plotted at the critical point in the log-log plot. The linear fit shows a time variation of the form $t^{-\delta}$ with $\delta = 1$, as is expected from Eq. (7.13)

in the long time limit, giving a time scale of relaxation $\tau \sim v^{-2} \sim (\lambda - \lambda_c)^{-2}$. We have fitted the relaxation of $O_a(t)$, obtained numerically, with an exponential decay and found τ. As can be observed from Fig. 7.4 it shows a clear divergence close to critical point with exponent 2.

We have obtained the relaxation time of $\rho_a(t)$ and it also shows similar divergence. Note that at $\lambda = \lambda_c$, $v = 0$ and it follows from Eq. (7.13) that $O_a(t) \sim t^{-1/2}$. This behavior is also confirmed numerically (see Fig. 7.5). The average condensation fraction $\rho_a(t)$ too follows this scaling, giving $\delta = 1/2$ (as order parameter relaxes as $t^{-\delta}$ at critical point).

We have also investigated the effect of having an external field linearly coupled with $O(t)$. In the multiagent scenario, this can have the interpretation of the influence of media. The map equation now reads,

$$O(t+1) = \lambda(1+\varepsilon_t)O(t) + hO(t), \tag{7.14}$$

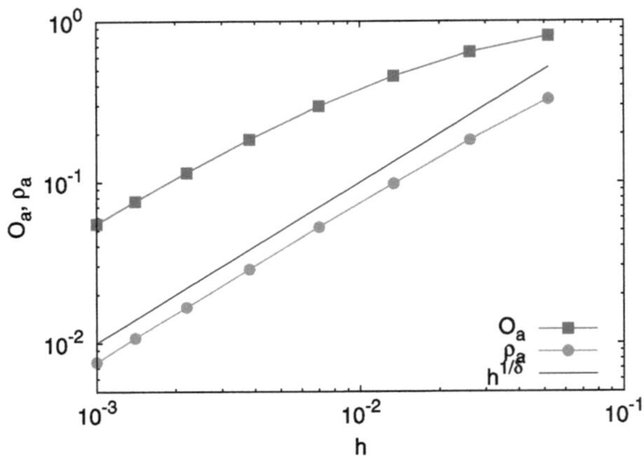

Fig. 7.6 The variation of O_a and ρ_a are plotted against the external field h at the critical point $\lambda = \lambda_C$. The linear fit in log-log scale shows $\delta' = 1$

where h is the field (constant in time). We have studied the response of O_a and ρ_a at $\lambda = \lambda_c$ due to application of small h. We find that (see Fig. 7.6) both grow linearly with h. One expects the order parameter to scale with external field at the critical point as $\rho_a \sim h^{1/\delta'}$. In this case $\delta' = 1$.

7.3.2 Random Walk with Discrete Step Size

One can simplify the random walk mentioned above and make it a random walk with discrete step sizes. This can be done by considering the distribution of ε to be a double delta function, i.e., $\varepsilon = 1$ or 0 with equal probability. This will make $\eta(t)$ in Eq. (7.2) to be $\log \lambda$ or $\log(2\lambda)$ with equal probability. Below critical point, both steps are in negative direction (away from reflector) and consequently taking the walker to $-\infty$. Exactly at critical point ($\lambda = \lambda_c$) the step sizes become equal and opposite i.e., $\log \lambda_c = -\log(2\lambda_c)$ giving $\lambda_c = 1/\sqrt{2}$. Above critical point, one of the steps is positive and the other is negative. However, the magnitudes of the steps are different. This unbiased walker (probability of taking positive and negative steps are equal) with different step sizes can approximately be mapped to a biased walker with equal step size in both directions. To do that consider the probability $p(x,t)$ that the walker is at position x at time t. One can then write the master equation

$$p(x,t+1) = \frac{1}{2}p(x+a,t) + \frac{1}{2}p(x+a+b,t), \tag{7.15}$$

where $a = \log \lambda$ and $b = \log 2$. Clearly,

$$\frac{\partial p(x,t)}{\partial t} = \left(a + \frac{b}{2}\right)\frac{\partial p(x,t)}{\partial x} + \left(\frac{a^2}{2} + \frac{ab}{2} + \frac{b^2}{4}\right)\frac{\partial^2 p(x,t)}{\partial x^2}. \tag{7.16}$$

Now the master equation for the usual biased random walker can be written as

$$p(x,t+1) = p' p\big(x+a',t\big) + q' p\big(x-a',t\big), \tag{7.17}$$

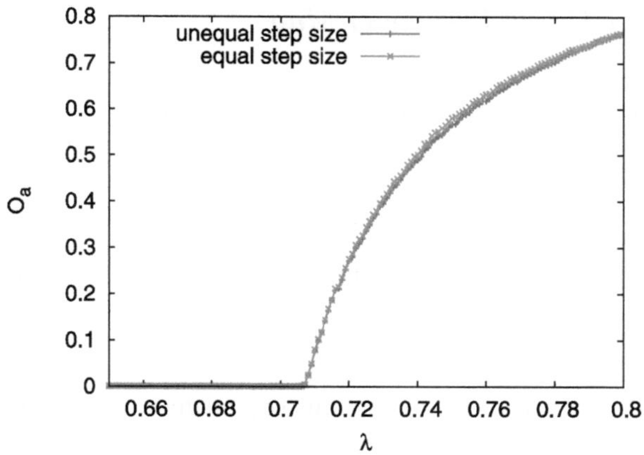

Fig. 7.7 The comparison of the biased walk with equal step size with the original walk is shown. Reasonable agreement is seen for a wide range of λ values

where p' and q' denote respectively the probabilities of taking positive and negative steps ($p'+q'=1$) and a' is the (equal) step size in either direction. The differential form of this equation reads

$$\frac{\partial p(x,t)}{\partial t} = (p'-q')a'\frac{\partial p(x,t)}{\partial x} + \frac{a'^2}{2}\frac{\partial^2 p(x,t)}{\partial x^2}. \tag{7.18}$$

Comparing these Eqs. (7.16) and (7.18), one gets

$$\begin{aligned} p' &= \frac{1}{2}\left[1+\frac{a+b/2}{a'}\right], \\ q' &= \frac{1}{2}\left[1-\frac{a+b/2}{a'}\right], \\ a' &= -\sqrt{\left(\log(\lambda)-\log(\lambda_c)\right)^2+\left(\log(\lambda_c)\right)^2}. \end{aligned} \tag{7.19}$$

Therefore, as $\lambda \to \lambda_c$, the bias $(p'-q') \sim (\lambda-\lambda_c)/a'$. These are consistent with the earlier calculations where we have taken the bias to be proportional to $(\lambda-\lambda_c)$. To check if this mapping indeed works, we have simulated a biased random walk with above mentioned parameters and found it to agree with the original walk (see Fig. 7.7). Similar to the approach taken for the continuous step-size walk, one can find the return time of the walker. This time the walker is exactly located at λ after it is reflected from the barrier. The return time again diverges as $(\lambda-\lambda_c)^{-1}$. Also $O_a(t)$ will have a similar form upto some prefactors. Condensation fraction will increase linearly with $(\lambda-\lambda_c)$ close to the critical point. All the other exponents regarding the relaxation time, time dependence at the critical point and dependence with external field are same as before. This shows that the critical behavior is universal with respect to changes in the distribution of ε.

7.3.3 Two Parameter Map

As the multi-agent model was generalized in a two-parameter model [16], one can also study the map version of that two-parameter model. It would read

$$O(t+1) = (\lambda + \mu\varepsilon)O(t). \tag{7.20}$$

As before, one can take log of both sides and in similar notations

$$X(t+1) = X(t) + \log(\lambda + \mu\varepsilon). \tag{7.21}$$

This can also be seen as a biased random walk. In fact, one can write the above equation as

$$X(t+1) = X(t) + \log\bigl[\lambda\bigl(1+\varepsilon'\bigr)\bigr], \tag{7.22}$$

where $\varepsilon' = (\mu/\lambda)\varepsilon$. This effectively changes the limit of the distribution of the stochastic parameter. Here ε' is distributed between 0 and μ/λ. One can then do the earlier exercise with this version as well. If one makes ε discrete, this is again a walk with unequal step sizes, which can again be mapped to a biased walk with equal step sizes. Therefore, in either case, some pre-factors will change, but the critical behavior will be the same as before. The critical behavior is, therefore, universal when studied in terms of the proper order parameter (condensation fraction). For uniformly distributed ε (in the range $[0:1]$), one can get the expression for the phase boundary from the equation

$$\int_0^1 \log(\lambda + \mu\varepsilon)d\varepsilon = 0, \tag{7.23}$$

which gives

$$\log(\lambda_c + \mu_c) + \frac{\lambda_c}{\mu_c}\log\left(\frac{\lambda_c + \mu_c}{\lambda_c}\right) = 1. \tag{7.24}$$

Of course, this gives back the $\lambda_c = e/4$ limit when $\lambda = \mu$. The phase boundary is plotted in Fig. 7.8. It also agrees with numerical simulations.

7.3.4 A Map with a Natural Bound

In the maps mentioned above, the upper (and lower) bounds are additionally provided with the evolution dynamics. Here we study a map where this bound occurs naturally. We intend to see its effect on the dynamics.

Consider the following simple map

$$O(t+1) = \tanh\bigl[\lambda(1+\varepsilon)O(t)\bigr]. \tag{7.25}$$

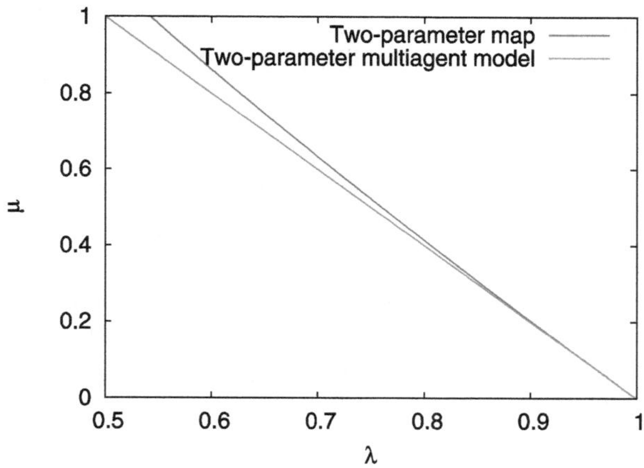

Fig. 7.8 The phase diagram for the two-parameter map as predicted from Eq. (7.24) (*upper line*). The phase diagram for the multi-agent version ($\mu_c = 2(1 - \lambda_c)$) is also plotted for comparison (*lower line*)

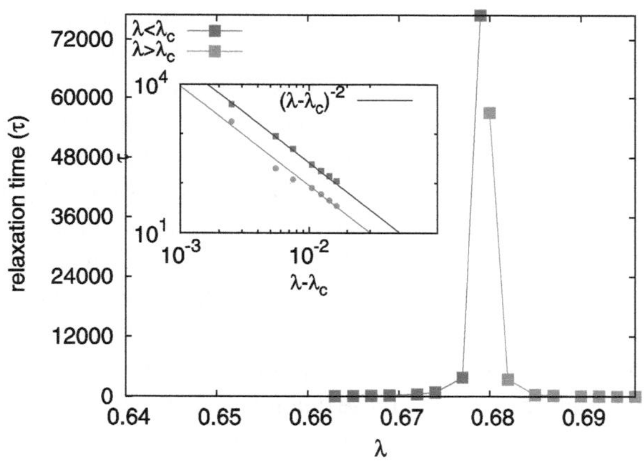

Fig. 7.9 The divergence of relaxation time is shown for the map described by Eq. (7.25) for both sides of the critical point $\lambda_c = e/4$. The *inset* shows that the exponent is 2 as is argued in the text

Due to the property of the function, the bounds in the values of $O(t)$ are specified within this equation itself. This map shows a spontaneous symmetry breaking transition as before. This function appears in mean field treatment of Ising model, but of course without the stochastic parameter.

Numerical simulations show that the O_a behaves as $(\lambda - \lambda_c)^{1/2}$. An analytical estimate of λ_c can be made by linearizing the map for small values of $O(t)$, which is valid only at critical point. Of course, after linearization the map is the same as the initial single parameter map, giving $\lambda_c = e/4$. The relaxation of $O_a(t)$ at $\lambda = \lambda_c$ behaves as $t^{-1/2}$ as before. These are also seen from numerical simulations. However, apart from the critical point, the map is strictly non-linear. Hence the results for its linearized version do not hold except for critical point. Also $\rho = 0$ here always.

To check the divergence of the relaxation time at the critical point, it is seen that it follows $\tau \sim (\lambda - \lambda_c)^{-2}$. Of course, in the deterministic version (mean-field Ising model), the exponent is 1. But as this map has stochasticity, the time exponent is doubled (see Fig. 7.9) (i.e., the relaxation time is squared as happens for random motion as opposed to ballistic motion).

7.4 Summary and Conclusions

In this paper we have studied the simplified map version of a recently proposed opinion dynamics model. The single parameter map was proposed in Ref. [15] and the critical point was estimated. Here we study the critical behavior in details as well as propose the two-parameter map motivated from Ref. [16]. The phase diagram is calculated exactly.

The maps can be cast in a random walk picture with reflecting boundary. Then using the standard random walk statistics, some steady state as well as dynamical behaviors are calculated and these are compared with the corresponding numerical results. It is observed that the usual order parameter of the system, i.e. the average value of the opinion (O_a) in the steady state, does not follow any power-law scaling (see Eq. (7.9)). In fact, it is the condensation fraction, or in this case the probability (ρ_a) that the opinion values touch the limiting value, turns out to be the proper order parameter, showing a power-law scaling behavior near critical point.

The dynamical behaviors of the two quantities ($O_a(t)$ and $\rho_a(t)$) are similar. We have calculated the power-law relaxation of these quantities at the critical point, which compares well with simulations. Also, the divergence of the relaxation time on both sides of criticality shows similar behavior. We have also studied the effect of an "external field" (representing media or similar external effects) in these models. At critical point, both the quantities grow linearly with the applied field. The average fluctuation in both of these quantities show a maximum near the critical point. These theoretical behavior fits well with the numerical results.

In summary, we develop an approximate mean field theory for the dynamical phase transition observed for the map Eq. (7.2) and find that the average condensation fraction $\rho_a \sim (\lambda - \lambda_c)^{\beta}$ with $\beta = 1$ behaves as the order parameter for the transition and it has typical relaxation time $\tau \sim (\lambda - \lambda_c)^{-z}$ with $z = 2$ and at critical point $\lambda = \lambda_c (= e/4)$ decays as $t^{-\delta}$ with $\delta = 1/2$.

References

1. Chakrabarti BK, Chakraborti A, Chatterjee A (eds) (2006) Econophysics and sociophysics. Wiley-VCH, Berlin
2. Stauffer D (2009) In: Meyers RA (ed) Encyclopedia of complexity and systems science. Springer, New York
3. Stauffer D (2003) Comput Sci Eng 5:71
4. Castellano C, Fortunato S, Loreto V (2009) Rev Mod Phys 81:591
5. Galam S, Gefen Y, Shapir Y (1982) J Math Sociol 9:1
6. Liggett TM (1999) Interacting particle systems: contact, voter and exclusion processes. Springer, Berlin

7. Sznajd-Weron K, Sznajd J (2000) Int J Mod Phys C 11:1157
8. Galam S (2008) Int J Mod Phys C 19:409
9. Biswas S, Sen P (2009) Phys Rev E 80:027101
10. Hegselman R, Krause U (2002) J Artif Soc Soc Simul 5:2
11. Deffuant G, Neau N, Amblard F, Weisbuch G (2000) Adv Complex Syst 3:87
12. Fortunato S (2005) Int J Mod Phys C 16:17
13. Fortunato S, Latora V, Pluchino A, Rapisarda A (2005) Int J Mod Phys C 16:1535
14. Toscani G (2006) Commun Math Sci 4:481
15. Lallouache M, Chakrabarti AS, Chakraborti A, Chakrabarti BK (2010) Phys Rev E 82:056112
16. Sen P (2011) Phys Rev E 83:016108
17. Biswas S, Chatterjee A, Sen P (2012) Physica A 391:3257
18. Biswas S (2011) Phys Rev E 84:056106
19. Crokidakis N, Anteneodo C (2012) Phys Rev E 86:061127
20. Chakrabarti AS (2011) Physica A 390:4370
21. Chandra AK (2011) Phys Rev E 85:021149
22. Sen P (2012) Phys Rev E 86:016115
23. Chakraborti A, Chakrabarti BK (2000) Eur Phys J B 17:167
24. Chatterjee A, Chakrabarti BK, Manna SS (2004) Physica A 335:155
25. Chatterjee A, Chakrabarti BK (2007) Eur Phys J B 60:135
26. Chatterjee A (2009) Eur Phys J B 67:593
27. Biswas S, Chandra AK, Chatterjee A, Chakrabarti BK (2011) J Phys Conf Ser 297:012004

Part II
Miscellaneous

Chapter 8
An Overview of the New Frontiers of Economic Complexity

Matthieu Cristelli, Andrea Tacchella, and Luciano Pietronero

Abstract The fundamental idea developed throughout this short overview on Economic Complexity is that a revolution of the revolution of Economics is needed to turn this field into a mature discipline. The first revolution of Economic Complexity (Bouchaud in Nature 455:1181, 2008) led to a conceptual paradigm shift and agent-based models have shown, from a qualitative point of view, the crucial role played by concepts like agent heterogeneity and herding behavior to understand the non-trivial features of financial time series. The second revolution must lead the paradigm shift from a conceptual and qualitative level to a quantitative and effective description of economic systems. This can be achieved through the introduction of new metrics and quantitative methods in Social Sciences (Economics, Finance, opinion dynamics, etc.). In fact, the concept of metrics is usually neglected by mainstream theories of Economy and Finance. Only in that way Economic Complexity can concretely affect the thinking of Economic mainstream and, in this sense, become a mature discipline. The large availability of datasets (the so-called Big Data Science) has recently revealed new promising path towards such perspectives and, as an example, we briefly discuss how archival data about export flows can be turned into a concrete tool to assess the competitiveness of countries and the complexity of products.

M. Cristelli (✉)
ISC-CNR, Sapienza University, Piazzale A. Moro 5, 00185 Rome, Italy
e-mail: matthieu.cristelli@roma1.infn.it

A. Tacchella · L. Pietronero
Sapienza University, Piazzale A. Moro 5, 00185 Rome, Italy

A. Tacchella
e-mail: andrea.tacchella@roma1.infn.it

L. Pietronero
e-mail: luciano.pietronero@roma1.infn.it

L. Pietronero
ISC-CNR, Via dei Taurini 19, 00185 Rome, Italy

F. Abergel et al. (eds.), *Econophysics of Agent-Based models*,
New Economic Windows, DOI 10.1007/978-3-319-00023-7_8,

8.1 Introduction

Why did physicists start to be interested in Economics, Finance and in general in Social Sciences? Why and how Physics can give an effective contribution to the description and the comprehension of these disciplines?

The answer can be traced back to the differences between Physics and Economics (and to some extent Social Sciences). Physics typically proceeds through a virtuous circle usually called Scientific Method: theoretical and experimental approaches are parallel, complementary, they progress by an ongoing reciprocal feedback and they are the *two sides of the same coin*. Therefore Physics is both an experimental and theoretical science.

On the other hand standard approaches in Social Sciences and Economics traditionally neglect the experimental side of the coin. Social sciences are indeed listed among those science defined as observational. Observational sciences are disciplines which operate in field where controlled observations cannot be used in order to study causes and effects. Unavoidably, social phenomena are usually characterized by a high degree of non-stationarity and a low level of controllability and reproducibility of experiments.

However, we argue that in those disciplines there is a systematic lack of data-oriented approaches and especially of *metrics* in order to give a quantitative description of social phenomena. By absence of a metrics, we mean that in Social Sciences and especially in Economics the introduction of the concept of *distance* to make a phenomenon quantitative is often neglected or does not appear as a crucial point. For instance, quantities like the systemic risk of financial markets, the mood of people, the quality/complexity of a product, the competitiveness of a country, etc., can be hardly measured by the tools provided by the mainstream of economic theories.

From this observation, we argue that economic mainstream cannot be defined as an experimental science and not even an observational one, given the actual state of the art. We are aware that there exist some works in Economics where the interplay between empirical evidences and models is not marginal but they are usually neglected or never incorporated in the mainstream of economical theories.

It clearly appears, from these considerations, which can be the contribution of Physics to Economics. Physics can contribute to make Economics an observational and experimental science. In particular methods and tools derived from Physics and Mathematics can be effective to provide and turn from qualitative to quantitative the description of social and economic phenomena. Thus by introducing effective metrics for economic systems, a twofold progress would be reached: on one hand, data oriented approaches and economic theories would have a common language to achieve a mutual feedback and on the other hand key issues for policy modeling and policy making could be properly addressed and suitable interventions could be undertaken.

The absence of a mutual feedback between data and theories in Economics has lead this discipline towards two problematic points with respect to the *desirable* forecast skills which should be required by this field given its societal impact:

1. On one hand Economics has basically developed only mathematically solvable models which are typically very far from the scenarios depicted by empirical analysis. If a similar approach was adopted in Physics, only the Hydrogen atom should exist!
2. On the other hand the very poor feedback between theoretical models and data-driven studies has lead to inadequate or partial measurement tools (i.e. mathematical tools) for Economics and Finance. In fact these tools are usually defined only on the basis of the model's results. The attempts to give a quantitative description of economical systems, to some extent, reflect only the theoretical *a priori* interpretation of these ones and never the empirical evidences. Tools are modeled to be optimal proxies only for what it is expected from models and not, as it happens in Physics, for what it is really observed. In summary, differently from Physics, very often in Economics, the only driving forces for the discipline are mainstream theories and the problem of the empirical validation is completely ignored. For instance, recent works [2], the standard forecast framework of linear regressions appears to be the wrong tool to quantify the forecast of the growth and the development of country economies.

We want to point out that we are not rejecting all economical theories, our criticisms are mainly towards the methodology of Economics with respect to empirical data. Economics needs a revolution [1] but this does not imply that economical theories are all and completely wrong. Physicists (and in general non-Economists) are not tempting to discard all current economical Theories, on the contrary we believe that some of the current theories can be seen as the zero order for new contributions which can be given to social sciences, especially Finance and Economics, by Physics. The current *mean field* theories of Economics cannot be the arrival point of this discipline but the starting one for the revolution which is now necessary given the powerlessness of traditional previsional tools in front of the last decade financial crisis and to the following economic stagnancy.

In the last 20–30 years the basis of a paradigm change in economic and in general social sciences has risen: new sources of data emerged and now permit to overcome, al least in some field of social sciences and economics, the lack of empirical data. The so-called Big Data Science and ICT (Information and Communications Technology) have disclosed such a huge amount of raw data about social and economic phenomena that now the invoked revolution of the methodology of Economics is possible.

This also explains why financial markets appeared as the first candidate for this interdisciplinary application of Physics because a systematic approach can be undertaken. In fact since two decades there exists a very huge amount of high frequency data from stock-exchanges which permits to perform *experimental procedures* as in Natural Sciences. Financial markets appear as a perfect playground where models can be tested and where repeatability of empirical evidences are well-established features. In addition the methods of Physics have been very effective in this field and have often given rise also to concrete (and sometimes profitable) financial applications.

Nevertheless, until now, Economic Complexity does not have led to a concrete change in economic mainstream. We believe this is due to the fact that the revolution invoked in [1] has taken place mainly at a qualitative and conceptual level. As an example, agent-based models have been a fundamental turning point for economic theories of financial markets. They allowed for the overcoming of the market description in terms of homogeneous agents, rational expectation, cause-effect relation in price evolution, price equilibrium and revealed the importance of concepts as belief and strategy heterogeneity, herding and non-rational behavior to explain the complex features of the dynamics of financial markets.

This new perspective unfortunately takes place only at a level of metaphor and the agent-based models, even if they are able to explain from a qualitative point of view the Stylized Facts, do not provide concrete tools to address key issues of markets such as risk.

In the following of this overview we are going to discuss which can the development perspectives of Economic Complexity towards a mature discipline. In particular, after a short summary of the state-of-the-art of agent-based modeling, we depict some possible paths to turn this class of models into concrete tools for the measure of market risks.

In the last section we give some insights on new perspectives of Economic Complexity beyond Financial markets. The availability of new datasets (the so called-Big Data Science) has recently opened new domains—traditionally characterized by an intrinsic lack of empirical data—towards which methodology from Physics and Complex Systems can be fruitfully applied. This can be achieved through the introduction of metrics and quantitative methods in Economics and as an example of this frontier of research in Economic Complexity we review how archival data about export flows can be turned into a concrete tool to assess the competitiveness of countries and the complexity of products

8.2 The Revolution of Economics Needs a Revolution

The major contributions of Economic Complexity to financial markets are twofold: phenomenological and effective description of the dynamics of markets at a micro-level and conceptual comprehension of Stylized Facts.

The former contribution provides fundamental insights in the non-trivial nature of the stochastic process performed by stock prices [3–7] and in the role of the interplay between agents to explain the behavior of the order impact on prices [7–14].

The latter approach instead has tried to overcome the traditional economical models based on concepts like price equilibrium and homogeneity of agents in order to investigate the role of heterogeneity of agents/strategy with respect to the price dynamics [15–23].

In the next sections we are going to focus on this second contribution to Finance and we discuss how to turn this approach into quantitative tools for key issues of financial markets.

8.2.1 Agent-Based Modeling: The State of the Art

The standard theory of Economics for Financial Markets is based on the following elements:

- Situation of equilibrium with equal (representative) agents which are (quasi) rational, have the same information and process it in the same way;
- Important price changes correspond to new information which arrives on the market. The fact that this information is random and independent leads to the famous random walk model and the corresponding Black and Scholes equations;
- This new information modifies the ratio between offer and demand and then also the price. This corresponds to a mechanical equilibrium of the market;
- These concepts also imply a Cause-Effect relation in which large price changes are due to the market reaction to the arrival of exogenous important news. Therefore a large price change is supposed to be associated to an equally large exogenous event.

It is too easy to argue that most of these assumptions have no basis at all. So one may wonder why they are so widely adopted. The real reason is that they permit an analytical treatment of the problem. This is a very different perspective from natural sciences in which very few realistic problems can be treated exactly. In addition these ideas are usually not tested against empirical data; a fact that strongly limits the scientific basis of the whole framework. The most reasonable of the above assumptions is the fact that indeed many external news are random and incoherent so the random walk appears reasonable as a simple modeling. On the other hand the assumptions made for the behavior of the agents are very far from reality. The agents can be very different from each other, their level of information and the way the use it is also very different. Finally they are not at all independent and, especially in situations of crisis the rationality hypothesis can be seriously wrong. In these situations fear and panic, as well as euphoria, lead to very strong herding behavior which is completely neglected in the standard model. This situation naturally calls for a possible description in terms of critical phenomena and complex systems. The study of complex systems refers to the emergence of collective properties in systems with a large number of parts in interaction among them. This implies a change of paradigm form the previous mechanical model to a complex model in which intrinsic instabilities can develop in a self-organized way without a cause-effect relation.

This change of paradigm is witnessed by the fact that in the past years there has been a large interest in the development of Agent Based Models aimed at reproducing and understanding the Stylized Facts observed in the financial time series. The Agent Based Models represent a broad class of models which have been introduced to describe the economic dynamics in a more realistic way. Their building blocks are:

- the agents are heterogeneous with respect to their various properties like strategies, wealth, time scale, etc.;
- the interaction between them is a fundamental element and, of course, it can have many different characteristics;

- price dynamics depends on the balance between offer and demand but the specific implementation can be different.

Usually the dynamics of this type of models is able to produce deviations from the Gaussian behavior related to the Stylized Facts for some specific range of parameters. The main criticisms towards Agent Based Models rely on the fact that up to now these models have not yet been able to produce concrete tools for measuring risk. They are somehow conceptual metaphors which showed that the mainstream of economical theories on financial markets were inadequate. They certainly revolutionized Economics but a new step is now required to make Agent Based Models a concrete tool to properly address issues like systemic risk and market instabilities. The path towards this new step is one of the new frontiers of the Economic Complexity.

8.2.2 Agent-Based Modeling: Perspectives

There exists a point which is usually neglected by Agent Based Models: the Stylized Facts usually correspond to a particular situation of the market. If the market is pushed outside such a situation, it will evolve to restore it spontaneously. The question is why and how all markets seem to self-organize in this special state and, even more important, when markets get less stable and can be more likely pushed far from this special state? The answer to this question conceptually represents a fundamental point in understanding the origin of the SF and from a practical point of view the starting point for the development of a quantitative measure for systemic risk.

In this respect, in a series of recent theoretical works [15–19], we show that Stylized Facts correspond to a finite size effect in the sense that they disappear in the limit of large N (total number of agents) or large time. Given the general nature of these results we may conjecture that these should apply to a broad class of agent based model and they are not a special property of a specific model. This situation has important implications both for the microscopic understanding of the Stylized Facts and for the analysis and interpretation of experimental data. In this series of works it is shown that a price which is very stable corresponds to a very large number of market participants. On the other hand a small number of agents leads to large movements in the price. In other words, when markets are in a low-volatility state agents tend to operate incoherently and the impact of an agents' action is negligible. Instead when the number of independent agents/strategies diminishes, the market is evolving towards a situation in which almost all the agents react in the same way. In such a scenario, on one hand the market susceptibility increases enormously giving rise to potential systemic instabilities. On the other hand the more the agents act coherently, the more the amplitude of the market fluctuations grow, triggering a self-reinforcing effect. From an empirical point of view a number of works confirms that markets oscillates among states characterized alternatively by mean reverting (stable) and mean escaping tendencies (large fluctuations, market instabilities). That being so the empirical comprehension of herding behaviors and the dynamics of the number of effective agents represent the second step to complete the paradigm change introduced by Agent-Based Models and to turn them into concrete predictive tools to help policy making in preventing global financial crashes.

The background of such a vision is twofolds:

- we believe that this dynamics (number of independent strategies) may anticipate the market instabilities, and a measure of the number of effectively independent strategies can be a concrete early warning to detect them. We want to stress that we do not aim at predicting when the next systemic crash will occur but when potential conditions for large instabilities arise;
- as discussed previously, a weakness of gent-Based Models is that the experimental framework defined by the Stylized Facts is rather limited and an improvement of this body of knowledge appears to be the bottle neck in the field. By the knowledge of the time series of the number of independent agents these models can be tuned, for the first time, on a quantity that have a direct and clear counterpart in the model: the number of agents. In such a way agent based models can be turned in powerful instruments to give a complete scenario in terms of their ingredients: herding parameters, trust, kind of strategies, liquidity and in general all the elements inserted in these models.

8.2.3 *Micro-structure of Financial Markets*

The availability of the high frequency records of stock-exchanges has led to a better description and comprehension of price dynamics, especially for price formation, price response and order impacts. This huge amount of data also improved the so-called financial engineering techniques. This data also provided the development of very effective and powerful phenomenological description of market activities which give some insights on market risk. However, systemic risk of financial is far being a solved problem. A complementary approaches to the analysis of micro-structure to address system risk is represented by network-based techniques [24] which looks at the robustness/fragility of credit and financial networks.

8.2.4 *Non-financial Proxies*

An interesting data-driven approach to the modeling of economic and financial systems is represented by the recent availability of large datasets which ranges from web activity of search engine and social networks to human activity records such as phone, car usage and international trade flows. The so-called Global Systems Science sees the systems in which we live as a series of interacting network. This networked reality has favored the emergence of a new data-driven research field where mathematical methods of computer science, statistical physics and sociometry provide quantitative insights on a wide range of disciplines [2, 25–32]. In this perspectives, Big Data may represents an important source of informative contents to improve the comprehension and the description of Economic Systems and social phenomena. The last 5–10 years delivered a series of pioneering works on this subject which showed the potential of this non-standard sources of data. We briefly list the main results in this very young field and in the next section we discuss in detail one of this recent work:

- user web activity can be turned into a useful tool to track flu and epidemics spreading and evolution;
- web search volumes can be predictive on trading volumes;
- web searches provide informative contents on customer habits, trends and can permit to track sales in various sectors such as phone activations, car sales and tourism;
- human mobility and human dynamics in highly populated area as metropolis can give key hints to project the so-called Smart Cities;
- import-export archival datasets collected by customs provides represents an important source of data to quantify the economic competitiveness of countries and the complexity of products.

8.3 Economic Complexity: From Finance to Economics

Differently from Finance, Economics is far from being an ideal candidate to export the methodology of Natural Sciences because of the lack of data since controlled (and repeatable) experiments are totally artificial while real experiments are almost incontrollable and non repeatable due to a high degree of non stationarity of economical systems. However, the application of method deriving from complexity theory to economic theory of growth is one the most important achievement of a stream of works emerged in the last five years [2, 31, 32].

A concrete example of how data-driven approaches can lead to new methods and tools for economics is the network defined by international trade flows [32]. It is indeed possible to introduce [2] a (non-monetary) metrics to measure the complexity and the competitiveness of the productive system of countries. In addition we are able to define a metrics for product complexity which overcomes traditional economic measure for the quality of products given in terms of hours of qualified labor needed to produce a good. The method developed provides some impressive results in predicting economical growth of countries and offers many opportunities of improvements and generalizations [2, 33].

8.3.1 Country Competitiveness and Products Complexity

Which is the country with the highest competitiveness in the ecosystem defined by the world trade web? Which one has the most complex productive system?

Some recent works [31, 32] have indicated new approaches to macroeconomic issues, inspired by network-based methods. The basic idea is to study the networks defined by international trade flows [34] and to gain information about the productive structure of countries and its future development. In this kind of analysis methods derived from physics have proven to be extremely effective.

According to mainstream theories on economical growth [35–39], of which Adam Smith is the father, specialization should provide an evolutionary advantage for a country. In fact, citing Adam Smith "it is the great multiplication of the productions [...] in consequence of the division of labor, which occasions [...] that universal opulence which extends itself to the lowest ranks of the people." This claim can easily be tested by looking at the binary Country-Product (C-P) matrix. In the C-P matrix each entry is 1 if a particular country C is a significative exporter of product P, 0 if it's not. Specialization would mean that when rows and columns of the C-P matrix are arranged properly, it should appear as a block-diagonal matrix. Observations disagree with this prediction and the C-P matrix is found to be approximately triangular.

It is worth noticing that for biological organisms and species, there is the evidence that diversification usually gives an evolutionary advantage with respect to specialization, in particular too specialized species are weaker when global changes occur and tend to become extinct. Therefore it appears that in a highly interconnected and dynamic environment as modern globalized economies are, country competitiveness resembles the biological fitness of species.

The existence of significative off-diagonal terms in the C-P matrix implies that specialization does not appear as a natural strategy for countries. Furthermore the triangular shape gives an additional and specific piece of information about the correlation between the composition of the export basket and the kind of products exported by a country. In fact we find that some countries have a large diversification of their production and consequently make almost all products, while scrolling down the rows of the matrix, the number of exported products decreases and countries become more and more specialized on a small subset of products which are exported by almost all countries. Furthermore the most diversified countries coincide with the richest ones, while the poorest ones are the ones specialized on those few products exported by almost every country. These empirical findings suggest that the diversification is more important than specialization for country growth as for biological systems. This is not surprising since specialization may be preferred or provide advantages only in a static and in equilibrium system. On the other hand in an out-of-equilibrium and dynamic playground where different productive systems compete, diversification becomes an evolutionary advantage in case of changes. The fact that the richest countries are also the most diversified ones confirms this hypothesis.

The question which now arises is how to measure the complexity of a productive system given the export basket. A recent work [32] proposes a new conceptual framework in order to explain why diversification is an optimal strategy and why the country-product matrix is basically triangular-shaped. The authors hypothesize that there exist some special endowments of the countries, called capabilities, which are non-exportable and stay localized in the nation where they are. The authors of [32] assume that these capabilities are the key element to grasp the complexity of a productive system because each product requires a specific set of capabilities which must be owned by a country in order to produce and then to export it. The first step of our work is then to define a consistent and solid way to infer about the level capabilities that each country has. We call this measure "Fitness". For products we define a somewhat symmetrical measure, the "Complexity".

8.3.1.1 Fitness and Complexity

The theory of capabilities suggests a strategy for using the information contained in the C-P matrix in order to determine Fitness and Complexity in a self-consistent, recursive way. At order 0 a natural guess for the fitness of a given country is to consider its diversification: the more products it exports, the more capabilities it should have. This is easily generalized when the diversification is weighted with the complexity of products, in a simple linear way.

Concerning products the measure has to be different because a strong nonlinearity is needed to take into account the following: considering a product of two very different countries (with respect to fitness), say Germany and Nigeria, its complexity cannot simply be obtained by averaging the level of fitness of the two countries, because the signal that Nigeria is able to produce it means that its capabilities (which are very few) are enough to enable the production of that product. Thus in the evaluation of the Complexity a strong non-linearity must be involved, in order to make the "worst" producer of a given product the most important in the determination of Complexity. This has to be achieved by still considering that an ubiquitous product is likely to be of low complexity. These two characteristics have to be consistently mixed. Given these qualitative definitions, each one involving the other, a recursive procedure (with a proper renormalization) leads to an asymptotic, stable distribution for Fitness and Complexity.

8.3.1.2 Fundamental Analysis of Countries and Complexity

Our measure allows to define a new kind of fundamental analysis of the growth potential of countries. First, combining information about the complexity of products with the explicit normalized volumes of export of each single country, it is possible to realize a "spectroscopy" of each productive system, and using information about the dynamics of such spectra infer about their future structure. Moreover it is possible to combine the information about the Fitness of a single country and its competitors on any single product, and consider that product's complexity in order to define measures of "competitiveness".

Through this new algorithm and the associated metrics it is possible to compare non-monetary factors of fitness and complexity to measures of economic intensity for instance the countries' GDPs per capita. We argue that countries that show both a high Fitness and a high Complexity, but a low GDP, are very likely to strongly boost their income in the next decades. This difference reveals a crucial information. When retrofitting its model with data from 1995 to 2010, it is possible to see that results reflected well what has occurred in the real world over that period.

This type of analysis can lead to a novel perspective for the study and industrial planning of countries, it also provides information on the GDP growth and an estimate of the global risk in relation to fundamental economics. It is just the beginning of a novel era for economics in which the large amount of high quality data and some original ideas open the way to more consistent methods and permit us to move from qualitative to quantitative considerations, shedding new light on important and unexpected properties.

8.4 Conclusions

Complex systems perspectives have revolutionized Economics and Financial Market theories. However, as argued throughout this paper, a second revolution is required to turn this interdisciplinary field in a mature discipline which can concretely affect the thinking of Economic mainstream. The first revolution somehow produced only a conceptual breaktrhough: for instance the metaphors of agent-based models have shown the importance of concept as heterogeneity, non rationality and herding behavior to understand the complex nature of the markets Stylized Facts.

In the last five years we are observing the first step towards the maturity of Economic Complexity. The paradigm shift now must be exported from conceptual and qualitative approaches to quantitative methods to assess key issue of Economic such as systemic risk of financial markets, risk rating, economic growth, technological development, etc.

Acknowledgements This work is support by Italian PNR project CRISIS-Lab.

References

1. Bouchaud J-P (2008) Economics needs a scientific revolution. Nature 455:1181
2. Tacchella A, Cristelli M, Caldarelli G, Gabrielli A, Pietronero L (2012) A new metrics for countries' fitness and products' complexity. Sci Rep Nat 2:723
3. Bouchaud J-P, Kockelkoren J, Potters M (2004) Random walks, liquidity molasses and critical response in financial markets. Quant Finance 6:115
4. Lillo F, Farmer J (2004) The long memory of the efficient market. Stud Nonlinear Dyn Econom 8:1226
5. Farmer JD, Gerig A, Lillo F, Mike S (2006) Market efficiency and the long-memory of supply and demand: is price impact variable and permanent or fixed and temporary. Quant Finance 6:107
6. Bouchaud J-P, Farmer JD, Lillo F (2008) How markets slowly digest changes in supply and demand. Elsevier/Academic Press, Amsterdam
7. Chakraborti A, Toke IM, Patriarca M, Abergel F (2011) Econophysics review, II: agent-based models. Quant Finance 11:1013–1041. doi:10.1080/14697688.2010.539249

8. Bouchaud J-P, Mezard M, Potters M (2002) Statistical properties of stock order books: empirical results and models. Quant Finance 2:251
9. Potters M, Bouchaud J-P (2003) More statistical properties of order books and price impact. Physica A 324:133
10. Wyart M, Bouchaud J-P, Kockelkoren J, Potters M, Vettorazzo M (2008) Relation between bid-ask spread, impact and volatility in order-driven markets. Quant Finance 8:41
11. Lillo F, Farmer JD, Mantegna RN (2002) Single curve collapse of price impact function for the New York stock exchange. arXiv:cond-mat/0207428
12. Farmer JD, Gillemot L, Lillo F, Mike S, Sen A (2004) What really causes large price changes? Quant Finance 4:383
13. Lillo F, Farmer JD, Mantegna RN (2003) Master curve for price-impact function. Nature 421:129
14. Weber P, Rosenow B (2006) Large stock price changes: volume and liquidity. Quant Finance 6:7
15. Cristelli M, Zaccaria A, Pietronero L (2011) Critical overview of agent-based models for economics. In: Proceedings of the school of physics E. Fermi, course CLXXVI, 2010, Varenna
16. Alfi V, Pietronero L, Zaccaria A (2009) Self-organization for the stylized facts and finite-size effects in a financial-market model. Europhys Lett 86:58003
17. Alfi V, Cristelli M, Pietronero L, Zaccaria A (2009) Minimal agent based model for financial markets, I: origin and self-organization of stylized facts. Eur Phys J B 67:385
18. Alfi V, Cristelli M, Pietronero L, Zaccaria A (2009) Minimal agent based model for financial markets, II: statistical properties of the linear and multiplicative dynamics. Eur Phys J B 67:399
19. Alfi V, Cristelli M, Pietronero L, Zaccaria A (2009) Mechanisms of self-organization and finite size effects in a minimal agent based model. J Stat Mech P03016
20. Samanidou E, Zschischang E, Stauffer D, Lux T (2007) Microscopic models of financial markets. Rep Prog Phys 70
21. Lux T, Marchesi M (1999) Scaling and criticality in a stochastic multi-agent model of a financial market. Nature 397:498
22. Giardina I, Bouchaud J-P (2003) Bubbles, crashes and intemittency in agent based market. Eur Phys J B 31:421
23. Caldarelli G, Marsili M, Zhang Y-C (1997) A prototype model of stock exchange. Europhys Lett 40:479
24. Battiston S, Puliga M, Kaushik R, Tasca P, Caldarelli G. DebtRank: too central to fail? Financial networks, the FED and systemic risk. Nature 2:541
25. Golub T (2010) Counterpoint: data first. Nature 464:679
26. Evans J, Rzhetsky A (2010) Machine science. Science 329:399
27. Lazer D et al (2009) Life in the network: the coming age of computational social science. Science 323:5915
28. Gonzalez M, Hidalgo C, Barabasi A-L (2008) Understanding individual human mobility patterns. Nature 453:479
29. Choi H, Varian H (2009) Predicting the present with Google trends. Technical report
30. Bordino I, Battiston S, Caldarelli G, Matthieu M, Ukkonen A, Weber I (2011) PLoS ONE 7(10):e47278. doi:10.1371/journal.pone.0047278
31. Hidalgo C, Klinger B, Barabási AL, Hausmann R (2007) The product space conditions the development of nations. Science 317:482–487
32. Hidalgo C, Hausmann R (2009) The building blocks of economic complexity. Proc Natl Acad Sci USA 106:10570–10575
33. Caldarelli G, Cristelli M, Gabrielli A, Pietronero L, Scala A et al (2012) A network analysis of countries' export flows: firm grounds for the building blocks of the economy. PLoS ONE 7(10):e47278. doi:10.1371/journal.pone.0047278
34. Gaulier G, Zignago S (2010) Baci: international trade database at the product-level. http://www.cepii.fr/anglaisgraph/workpap/pdf/2010/wp2010-23.pdf

35. Smith A (1776) The wealth of nations. Strahan and Cadell, London
36. Ricardo D (1817) On the principles of political economy and taxation. Murray, Sydney
37. Romer PM (1990) Endogenous technological change. J Polit Econ 98:71–102
38. Grossman GM, Helpman E (1991) Quality ladders in the theory of growth. Rev Econ Stud 58:43–61
39. Flam H, Flanders MJ (1991) Heckscher-Ohlin trade theory. MIT Press, Cambridge

Chapter 9
Jan Tinbergen's Legacy for Economic Networks: From the Gravity Model to Quantum Statistics

Tiziano Squartini and Diego Garlaschelli

Abstract Jan Tinbergen, the first recipient of the Nobel Memorial Prize in Economics in 1969, obtained his PhD in physics at the University of Leiden under the supervision of Paul Ehrenfest in 1929. Among many achievements as an economist after his training as a physicist, Tinbergen proposed the so-called Gravity Model of international trade. The model predicts that the intensity of trade between two countries is described by a formula similar to Newton's law of gravitation, where mass is replaced by Gross Domestic Product. Since Tinbergen's proposal, the Gravity Model has become the standard model of non-zero trade flows in macroeconomics. However, its intrinsic limitation is the prediction of a completely connected network, which fails to explain the observed intricate topology of international trade. Recent network models overcome this limitation by describing the real network as a member of a maximum-entropy statistical ensemble. The resulting expressions are formally analogous to quantum statistics: the international trade network is found to closely follow the Fermi-Dirac statistics in its purely binary topology, and the recently proposed mixed Bose-Fermi statistics in its full (binary plus weighted) structure. This seemingly esoteric result is actually a simple effect of the heterogeneity of world countries, that imposes strong structural constraints on the network. Our discussion highlights similarities and differences between macroeconomics and statistical-physics approaches to economic networks.

9.1 Introduction

Over the last fifteen years, there has been an ever-increasing interest in the study of networks across many scientific disciplines, from physics to biology and the social sciences [1]. Economics is no exception. Empirical [2, 3] and theoretical [4, 5]

T. Squartini (✉) · D. Garlaschelli
Instituut-Lorentz for Theoretical Physics, Leiden Institute of Physics, University of Leiden, Niels Bohrweg 2, 2333 CA Leiden, The Netherlands
e-mail: squartini@lorentz.leidenuniv.nl

D. Garlaschelli
e-mail: garlaschelli@lorentz.leidenuniv.nl

F. Abergel et al. (eds.), *Econophysics of Agent-Based models*, New Economic Windows, DOI 10.1007/978-3-319-00023-7_9,

analyses of economic networks have been growing steadily, gradually encompassing different scales: from 'microscopic' networks of individual agents and financial assets [4, 6], through 'mesoscopic' networks of firms, banks and institutions [2, 7, 8], to 'macroscopic' networks of world countries and economic sectors [9–11]. An unprecedented element of continuity across these different economic scales has been the search for empirical laws characterizing real-world networks and the subsequent introduction of simple models aimed at reproducing the observed 'stylized facts'. Placing observations, rather than theoretical postulates, at the starting point of scientific investigations is probably the main positive outcome of the interaction between economists and physicists, an interaction that—over the last two decades—has given rise to the controversial field of 'Econophysics'. The interdisciplinary study of economic networks is another very fruitful result of this interaction. The added value of using the network approach to economic problems is the possibility to investigate indirect effects arising as the combination of many pairwise interactions between economic agents or units. The prototypical example is the study of *systemic risk*, i.e. the risk of a system-wide cascade of defaults of banks or institutions connected to each other in a financial network, as opposed to traditional measures of risk for single financial entities.

Despite the 'network approach' is relatively recent, much earlier studies in Economics already recognized the importance of (what we now call) socio-economic networks, even if this knowledge was more or less dispersed across sub-fields that used to be largely disjoint. An important example is the so-called Gravity Model [12]. The name originates from a loose analogy with Newton's law of gravitation, which states that the gravitational force between two objects is proportional to the product of their masses, and inversely proportional to the square of the distance between them. Strictly applying the analogy to the economic setting, the Gravity Model (see [12] for an excellent review) assumes that a 'mass' S_i of goods (or services, or factors of production such as labor) supplied at an origin i is 'attracted' to a mass D_j of demand for such goods located at a destination j. This attraction generates a flow F_{ij} of goods, but the flow is reduced by the geographic distance d_{ij} between origin and destination as follows

$$F_{ij} = K\frac{S_i D_j}{d_{ij}^2} \tag{9.1}$$

where K is a global free parameter to be fitted to real data. The Gravity Model predicts larger fluxes between closer and 'bigger' (in terms of the size of supply and demand) locations, exactly in the same way as the gravitational force is stronger between closer and more massive objects.

The use of the Gravity Model was pioneered by Ravenstein [13] in studies of migration patterns, where flows represent movements of people, and S_i and D_j are mainly determined by the sizes of the two populations located at the origin and destination. Jan Tinbergen, the first recipient of the Nobel Memorial Prize in economics, was instead the first to use the Gravity Model to explain international trade flows [14]. In this case, flows represent movements of goods among world countries, and S_i and D_j are expected to be determined by the values of the Gross Domestic Product (GDP) of the countries of origin and destination, i.e. $S_i = D_i = GDP_i$.

Indeed, it is precisely as the first data-driven model of trade that the Gravity Model acquired its great popularity, presumably because of the accuracy with which it predicts observed trade fluxes. On the other hand, international trade is also one of the best examples of economic systems that have been intensively studied, using a completely different 'network' approach, over the last decade. This coincidence makes the International Trade Network a useful example to compare traditional (economic) and recent (network) approaches when applied to the same system, which in this case is also an empirically well documented one. For this reason, in what follows we will focus on Jan Tinbergen's Gravity Model of trade, its successes and limitations, and the more recent approaches that overcome some of these limitations. As another curious coincidence, these very different frameworks have a common element: the modeling of an economic network in close analogy with physical laws, from gravitation to statistical physics and quantum statistics.

9.2 Jan Tinbergen and the Gravity Model of Trade

In a slightly (and not fully) generalized form, the Gravity Model of international trade states that the expected amount of trade from country i to country j is

$$\langle w_{ij} \rangle = K \frac{GDP_i^{\alpha} GDP_j^{\beta}}{d_{ij}^{\gamma}} \tag{9.2}$$

where d_{ij} is the geographic distance between countries, and α, β and γ are additional (besides K) free parameters. The angular brackets in Eq. (9.2) denote an expected value: this means that the model is not intended to be a deterministic one, since real data will obviously deviate from the postulated expression. So, strictly speaking, the model predicts that the realized amount of trade is $w_{ij} = \langle w_{ij} \rangle + \varepsilon$ where ε is an error term with zero mean (if a linear regression to the observed trade flows is used), or alternatively $w_{ij} = \langle w_{ij} \rangle \cdot \eta$ where η is an error term with unit mean (if a linear regression to the *logarithm* of the observed trade flows is used). In both cases, the fitted values of the parameters are usually around $\alpha \approx \beta \approx \gamma \approx 1$ [12]. Further extensions of Eq. (9.2) include additional factors either favouring or suppressing trade.[1] Despite the inclusion of these additional factors improves the fit, the main factor determining trade flows remains the GDP, followed by geographic distances. So Eq. (9.2), with exponents $\alpha \approx \beta \approx \gamma \approx 1$, captures the basic lesson

[1]Examples of favouring factors are: trade agreements, membership to common economic groups, shared geographic borders, common currency, etc. Examples of suppressing factors are: embargoes, trade restrictions, and other factors representing a trade friction.

learnt from real trade data and makes the Gravity Model closer to the expression for the gravitational energy ($\gamma = 1$) than to the one for the gravitational force ($\gamma = 2$).

Obviously, there is absolutely nothing fundamental in the formal analogy between the empirical laws of trade (or any other economic flux) and gravity, and no profound reason why these laws should bear any mathematical similarity at all. Rather, the deep similarities must be looked for at different levels:

- A first analogy involves the implicit use of *symmetry* in both cases: both Eq. (9.2) and Newton's law, state that, *all else being equal*, only mass/GDP and distance determine the amplitude of the interaction. In physics, this means that Newton's law holds *in vacuum*, i.e. in absence of anything else that can interact gravitationally with the two objects. In macroeconomics, this means that Eq. (9.2) holds in absence of other factors affecting trade, such as the additional regressors we mentioned in footnote 1.
- A second similarity is the *qualitative dependence* on the key quantities: both laws assume that interaction amplitudes increase with increasing mass/GDP, and decrease with increasing distance. In principle, there is an infinity of quantitative ways (functional forms) to implement this qualitative idea. Accidentally, the functional forms describing gravitation and trade turn out to be very similar, but their qualitative analogy would have held even if the two mathematical expressions were different. In some sense, this makes the qualitative analogy more fundamental than the mathematical one.
- The above consideration leads us to a third analogy, i.e. the *phenomenological* character common to Eq. (9.2) and Newton's formula. In both laws, the particular functional form that implements the previous theoretical arguments is established on the basis of its success in reproducing real data, and thus *a posteriori*. Other functional forms, while possible *a priori* on the basis of the above two points, must be discarded if they do not explain observations. Only after they were widely accepted as powerful empirical laws explaining observations, Netwon's law and the Gravity Model became the 'target' of more general and abstract theories. For instance, to be acceptable, Einstein's theory of General Relativity *must* reduce to Netwon's law in the appropriate circumstances, and micro-founded economic models must generate the Gravity Model when aggregated at the macro level [12].

In our view, the above epistemological analogies are even more fundamental than the (accidental) mathematical analogy between Eq. (9.2) and Netwon's law. Another deep connection between physics and economics exists at a personal level: Jan Tinbergen, the founder of the Gravity Model of trade, was a physicist before becoming an economist.

Born in Den Haag, the Netherlands, in 1903, Jan Tinbergen started his studies in mathematics and the natural sciences at the University of Leiden, soon after graduating from high school in 1921 with the highest honors. In Leiden, he later started a PhD in physics under the supervision of Paul Ehrenfest, who was then professor in Theoretical Physics (see Fig. 9.1). Tinbergen became Ehrenfest's assistant, the private tutor of Ehrenfest's son, and a frequent visitor of Ehrenfest's house, that was regularly visited also by Einstein, Bohr, Heisenberg, Fermi and Pauli. Tinbergen had

Fig. 9.1 Ehrenfest's students, Leiden 1924. Left to right: Gerhard Heinrich Dieke, Samuel Abraham Goudsmit, Jan Tinbergen, Paul Ehrenfest, Ralph Kronig, and Enrico Fermi (copyright © Chicago University Press)

always been attracted by economics, and Ehrenfest was interested in the analogies between economics and physics. This resulted in Tinbergen's PhD thesis, entitled 'Minimum Problems in Physics and Economics' and defended in 1929. Shortly after, despite Ehrenfest had repeatedly tried to convince him to remain a physicist, Tinbergen started a brilliant career as an economist. His pioneering views led him to introduce Econometrics, a synthesis between mathematics, economic theory and statistics. In Tinbergen's view, economic theory should formulate hypotheses translated into mathematical relations that are then statistically tested on empirical data. This distinctive quantitative approach was almost surely due to Tinbergen's graduate training as a physicist. His idea of introducing a quantitative model of international trade flows is clearly in line with this approach. Jan Tinbergen's career culminated in 1969 when he received the first *Bank of Sweden Prize in Economic Sciences in Memory of Alfred Nobel* or shortly *Nobel Memorial Prize in Economics*, often mistakenly referred to as the 'Nobel Prize in Economics' (that, strictly speaking, does not exist).

As a physicist, Jan Tinbergen of course knew Netwon's law very well, a knowledge that might have facilitated making a mathematical connection to the study of international trade. But, we believe, his idea of using the Gravity Model in his research as an economist was most probably triggered by the deeper similarities, discussed above, at the epistemological level. Jan Tinbergen's familiarity with the scientific method universally used in physics is probably the reason why his many achievements as an economist are all characterized by a strong quantitative approach and a clear focus on empirical data. Without trying to distort scientific and personal history, we might therefore presume that Tinbergen's view was not too far from what, in modern jargon, are the inspiring concepts of 'Econophysics'. His Gravity Model of trade can also be regarded as the first model of the system that, in the more recent Complex Networks literature, has been intensively studied under the names of 'International Trade Network' (ITN) or 'World Trade Web' (WTW) [9, 15–19]. Therefore, in our view, Jan Tinbergen's pioneering work deserves full attention from the scientific community active in Econophysics and Network Science. Along these lines, the Gravity Model and the ITN, both still under intense investigation, can be considered two brilliant examples of how ideas from physics can fruitfully interact with economic problems.

9.3 The Network Approach

At the time of Tinbergen's analyses, data about international trade flows were of course much less accurate than today. Missing data were the rule rather than the exception, so it was practically impossible to distinguish between the absence of data documenting an existing trade relationship and a 'true' absence of the relationship itself in the real world. While this confusion still cannot be completely eliminated, in modern databases [20, 21] it only affects a few percent of the data. A simple analysis of such databases yields to a systematic result: in each yearly snapshot of the ITN from the 50's until now, only 50–60 % of the total pairs of world countries are found to be connected by trade relationship [9, 16]. With little error, pairs of countries that do not trade at all are the remaining fraction.

If we look again at Eq. (9.2), we immediately see that the observation of a 'half-connected world' is in contrast with the predictions of the Gravity Model, as it cannot predict zero trade flows.[2] Exactly as the gravitational force between any two masses (no matter how small or distant) is never zero according to Newton's law, the Gravity Model predicts that trade exchanges between any two countries (no matter how poor or distant) are always positive. However, while any two massive objects are indeed found to be attracted over cosmological distances in our Universe, the observation of an economically half-connected world implies that *the Gravity Model fails in reproducing the missing links of the world trade network.*[3] In other words, if the set of existing connections (i.e. the *topology*) of the ITN is preliminarly specified, then the Gravity Model succeeds in reproducing the magnitude of trade connections. But in general, it fails in reproducing the observed topology of the network.

It is interesting to notice that the awareness of the importance of the *binary* topology of a network, besides that of *weighted* structural properties, is a recent conquest of Network Science—clearly absent at the time of Tinbergen. Motivated by this

[2]Strictly speaking, the introduction of an error term into Eq. (9.2) allows to have zero or even negative values. However, after fitting the model to the data, or simply in order to avoid the generation of precisely those unrealistic negative values, the variance of the error term is so small that zero trade flows have a vanishing probability.

[3]In principle, also this limitation can be overcome if the Gravity Model is extended into the so-called *zero-inflated* models [22, 23] that use Eq. (9.2) (or its generalizations) in a two-step procedure: first in order to estimate the probability of a trade connection, and then in order to estimate the intensity of the connection. However, recent analyses [23] have shown that this procedure provides a bad fit to the observed network: when used in order to estimate link probabilities and link intensities simultaneously, the Gravity Model turns out to be a very bad model.

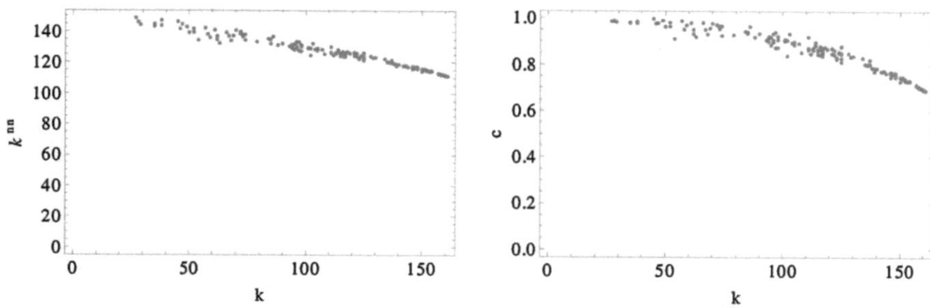

Fig. 9.2 The heterogeneous topology of the binary World Trade Web. *Left*: average nearest neighbour degree k^{nn} as a function of the degree k, for all vertices. *Right*: clustering coefficient c as a function of the degree k. Data: UNCOMTRADE database [21], year 2001 ($N = 162$ countries)

awareness, the analyses of the ITN carried out over the last decade have documented an intricate and heterogeneous topology. Let us for instance consider the *undirected* version of the ITN, where two countries (the nodes, or *vertices*, of the network) are connected by a *link* (or *edge*) if there exists at least a trade relationship (in any direction) between them. In this network, the number of connections (the so-called *degree*, denoted by k) of world countries is found to be very broadly distributed, with poor countries having only one or two connections (typically including the USA) and rich countries being connected to a significant number of partners, up to the total number of countries in the world. This result is very robust, since the degree is found to systematically increase with the GDP [9]. Moreover, (anti)correlations between the degrees of two trading partners are significant: the average degree of trade partners (the *average nearest neighbour degree*, denoted by k^{nn}) is smaller for countries with larger degree [9, 15] (see Fig. 9.2). This means that more connected (richer) countries trade with countries having on average a smaller number of partners, and less connected (poorer) countries trade with countries having on average a larger number of partners. A similar result holds for the so-called *clustering coefficient* (denoted by c) of a country, defined as the realized fraction of links (local link density) among the partners of that country. Just like k^{nn}, c is found to decrease as k increases, meaning that more connected countries have a less interconnected neighbourhood, and less connected countries have a more interconnected neighbourhood [9, 15] (see Fig. 9.2). All these topological properties can be generalized to the *directed* version of the network (where links follow the direction of, say, exports), and similar results are found [16, 18].

Even when weighted properties of the network are studied, the importance of the underlying topology is still manifest, e.g. when local averages of weighted quantities are performed. For instance, let us consider the weighted analogue of the degree, i.e. the *strength* (denoted by s) defined as the total weight of the links of a country (the total value of imports and exports for that country). As in the binary case, the average strength of the partners of a country (the *average nearest neighbour strength*, denoted by s^{nn}) is found to decrease as the strength of that country increases [17, 18] (see Fig. 9.3). Being a local average over the partners of each country, s^{nn} is strongly

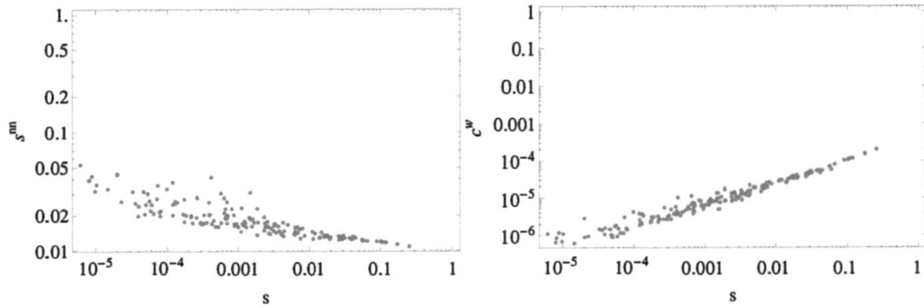

Fig. 9.3 The nontrivial structure of the weighted World Trade Web. *Left*: average nearest neighbour strength s^{nn} as a function of the strength s, for all vertices. *Right*: weighted clustering coefficient c^w as a function of the strength s. Data: UNCOMTRADE database [21], year 2001 ($N = 162$ countries). The values are rescaled by the total weight

influenced by the degree, which is a binary property. Similarly, a weighted generalization of the clustering coefficient (denoted by c^w) is also influenced by the binary structure, since it is still defined on the local neighbourhood of countries. Unlike its binary counterpart, c^w is found to increase as the strength increases [17, 18] (see Fig. 9.3).

9.4 Statistical Physics and Maximum-Entropy Models

Taken together, the above findings highlight that the topology of the ITN is nontrivial and very different from the complete network predicted by the Gravity Model. In the previous section we discussed some coincidences and deeper similarities behind the use of the gravity law in physics and economics. As another interesting similarity, recent results [16–18, 24] suggest that the limitations of the Gravity Model can only be overcome after a change of paradigm which is not dissimilar from the one that accompanied two revolutions in physics, namely the advent of statistical mechanics and that of quantum physics. The new paradigm assumes that, in order to predict the presence of a link (and not only its weight), *probabilistic models* of networks need to be considered. The great conceptual jump consists in assuming that the observed network is not unique, but one of many possible realizations, each of which has a probability P to occur. This probability must be determined by establishing which of the properties of the network are somehow 'unavoidable', and assuming that all possible networks displaying those properties (including the real network) are equally probable. This change of approach is equivalent to the one leading to the introduction of statistical physics: if the detailed *microscopic* configuration of a large system is unknown (as is generally the case), and only a few *macroscopic* quantities are known (e.g. the total energy), then some properties of the system can be inferred by averaging over all possible configurations compatible with the known macroscopic quantities. The probability of each configuration therefore depends on the choice of the macroscopic quantities to be reproduced.

As Jaynes pointed out in his work devoted to the connections between statistical mechanics and information theory [25], the fundamental problem of statistical physics can be regarded as a particular case of a more general class of problems of inference from partial information. In the general case, one looks for the probability distribution that maximizes the uncertainty about the system, given the partial knowledge of the latter. Mathematically, if C denotes a possible (microscopic) configuration of the system, the solution of the problem is obtained maximizing Shannon's entropy

$$S \equiv -\sum_{C} P(C) \ln P(C) \tag{9.3}$$

subject to a set of *constraints*, representing what is known about the system [25]. The result of this constrained maximization problem is the probability

$$P(C) = \frac{e^{-H(C)}}{Z} \tag{9.4}$$

where $H(C)$ is a linear combination of the constraints (where each constraints is coupled to its Lagrange multiplier) and

$$Z \equiv \sum_{C} e^{-H(C)} \tag{9.5}$$

is the normalization factor. The ensemble of configurations generated by Eq. (9.4) is the *maximum-entropy ensemble* specified by the chosen constraints.

Jaynes noticed that, if the system under consideration is a physical one, and the only constraint is the total (macroscopic) energy $E(C)$, then $H(C) = \beta E(C)$ where β is the Lagrange multiplier ensuring that (the ensemble average of) $E(C)$ can be set equal to its observed value. This leads to the identification of Eq. (9.4) with the Gibbs-Boltzmann distribution, if β is identified with the inverse temperature through $\beta = (kT)^{-1}$ (k being Boltzmann's constant). Automatically, this also shows that Eq. (9.5) can be identified with the *partition function*, and Eq. (9.3) with the Gibbs-Boltzmann entropy.

Coming to the case of networks, it has been shown [26–29] that the class of network models known in the social sciences under the name of *Exponential Random Graphs* or p^* models [26] is also a particular case of the above maximum-entropy problem. In these models, the constraints are (not necessarily macroscopic) topological properties that one wants to control for. Consider for instance the case of *binary graphs* with a given number N of vertices. In undirected binary graphs, each pair of vertices is either connected or not, with no possible variation in the direction and intensity of the connection. Each configuration C is uniquely specified by the *adjacency matrix* A of the graph, defined as a symmetric $N \times N$ matrix with entries $a_{ij} = 1$ if a link exists between the vertices i and j, and $a_{ij} = 0$ otherwise. Therefore, we can label each configuration with A rather than C, and the corresponding probability with $P(A)$. The simplest example is when the only constraint is the total number L of links. It has been shown [26, 28] that this particular case reduces to the popular Erdős-Rényi random graph model, where all pairs of vertices are connected independently of each other and with the same probability p (which can be viewed as the only Lagrange multiplier, uniquely specified by the observed value of L). In this model, Eq. (9.4) translates into the following probability $P(A)$, that simply factorizes over all pairs of vertices i, j:

$$P(A) = \prod_{i<j} p^{a_{ij}}(1-p)^{1-a_{ij}} = p^{L(A)}(1-p)^{N(N-1)/2-L(A)}. \tag{9.6}$$

This expression shows that the generation of an entire graph A is the combination of $N(N-1)/2$ independent *Bernoulli trials*,[4] each corresponding to the creation of a single link and characterized by the same success probability p.

The above simple example shows that individual links naturally inherit, from the maximum-entropy structure of the overall model, the character of random variables, to be described by probability distributions. If weighted networks are considered, maximum-entropy models can still be defined [26, 28] and lead again to a probabilistic description of the weight of all links, *including the possibility of zero weights which correspond to missing links* (we will discuss explicit cases later). Thus, in both binary and weighted descriptions, the probabilistic character of link creation characterizing the maximum-entropy approach eliminates the need to specify the topology of the network as 'given', overcoming the limitation encountered when using the Gravity Model. This makes maximum-entropy graph ensembles a potentially successful approach to the analysis of the ITN and economic networks in general. However, two aspects remain to be discussed:

- one needs of course to check whether a suitable choice of constraints can indeed reproduce the empirical properties of the ITN: this requires the identification of topological constraints that are both reasonable (i.e. they can be *a priori* justified as a meaningful choice) and effective (i.e. they are *a posteriori* successful in replicating the ITN);
- even if the specification of appropriate *topological* constraints turns out to satisfactorily reproduce the observed network, one needs to understand whether this result can be reconciled with, or at least related to, the main idea of the Gravity Model: the assumption that trade strongly depends on *non-topological* quantities such as GDP and distances.

[4] A *Bernoulli trial* (or *Bernoulli process*) is the simplest random event, i.e. one characterized by only two possible outcomes. One of the two outcomes is referred to as the 'success' (in this case, the creation of a link) and is assigned a probability p. The other outcome is referred to as the 'failure', and is assigned the complementary probability $1-p$. Equation (9.6) is indeed the product of the probability $p^{L(A)}$ of $L(A)$ successful events of link creation times the probability $(1-p)^{N(N-1)/2-L(A)}$ of the complementary number of failures, where $L(A)$ is the number of links in the particular graph A. Note that $N(N-1)/2$ is the total number of pairs of N vertices: we are uninterested in self-loops, so the diagonal matrix entries are $a_{ii}=0$, which leaves us with only $N(N-1)/2$ degrees of freedom in a symmetric $N \times N$ adjacency matrix. For the same reason, the sum in Eq. (9.6) runs over pairs with $i<j$, i.e. only over the upper triangle of the matrix A.

In Sects. 9.5–9.8 we will address the first point in detail, while in Sect. 9.9 we will deal with the second one.

So far, our general discussion has highlighted that the transition from the Gravity Model to maximum-entropy models is analogous, both conceptually and mathematically, to the paradigm shift that led to the introduction of Statistical Physics at the beginning of the twentieth century. The common aspect in both cases is the *probabilistic* description of the system. As we now show, another common aspect leads to a further formal similarity: the *discreteness* of both economic quantities and microscopic particles implies that, when a specific choice of constraints is made, Eq. (9.4) leads to the same mathematical expressions that are encountered in Quantum Physics. These expressions are the so-called *Fermi-Dirac* and *Bose-Einstein statistics*.[5]

9.5 Fermi-Dirac Statistics

Let us first consider binary networks, i.e. let us focus only on the presence/absence of links. In order to simplify the discussion, let us also consider undirected networks (the results that follow can be straightforwardly extended to directed configurations). We have already mentioned the example of the random graph model, obtained when the only constraint is the total number of links. That model is very simple, but severely limited by its complete homogeneity: all vertices have approximately the same topological properties, narrowly distributed around a common average value. This is in stark contrast with the strong heterogeneity of most real-world economic networks, including the ITN as we already discussed in Sect. 9.3. If we want to build a maximum-entropy model of the ITN whose topology is a real improvement over the Gravity Model, we need to reproduce the observed heterogeneity of the network. To this end, it is necessary to enforce different constraints that lead to more complicated models. One of the widespread choices in network theory is to consider an ensemble of networks where each vertex i has the same degree k_i as in the real network. This choice is justified by the fact that, being an entirely local topological property, the degree is expected to be directly affected by some intrinsic (non-topological) property of vertices. For instance, we already anticipated that in the ITN the degree of a country increases with the GDP of the latter [9]. It would of course not make sense to compare the real ITN with a randomized counterpart

[5]In quantum physics, fundamental particles are believed to be of two types: *fermions* or *bosons*, depending on the value of their *spin* (an intrinsic 'angular moment' of the particle). Fermions have half-integer spin and cannot occupy a quantum state (a configuration with specified microscopic degrees of freedom, or *quantum numbers*) that is already occupied. In other words, at most one fermion at a time can occupy one quantum state. The resulting probability that a quantum state is occupied is known as the *Fermi-Dirac* statistics. Bosons have integer spin and can occupy states with no restriction: any non-negative integer number of bosons can occupy the same quantum state. The resulting expected number of particles occupying a given quantum state is described by the so-called *Bose-Einstein* statistics.

where the degree of a country no longer corresponds to a realistic value of its GDP (for instance, where the USA have only one or two connections). This leads us to interpret the observed degrees of countries as 'unavoidable' topological constraints, in the sense that the violation of the observed values would lead to an 'impossible', or at least very unrealistic, world trade network.

The resulting model is known as the *Configuration Model*, and is defined as a maximum-entropy ensemble of graphs with given *degree sequence* [26, 28]. The degree sequence, which is the constraint defining the model, is nothing but the ordered vector $\mathbf{k}$ of degrees of all vertices (where the ith component k_i is the degree of vertex i). The ordering preserves the 'identity' of vertices: in the resulting network ensemble, the expected degree $\langle k_i \rangle$ of each vertex i is the same as the empirical value k_i for that vertex. In the Configuration Model, Eq. (9.4) translates into the graph probability

$$P(A) = \prod_{i<j} q_{ij}(a_{ij}) = \prod_{i<j} p_{ij}^{a_{ij}} (1 - p_{ij})^{1-a_{ij}} \tag{9.7}$$

where $q_{ij}(a) = p_{ij}^{a}(1 - p_{ij})^{1-a}$ is the probability that particular entry of the adjacency matrix A takes the value $a_{ij} = a$. The above expression shows that the creation of a link has still the form of a Bernoulli process (see footnote 4), but now, unlike the random graph model described by Eq. (9.6), different pairs of vertices are characterized by different connection probabilities p_{ij}. These probabilities read [28]

$$\langle a_{ij} \rangle = p_{ij} = \frac{x_i x_j}{1 + x_i x_j} \tag{9.8}$$

where x_i is the Lagrange multiplier obtained by ensuring that the expected degree of the corresponding vertex i equals its observed value: $\langle k_i \rangle = k_i \ \forall i$ [28]. Note that, as always happens in maximum-entropy ensembles described by Eq. (9.4), the probabilistic nature of configurations implies that the constraints are valid only on average (the angular brackets indicate an average over the ensemble of realizable networks). Also note that p_{ij} is a monotonically increasing function of x_i and x_j. This implies that $\langle k_i \rangle$ is a monotonically increasing function of x_i. An important consequence is that two countries i and j with the same degree $k_i = k_j$ must have the same value $x_i = x_j$.

Equation (9.8) provides an interesting connection with quantum physics, and in particular the statistical mechanics of the microscopic particles known as *fermions* (see footnote 5). The 'selection rules' of fermions dictate that only one particle at a time can occupy a single-particle state, exactly as each pair of vertices in binary networks can be either connected or disconnected. In this analogy, every pair i, j of vertices is a 'quantum state' identified by the 'quantum numbers' i and j. So each link of a binary network is like a fermion that can be in one of the available states, provided that no two objects are in the same state. Equation (9.8) indicates the expected number of particles/links in the state specified by i and j. With no surprise, it has the same form of the so-called *Fermi-Dirac statistics* describing the

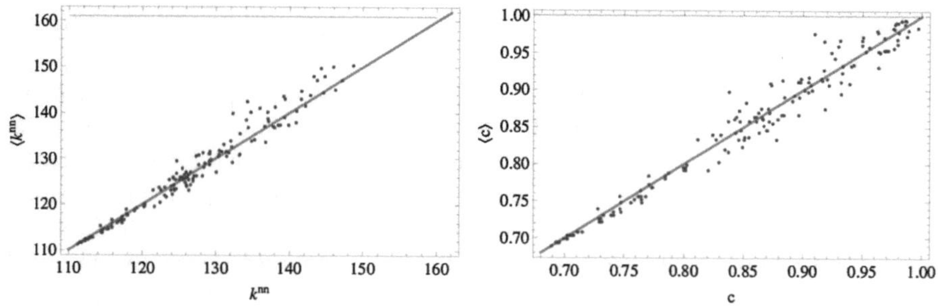

Fig. 9.4 The topology of the World Trade Web is well reproduced by specifying the number of trade partners of each country (binary Configuration Model). *Left*: observed VS expected average nearest neighbour degree k^{nn}, for all vertices. *Right*: observed VS expected clustering coefficient c. The *red curves* are identity lines (perfect agreement). The *green curves* represent the prediction for the same quantities under the Gravity Model, which instead predicts a completely connected network with all vertices characterized by the same value. Data: UNCOMTRADE database [21], year 2001 ($N = 162$ countries)

expected number of fermions in a given quantum state [26, 28, 29]. As we already discussed, the probabilistic nature of links allows also for the presence of empty states, whose occurrence is now regulated by the probability coefficients $(1 - p_{ij})$.

We now come to the application of the model to the topology of the ITN. Unlike the Gravity Model, the Configuration Model allows the whole degree sequence of the observed network to be preserved (on average), while randomizing other (unconstrained) network properties [28]. In order to check whether the model successfully reproduces the ITN, one needs to compare the higher-order (unconstrained) observed topological properties with their expected values calculated over the maximum-entropy ensemble. This automatically indicates whether the degree sequence is informative in explaining the rest of the topology. This can be done analytically, by means of the probabilities appearing in Eq. (9.8) [28]. The effectiveness of the degree sequence in reproducing other topological properties of the ITN is shown in Fig. 9.4, where we compare the observed values of the average nearest neighbour degree k_i^{nn} and clustering coefficient c_i (defined in Sect. 9.3) with the corresponding expected values $\langle k_i^{nn} \rangle$ and $\langle c_i \rangle$, for all vertices. In this type of plot, the agreement between model and observations can be simply assessed as follows: the less scattered the cloud of points around the identity function, the better the agreement between model and reality. In principle, a broadly scattered cloud around the identity function would indicate the little effectiveness of the chosen constraints in reproducing the unconstrained properties, signalling the presence of genuine higher-order patterns of self-organization, not simply explainable in terms of the degree sequence alone. However, the results in Fig. 9.4 indicate that the World Trade Web is well reproduced by the Configuration Model. This result is very robust, as documented by recent analyses that have confirmed it for different temporal snapshots, different levels of aggregation (up to individual commodities), and different datasets [16, 18]. With the appropriate generalizations, this conclusion also holds when the ITN is analysed as a *directed* network, still well described by the Fermi-Dirac statistics [16, 19].

For completeness, Fig. 9.4 also shows the Gravity Model's prediction of a complete network, which is dramatically different from the observed one. Thus the maximum-entropy approach, and in particular the Configuration Model, represents a significant advantage with respect to the Gravity Model. An unexpected implication is that the degrees of world countries are maximally informative about the ITN as a whole: if the empirical degree sequence of the ITN is not reproduced, the observed topology of the network as a whole will not be reproduced either [16]. Unfortunately, current micro-founded models in the economics literature do not attempt at replicating or explaining the particular value of the number of trade partners of a country. Rather, they aim at reproducing the Gravity Model, inspired by the success of the latter at the level of non-zero flows [12]. In so doing, these models are destined to fail in explaining the heterogeneous topology of the ITN. To overcome this limitation, future models should aim at replicating the degree sequence explicitly. As we anticipated at the end of Sect. 9.4, one should also investigate the relationship between the 'empirically informative constraints' in the maximum-entropy approach (the degree sequence in this case) and the 'macroeconomic explanatory factors' in the Gravity Model approach (such as GPD and distances). We will discuss this important point in Sect. 9.9.

We conclude this section by stressing again that the 'fermionic' character of the ITN, when treated as a binary network, is the mere result of the restriction that no two binary links can be placed between any two vertices, leading to a mathematical result which is formally equivalent to the one of quantum statistics. Clearly, there is nothing really 'quantum' in trade connections being described by the Fermi-Dirac statistics, exactly as there is nothing really 'gravitational' in non-zero trade flows being described by the Gravity Model. Still, the deep epistemological analogies leading to similar laws in physics and economics (the ones we discussed in Sect. 9.2) remain, and are now translated into a more sophisticated formalism that allows for the probabilistic and discrete nature of the system:

- In both physical and economic applications, the Fermi-Dirac statistics has the following *symmetry*: the probability p_{ij} only depends on the combination $x_i x_j$. In quantum physics, $x_i x_j$ in turn depends only on the energy of the quantum state, while in the Configuration Model it depends only on the end-point degrees k_i and k_j. This means that, all else being equal (e.g. given the same energy, or the same values of the end-point degrees), the occupation probability of two different states (i, j) and (m, n) is the same.
- In both applications, the Fermi-Dirac statistics implements the *qualitative* idea that, the larger the value of $x_i x_j$, the higher the probability that the state (i, j) is occupied.
- Finally, the validity of the Fermi-Dirac statistics is in both cases established *a posteriori*, by the fact that it reproduces empirical observations. This phenomenological agreement confirms that the postulated quantum numbers/topological constraints, which uniquely specify the values $\{x_i\}$, are indeed the (only) relevant ones for the problem under investigation.

As we will discuss in Sect. 9.9, in the ITN the value of $x_i x_j$ can also be related to the GDP of the two countries i and j, and to the geographic distance separating them. This restores a tight correspondence between the three points listed above and the three ones discussed in Sect. 9.2.

9.6 Bose-Einstein Statistics

We started this chapter stressing the importance of the ITN as a complex weighted network, while in the previous section we restricted ourselves to the description of its purely binary topology. From this point onwards, we go back to the full weighted level and discuss whether it is possible to reproduce the topology and weights of the ITN simultaneously. Naively, the results shown so far suggest that a first attempt in this direction could be the introduction of a two-step process where the topology is first established using the Configuration Model, and the realized link weights are then estimated using the Gravity Model. However, besides being disappointingly inelegant, this approach would result in a hybrid combination of maximum-entropy and economically inspired expectations, leaving the final results without a clear interpretation. A more satisfactory way to proceed is expanding the maximum-entropy formalism into one valid for weighted networks, automatically closer to the macroeconomic reasoning.

To many economists, the finding that the number of trade partners of a country is a particularly informative quantity might appear weak or misleading, given the expectation that the monetary value of imports and exports is in principle much more informative: common sense suggests that knowing how much (in dollars) a country trades with the rest of the world should be more informative than just knowing how many partners trade with that country. This leads to the expectation that the strength s should be more informative than the degree k. One might therefore suspect that an even better model of the ITN, still preserving the observed heterogeneity of countries, is one where the strengths, rather than the degrees, are enforced as constraints. In this section, we take this approach and show that it actually leads to a counter-intuitive result: unlike what we found for the degree sequence, knowing the strength of each world country turns out to be very poorly informative about the structure of the ITN as a whole.

The theoretical framework introduced in the previous section allowed us to treat links as probabilistic entities in order to overcome the Gravity Model's prediction of a completely connected network. Since we were only interested in the prediction of the presence or absence of links, the selection rules were formally analogue to the fermionic ones: $a_{ij} = 0, 1$. However, the formalism can be generalized in order to analyze *weighted networks* where links can have non-negative integer weights. If we keep considering undirected graphs for simplicity, each network is now specified by a $N \times N$ symmetric *weight matrix* W whose entry w_{ij} equals the weight of the link between the vertices i and j. Therefore, we can now label each configuration with W, and the corresponding probability with $P(W)$. If we define the *strength sequence* $\mathbf{s}$ as the ordered vector of strength values (with components s_i, $i = 1, \dots, N$), the *Weighted Configuration Model* can be introduced as the ensemble of *weighted* networks with given strength sequence. If one allows each w_{ij} to take non-negative integer numbers, Eq. (9.4) now becomes [28]

$$P(W) = \prod_{i<j} q_{ij}(w_{ij}) = \prod_{i<j} p_{ij}^{w_{ij}}(1 - p_{ij}). \tag{9.9}$$

Where $q_{ij}(w) = p_{ij}^{w}(1 - p_{ij})$, which has now the form of a *geometric distribution*,[6] is the probability that the vertices i and j are connected by a link of weight w. The outcome $w = 0$, corresponding to a missing link, occurs with probability $1 - p_{ij}$. Therefore p_{ij} still denotes the probability that i and j are connected, irrespective of the weight of this connection. This probability now reads

$$p_{ij} = y_i y_j \tag{9.10}$$

where y_i is the Lagrange multiplier required in order to ensure that the expected strength $\langle s_i \rangle$ of each vertex i equals the empirical value s_i [28]. This time, two countries i and j with the same strength $s_i = s_j$ (independently of their degrees) must have the same value $y_i = y_j$ [17].

As in the previous case, a connection with another well-known quantum statistics emerge. The 'selection rules' have now allowed us to treat link weights as formally analogue to *bosons* (see footnote 5), admitting multiple and unlimited occupations of the same 'quantum state' ($w_{ij} = 0, 1, 2, \dots, +\infty$). Indeed, the expected occupation number of a quantum state, which is the expected weight of the link between vertices i and j, is now formally identical to the so-called *Bose-Einstein statistics* [26, 28]:

$$\langle w_{ij} \rangle = \frac{p_{ij}}{1 - p_{ij}} = \frac{y_i y_j}{1 - y_i y_j} \tag{9.11}$$

As before, there is nothing fundamental in the mathematical analogy with quantum statistics, the only common element being the postulated *discreteness* of the numbers w_{ij}.[7] The deeper similarity involves again the concept of symmetry, which in this case refers to the assumption that, all else being equal, in the Bose-Einstein statistics the expected value $\langle w_{ij} \rangle$ only depends on $y_i y_j$. Similarly, the common qualitative aspect shared by the physical and economic applications is that $\langle w_{ij} \rangle$ is expected to increase with $y_i y_j$. However, as we now show, this time the phenomenological analogy between the Weighted Configuration Model of trade and quantum

[6] In one of its possible formulations, the *geometric distribution* describes the probability that, in a sequence of repeated Bernoulli trials (see footnote 4) with success probability p, the first w trials are all successful and the following one is unsuccessful. This happens with probability $p^{w}(1 - p)$.

[7] In quantum physics, the discreteness is implied by the fact that particles can only exist in integer number. In economic networks, the discreteness is implied by the fact that money can only exist in integer multiples of a fundamental, indivisible unit of currency (such as one Eurocent).

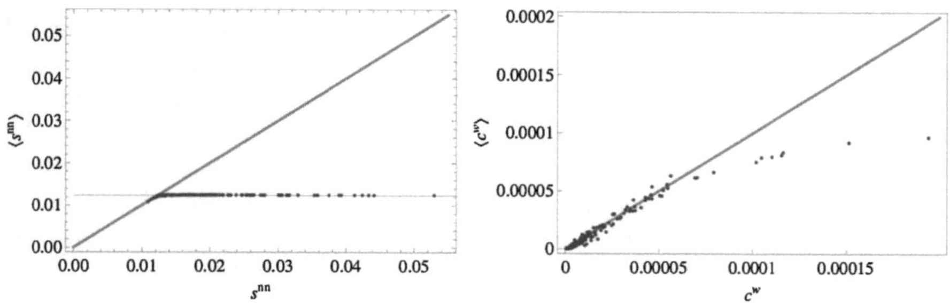

Fig. 9.5 The topology and weights of the World Trade Web are NOT well reproduced by specifying the total import and export value of each country (weighted Configuration Model). *Left*: observed VS expected average nearest neighbour strength s^{nn}, for all vertices. *Right*: observed VS expected weighted clustering coefficient c^w. The *red curves* are identity lines (perfect agreement). The *green curve* represents the prediction for the same quantities under the Gravity Model, which instead predicts a completely connected network with all vertices characterized by the same value. Data: UNCOMTRADE database [21], year 2001 ($N = 162$ countries). The values are rescaled by the total weight

statistics (i.e. the agreement of both with real data) breaks down: while the Bose-Einstein distribution describes the microscopic world of bosons remarkably well, Eqs. (9.9)–(9.11) fail miserably in reproducing the observed ITN.

To see this, one can again compare the observed and the expected values of higher-order (unconstrained) topological properties. In Fig. 9.5 we show the average nearest neighbour strength s^{nn} and weighted clustering coefficient c^w (see Sect. 9.3). The results now indicate how bad the accordance between the Weighted Configuration Model and the real network is. Interestingly enough, the model's prediction for the average nearest neighbors strength is almost identical to the Gravity Model's prediction for the same quantity. This unambiguously indicates that the two models suffer from the same limitation: their incapability to reproduce the topology of the observed network and, in particular, the fact that the Weighted Configuration Model generates an extremely dense network [17], not too different from the completely connected topology predicted by the Gravity Model (for a fully connected network whose weights are rescaled by the total weight, it is easy to estimate $s^{nn} \simeq \sum_i s_i/N = 2/N$). Similarly, even if the smaller values of the weighted clustering coefficient seem to partially agree with the model's prediction, the behaviour for large values indicates that major refinements are needed in order to improve the model performance [17]. The disagreement between the Weighted Configuration Model and the real ITN has been confirmed on different data, different temporal snapshots, and different commodities [17, 18].

The above result contradicts the intuitive expectation that, by taking a weighted quantity (the strength sequence **s**) as input, the Weighted Configuration Model should be more informative than its binary counterpart. Indeed, while the *complete* knowledge of all the weights of the network is of course more informative than the knowledge of the binary topology alone, it turns out that the *partial* knowledge of the weighted network (the strength sequence in this case) is *less* informative than the knowledge of the corresponding binary quantity (the degree sequence). A somewhat puzzling consequence for macroeconomic modeling is that, even if a micro-founded model of international trade successfully reproduces the observed total imports and exports of all world countries, this is definitely not enough in order to reproduce the structure of the ITN as a whole. If combined with the previous result about the extreme informativeness of the degree sequence, this finding strengthens the unconventional conclusion that *satisfactory models of international trade should aim at primarily reproducing the binary properties of countries (number of trade partners) rather than their weighted ones (total import and export)* [16–18].

9.7 Generalized Quantum Distributions

One therefore needs to look for a better model of the International Trade Network, able to reproduce the binary topology and the weighted structure of the network simultaneously. Since microscopic particles are either fermions or bosons, the Fermi-Dirac and Bose-Einstein distributions are the only two types of quantum statistics traditionally used in physics.[8] However, one can formally introduce generalized distributions that reduce to the Fermi-Dirac and the Bose-Einstein statistics as particular cases. While the physically realizable quantum systems might correspond only to the fermionic and bosonic extremes, it might well be that other systems, such as economic networks, can instead realize other non-trivial limits of those generalized distributions. Therefore, in this and in the next section we discuss two possible generalized 'quantum' statistics and their relationship with the structure of economic networks, and the ITN in particular.

The case we consider in this section is just a pedagogical example, while the one we discuss in the next section leads to a very important model that reproduces the observed ITN in great detail. As we showed, the only mathematical ingredient needed to generate the Fermi-Dirac statistics in maximum-entropy network ensembles is the $0/1$ character of binary links ($a_{ij} = 0, 1$), while the only ingredient needed to generated the Bose-Einstein statistics is the non-negative integer character of weighted links ($w_{ij} = 0, \dots, +\infty$). For non-physical systems, there is no reason why the only two allowed values for the maximum weight should be one and infinite. In general, we can consider a general family of distributions, obtained when the occupation number can range from 0 to a *finite* maximum value w_{max}. All the distributions within this family share the same discrete character, due to the integer occupation numbers of 'quantum states'. However, they only reduce to the Fermi-Dirac and Bose-Einstein distributions in the extreme cases $w_{max} = 1$ and $w_{max} \to +\infty$ respectively.

In the general case, with $w_{ij} = 0, 1, 2, \dots, w_{max}$, the maximum-entropy ensemble of networks with given strength sequence $\mathbf{s}$ is characterized by the probability

[8] Excluding the case of *anyons* [30], particles that can only exist in two dimensions and that can be described by a generalized 'fractional statistics' [31], which is however unrelated to the extensions we discuss in this section and in the next one.

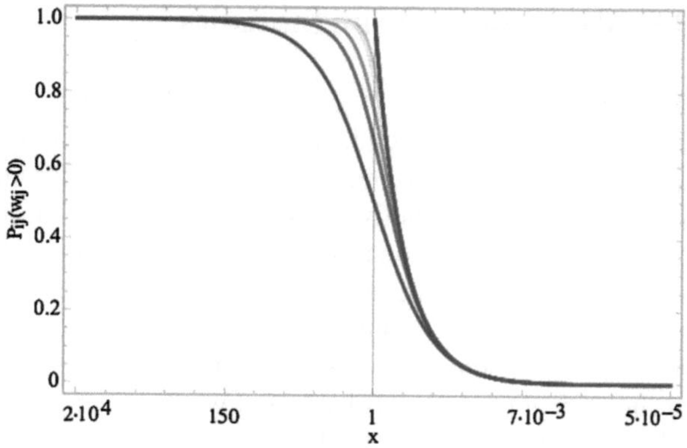

Fig. 9.6 Single-link cumulative distribution functions of the generalized statistics, defined in Eq. (9.13), ranging from the Fermi-Dirac (the *blue* one on the *left*, with $w_{max} = 1$) to the Bose-Einstein (the *blue* one on the *right*, with $w_{max} \to \infty$). The *x-axis* is logarithmically scaled

distribution

$$P(W) = \prod_{i<j} q_{ij}(w_{ij}) = \prod_{i<j} \left[\frac{y_{ij}(1-y_{ij})}{1-y_{ij}^{(w_{max}+1)}} \right]^{w_{ij}} \left[\frac{1-y_{ij}}{1-y_{ij}^{(w_{max}+1)}} \right]^{1-w_{ij}} \quad (9.12)$$

where for simplicity we have defined $y_{ij} \equiv y_i y_j$, if y_i still denotes the Lagrange multiplier needed to enforce the constraint $\langle s_i \rangle = s_i$. Like Eqs. (9.7) and (9.9), the above probability is still a product over single-link distributions, each characterized by the same, bounded range of values. In order to better visualize the functional form of such distributions, the corresponding single-link cumulative distribution functions can be plotted, as shown in Fig. 9.6. The latter can be computed quite easily as

$$P_{ij}(w_{ij} > 0) \equiv 1 - q_{ij}(0) = \frac{y_{ij}(1-y_{ij}^{w_{max}})}{1-y_{ij}^{(w_{max}+1)}}. \quad (9.13)$$

As w_{max} increases from 1 to $+\infty$, the intersections of these distributions with the y-axis form an interesting numerical succession, whose generic term is

$$P_{ij}(w_{ij} > 0 | y_{ij} = 1) = \frac{w_{max}}{w_{max}+1} = \frac{1}{2}, \frac{2}{3}, \frac{3}{4}, \ldots \quad (9.14)$$

and whose limit when $w_{max} \to +\infty$ is 1.

The generalized distribution considered above is an example showing how it is possible to gradually interpolate between the Fermi-Dirac and Bose-Einstein through the introduction of an extra parameter (w_{max}). In principle, intermediate values of the parameter can lead to different results than the ones we showed in Sect. 9.6, and potentially to an improvement over them. However, since the observed weights in the ITN are extremely large, the value of w_{max} required in order to generate realistic weights will be so large that the results are practically indistinguishable from those we have already discussed using the Bose-Einstein distribution. So this model does not represent a real improvement.

9.8 Mixed Bose-Fermi Statistics

A fundamentally different way to unify the Fermi-Dirac and Bose-Einstein distributions, other than introducing an extra parameter while keeping the strength sequence **s** as the constraint, is adopting a different choice of the constraint themselves. Specifically, motivated by the success of the binary Configuration Model discussed in Sect. 9.6, one can introduce a maximum-entropy ensemble of networks with given degree sequence **k** and strength sequence **s**, i.e. one where both constraints are enforced simultaneously [32]. To this end, the maximum allowed value of weight is again $w_{max} = +\infty$ as in the Bose-Einstein case. However, in terms of theoretical models, this leads to a whole new family of probability distributions, whose functional form is [32]

$$P(W) = \prod_{i<j} q_{ij}(w_{ij}) = \prod_{i<j} \left[\frac{(x_i x_j)^{a_{ij}} y_i y_j (y_i y_j)^{w_{ij}-1}(1 - y_i y_j)}{1 - y_i y_j + x_i x_j y_i y_j} \right] \tag{9.15}$$

where a_{ij}, the element of the adjacency matrix, is 0 if $w_{ij} = 0$ and 1 if $w_{ij} > 0$. In the above expression, the **x** vector controls for the degrees and the **y** vector controls for the strengths. This double set of constraints implies that, while different pairs of vertices are still independent, the creation of a link of given weight between two vertices is neither a Bernoulli nor a geometric process (see footnotes 4 and 6), but a combination of the two.

As in the previous example, the (now generalized) 'quantum' or discrete character of the statistics becomes evident as soon as the expected occupation numbers are computed [32]:

$$\langle a_{ij} \rangle = \frac{x_i x_j y_i y_j}{1 - y_i y_j + x_i x_j y_i y_j}, \qquad \langle w_{ij} \rangle = \frac{x_i x_j y_i y_j}{(1 - y_i y_j + x_i x_j y_i y_j)(1 - y_j y_j)}. \tag{9.16}$$

In this case, the unification of the Fermi-Dirac and Bose-Einstein distributions is achieved by combining binary and weighted constraints 'as a block', i.e. in a single big step. As a result, one cannot gradually interpolate between the two ordinary statistics: for instance, in order to retrieve the Bose-Einstein distribution, one must drop the entire degree sequence **k** from the set of constraints (mathematically, this corresponds to set **x** equal to the unit vector **1**).[9] The fundamental difference with respect to the intermediate distributions defined by Eq. (9.12) is that now the occupation probability of an empty 'single-link state' differs from the occupation probability of an already occupied state. In fact, this family of distributions can be intuitively described by saying that the 'first' appearance of a link of unit weight between two disconnected vertices is regulated by the Fermi-Dirac statistics, while the

[9] Note that the dual operation, i.e. dropping the entire strength sequence **s** from the set of constraints (mathematically corresponding to $\mathbf{y} = \mathbf{1}$), leads to an undefined model and does not correspond to the Fermi-Dirac statistics. The reason is that the maximum weight is still $w_{max} = +\infty$ and not $w_{max} = 1$: without constraints on the weighted properties, the expected weights become infinite.

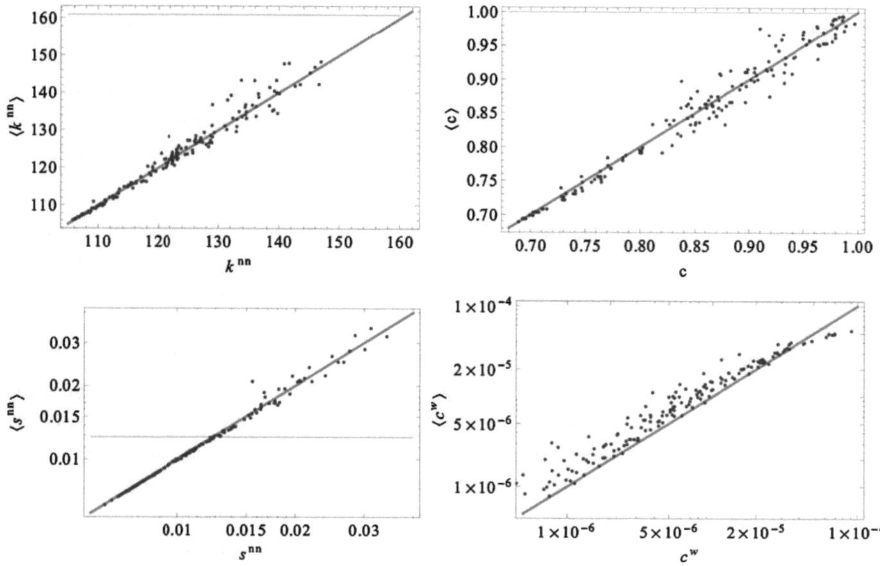

Fig. 9.7 The topology and weights of the World Trade Web are simultaneously well reproduced by specifying the number of partners of each country, as well as the total import and export value of each country (mixed Configuration Model). *Top left*: observed VS expected average nearest neighbour degree k^{nn}, for all vertices. *Top right*: observed VS expected clustering coefficient c. *Bottom left*: observed VS expected average nearest neighbors strength s^{nn}. *Bottom right*: observed VS expected clustering coefficient c^w. The *red curves* are identity lines (perfect agreement). The *green curves* represent the prediction for the same quantities under the Gravity Model, which instead predicts a completely connected network with all vertices characterized by the same value. Data: UNCOMTRADE database [21], year 2001 ($N = 162$ countries). The values are rescaled by the total weight

'subsequent' appearance of units of weights between two already connected vertices is regulated by the Bose-Einstein statistics. For this reason, the statistics defined by Eq. (9.16) is called the *mixed Bose-Fermi statistics* [32].

The Bose-Fermi statistics reproduces with great accuracy all the four higher-order structural quantities (both binary and weighted) of the ITN considered so far. This is shown in Fig. 9.7, where we compare the expected and observed values of k_{nn}, c, s^{nn}, and c^w for all vertices. For the first time, we find a very close agreement for all these topological properties simultaneously. This result is very robust, as it holds for different snapshots and different commodities [24]. Two main conclusions can be drawn:

- on one hand, the addition of weighted constraints to the binary ones does not affect the effectiveness of the mixed model in reproducing the purely topological properties: the two upper panels of Fig. 9.7 look approximately the same as the two panels of Fig. 9.4;
- on the other hand, the addition of binary constraints to the weighted ones dramatically improves the performance of the model in predicting purely weighted

quantities. This is evident by comparing Fig. 9.5 with the two bottom panels of Fig. 9.7. The latter show that now both the average nearest neighbour strength and the weighted clustering coefficient closely follow the identity function. In a sense, the addition of purely binary constraints compensates the incapability of the purely bosonic model in reproducing the network structure and brings the model back to high levels of performance at the topological level.

The family of mixed Bose-Fermi statistics not only represents a powerful model in order to explain *both* the binary and the weighted quantities of interest of a given, observed network; it also points out the strong effects of the underlying topology on the weighted structural patterns.

In economic terms, our discussion leads to the conclusion that the knowledge of monetary/weighted structural properties (such as total imports and exports) is informative only if in combination with non-monetary/binary properties (such as the number of trade partners). The expectation that monetary quantities are *per se* more informative than the corresponding binary ones turns out to be incorrect. For this reason, we believe that the Bose-Fermi statistics is a very useful tool in the understanding of economic networks in general. The fact that both strengths and degrees are enforced allows to study the interplay between the topological and monetary levels of organization, while still keeping the model parsimonious: only *local* (country-specific) structural properties, the ones that we discussed as the somewhat irreducible and 'unavoidable' level of heterogeneity, are enforced.

9.9 The Role of GDP and Distance

The results presented so far show that the attempt to model the International Trade Network using analogies with physical laws, initiated by Jan Tinbergen with the introduction of the Gravity Model, turns out to be extremely successful, even if the appropriate formal expressions are very different from Tinbergen's original idea. However, to complete our discussion, we need to address the final point anticipated in Sect. 9.4, i.e. how to reconcile maximum-entropy graph ensembles (that take structural properties as input) with the Gravity Model's expectation that international trade strongly depends on non-structural macroeconomic properties, such as GDP and geographic distances.

We start with a discussion about the role of GDP. As we anticipated in Sects. 9.3 and 9.5, the GDP of a country turns out to be highly correlated with the degree of that country in the ITN. Interestingly, the functional dependence of the degree on the GDP can be adequately characterized by Eq. (9.8). In more detail, it was shown [9] that the connection probability

$$p_{ij} = \frac{zGDP_i GDP_j}{1 + zGDP_i GDP_j} \tag{9.17}$$

(where z is a global free parameter) defines a model that reproduces the properties of the binary topology of the ITN very well, just like the Configuration Model defined by Eq. (9.8) does. The value of z is fitted to the data by requiring that the expected number of links $\langle L \rangle$ equals the observed number L [9].[10] This result shows that the parameter x_i is approximately proportional to GDP_i, as can be confirmed explicitly [33]. In terms of the network formation process, this means that the results discussed in Sect. 9.5 can be almost entirely rephrased as follows. The GDP is found to determine *directly* the number of trade partners of each country (because GDP_i has the same role of the Lagrange multiplier x_i determining k_i), and *indirectly* the whole topology of the International Trade Network (because the functional form of the connection probability is that of the Configuration Model, where the higher-order topological properties are entirely determined by the degree sequence). The only topological quantity we need to know about the real network is the total number of links specifying the parameter z.

In an only slightly more complicated way, it is also possible to incorporate distances into Eq. (9.17) [34]. This leads to the probability

$$p_{ij} = \frac{zGDP_i GDP_j e^{-\gamma f(d_{ij})}}{1 + zGDP_i GDP_j e^{-\gamma f(d_{ij})}} \tag{9.18}$$

where $f(d_{ij})$ is some increasing function of the geographic distance between countries i and j. The simplest choice for this function is $f(d_{ij}) = d_{ij}$ [35]. The model has now two parameters, which can be fixed simultaneously by imposing $\langle L \rangle = L$ and $\langle F \rangle = F$, where $F \equiv \sum_{ij} a_{ij} f(d_{ij})$ is a measure of the *filling* of space by the network [35]. A variant of this model has been recently used to analyse the directed version of the ITN [34]. The result one finds is that the addition of spatial information moderately improves the fit to the data. However, alternative models that include information about the *reciprocity* of trade [19, 36], rather than geographic distances, systematically outperform the spatial model [34].

We note that, along the same lines as above, it is straightforward to introduce GDPs and distances also in the weighted models defined by Eqs. (9.11) and (9.16), by simply replacing $y_i y_j$ and $x_i x_j$ with $zGDP_i GDP_j e^{-\gamma f(d_{ij})}$. Even if these weighted models have not yet been used in empirical analyses, the above discussion shows that maximum-entropy ensembles are not *per se* incompatible with the Gravity Model's approach of explaining trade patterns in terms of macroeconomic quantities such as GDP and distances. On the contrary, we believe that maximum-entropy models are a very promising tool to understand economic networks. Identifying the most informative properties explaining the topology and weights of real economic networks is extremely important in order to identify the most relevant 'targets' of theoretical models. The finding that the observed trade patterns cannot be adequately understood unless one is able to reproduce the degree sequence, and that the latter is directly determined by the GDP, could only be established using a maximum-entropy model. More in general, the important role played by binary properties even in weighted analyses is a highly nontrivial result.

---

[10] This choice for the parameter z corresponds to the maximization of the likelihood of the model defined by Eq. (9.17) [33], exactly like the values of $\{x_i\}$ that realize the conditions $\langle k_i \rangle = k_i \ \forall i$ maximize the likelihood of the Configuration Model defined by Eq. (9.8) [28, 33].

9.10 Conclusions

In 1962, in what we would now call a pioneering attempt to model economic networks, the physicist and first Nobel Memorial Prize laureate Jan Tinbergen introduced the Gravity Model of trade mimicking Newton's gravitation law. This very intuitive and elegant proposal aimed at explaining trade exchanges in terms of a few macroeconomic quantities (GDP and geographic distance) by combining them in the same way as nature combines gravitational masses and spatial distances.

The success of the Gravity Model is due to the fact that it reproduces well the observed (non-zero) trade flows between countries. Minor refinements to the model, such as the inclusion of additional factors either favouring or suppressing trade, are relatively simple to make and further improve the fit to the data. Therefore, for half a century the Gravity Model has been used more and more extensively in macroeconomic analyses, and it has become the standard model of international trade in the economics literature. However, the most serious and in some sense irreducible limitation of the Gravity Model emerged only relatively recently, after the publication of several empirical analyses documenting the topology of the International Trade Network in the statistical physics literature. While the Gravity Model predicts a completely connected network where every country trades with all other countries, the observed ITN is much more heterogeneous and hierarchical.

We have shown that this limitation can be overcome by adopting a probabilistic view, in exactly the same way as classical physics escaped its crisis at the end of XIX century by adopting the quantum-mechanical paradigm. In network theory, this amounts to consider the adjacency matrix entries as probabilistic entities and the node pairs as single-link states whose occupation numbers are regulated by the same selection rules that apply to fermions and bosons in quantum physics. In this way, various probability distributions can be defined in order to explain the observed structural patterns. On one hand, purely fermionic selection rules excellently reproduce the binary topology of the ITN, but are intrinsically limited by the fact that they give no information about the weights of links in the network. On the other hand, bosonic selection rules are suitable for weighted analyses but suffer from the same limitations of the Gravity Model, since they lead to the prediction of an almost completely connected network. Interestingly enough, the most effective probabilistic models are those combining fermionic and bosonic selection rules. In this combination, the limitations encountered when the two quantum statistics are considered as separate are overcome simultaneously, and both the topology and weights of the observed ITN are nicely reproduced.

The main conclusion we can draw is the fundamental role played by topology in explaining the observed patterns of real world networks: in contrast with the 'mainstream' economic thinking, purely weighted information (such as that encoded into the strength sequence) is *not* enough to reproduce all the observed properties and, in particular, the purely binary ones (such as the degree sequence). A genuine, purely binary information is also needed from the very beginning, as confirmed by the successful family of mixed Bose-Fermi probabilistic distributions. This shows that the naive expectation that weighted/monetary quantities are *per se* more informative than the corresponding binary/non-monetary ones is incorrect. The counter-intuitive nature of this finding shows that it is very important to further develop an appropriate information-theoretic formalism, based on maximum-entropy statistical ensembles, aimed at identifying the key structural properties of economic networks.

Curiously, the road taken by Jan Tinbergen appears to lead to 'physical' laws that are quite different from the ones originally postulated, and more similar to quantum statistics than gravitation. However, the deep epistemological reasons underlying Tinbergen's idea of introducing the Gravity Model of trade appear to be very appropriate, and persist throughout the more recent approaches. This is, we believe, the most important legacy that Jan Tinbergen left us for the modern understanding of economic networks.

Acknowledgements D.G. acknowledges support from the Dutch Econophysics Foundation (Stichting Econophysics, Leiden, the Netherlands) with funds from beneficiaries of Duyfken Trading Knowledge BV, Amsterdam, the Netherlands.

References

1. Caldarelli G (2007) Scale-free networks: complex webs in nature and technology. Oxford University Press, Oxford
2. Aoyama H, Fujiwara Y, Ikeda Y, Iyetomi H, Souma W (2010) Econophysics and companies: statistical life and death in complex business networks. Cambridge University Press, Cambridge
3. Battiston S, Glattfelder JB, Garlaschelli D, Lillo F, Caldarelli G (2010) In: Network science: complexity in nature and technology. Springer, London, pp 131–163
4. Jackson MO (2008) Social and economic networks. Princeton University Press, Princeton
5. Goyal S (2007) Connections: an introduction to the economics of networks. Princeton University Press, Princeton
6. Tumminello M, Aste T, Di Matteo T, Mantegna RN (2005) Proc Natl Acad Sci USA 102(30):10421–10426
7. Garlaschelli D, Battiston S, Castri M, Servedio VDP, Caldarelli G (2005) Physica A 350(2):491–499
8. Squartini T, van Lelyveld I, Garlaschelli D (2013) arXiv:1302.2063
9. Garlaschelli D, Loffredo MI (2004) Phys Rev Lett 93:188701
10. Hidalgo CA, Klinger B, Barabási A-L, Hausmann R (2007) Science 317:482–487
11. Tacchella A, Cristelli M, Caldarelli G, Gabrielli A, Pietronero L (2012) Sci Rep 2:723
12. Anderson JE (2010) NBER working paper 16576
13. Ravenstein EG (1889) J R Stat Soc 52(2):241–305
14. Tinbergen J (1962) Shaping the world economy: suggestions for an international economic policy. The Twentieth Century Fund, New York
15. Serrano MA, Boguná M (2003) Phys Rev B 68:015101
16. Squartini T, Fagiolo G, Garlaschelli D (2011) Phys Rev E 84:046117
17. Squartini T, Fagiolo G, Garlaschelli D (2011) Phys Rev E 84:046118
18. Fagiolo G, Squartini T, Garlaschelli D (2012) J Econ Interact Coord 8(1):75–107
19. Squartini T, Garlaschelli D (2012) Lect Notes Comput Sci 7166:24–35
20. Gleditsch KS (2002) J Confl Resolut 46:712–724
21. UNCOMTRADE database. http://comtrade.un.org/
22. Linders G-JM, de Groot HL (2006) Tinbergen Institute discussion papers 06-072/3
23. Duenas M, Fagiolo G (2013) J Econ Interact Coord 8(1):155–178

24. Mastrandrea R, Squartini T, Fagiolo G, Garlaschelli D (2013) arXiv:1307.2104
25. Jaynes ET (1957) Phys Rev 106:620
26. Park J, Newman MEJ (2004) Phys Rev E 70:066117
27. Fronczak A, Fronczak P, Holyst JA (2006) Phys Rev E 73:016108
28. Squartini T, Garlaschelli D (2011) New J Phys 13:083001
29. Fronczak A (2012) arXiv:1210.7828
30. Arovas DP, Schrieffer R, Wilczek F, Zee A (1985) Nucl Phys B 251:117–126
31. Ramanathan R (1992) Phys Rev D 45:4706–4707
32. Garlaschelli D, Loffredo MI (2009) Phys Rev Lett 102:038701
33. Garlaschelli D, Loffredo MI (2008) Phys Rev E 78:015101
34. Picciolo F, Squartini T, Ruzzenenti F, Basosi R, Garlaschelli D (2012) arXiv:1210.3269
35. Ruzzenenti F, Picciolo F, Basosi R, Garlaschelli D (2012) Phys Rev B 86:066110
36. Garlaschelli D, Loffredo MI (2006) Phys Rev E 73:015101

Chapter 10
A Macroscopic Order of Consumer Demand Due to Heterogenous Consumer Behaviors on Japanese Household Demand Tested by the Random Matrix Theory

Yuji Aruka, Yuichi Kichikawa, and Hiroshi Iyetomi

Abstract So far the consumer theory was microscopically too restrictive to overlook many important scenes of the whole consumption activities. This view is deliberately dropping the inter-correlated factors between different income classes and household demands. The household demands must have a certain bias toward either common or different directions among different income classes. In some sense, the traditionally narrow interest may be dangerous because other decisive factors contributing to the consumption activities may be missed. This article argued to choose a particular scene where some natural or social correlative relations i.e., some dominant forces, may work in the consumption activities over the different income classes. By introducing the different income classes, we can just analyze a new facet of interactive correlations among the heterogeneous consumers. Here we can find any correlative relation, irrespective of price variations. Such a way of thinking may lead us observing another hidden force of the consumption activities.

10.1 A New Empirical Examination on Household Demand and the Analytical Targets

In view of historical development of economic theories, we identify the most essential constituents of economics with the following issues: demand law, utility function, production function, and general equilibrium. These issues were argued

Y. Aruka
Faculty of Commerce, Chuo University, 742-1 Higashinakano, Hacioji-shi, Tokyo 192-0393, Japan
e-mail: aruka@tamacc.chuo-u.ac.jp

Y. Kichikawa
Graduate School of Science and Technology, Niigata University, Ikarashi-2-cho, Nishi-ku, Niigata 950-2181, Japan
e-mail: f11a011f@mail.cc.niigata-u.ac.jp

H. Iyetomi (✉)
Faculty of Economics, University of Tokyo, 7-3-1 Hongo, Bunkyo-ku, Tokyo 113-0033, Japan
e-mail: hiyetomi@gmail.com

F. Abergel et al. (eds.), *Econophysics of Agent-Based models*,
New Economic Windows, DOI 10.1007/978-3-319-00023-7_10,

professional-mathematically in 1930s to 1950s, mainly by mathematicians and physicists.

- A design of demand law and a new form
  - We had a splendid work of Pareto-Slutsky equation. However, Abram Wald was the first scholar who tackled the demand law in a scientific way.
- The law of demand
  - Income effect in parallel with substitution effect are the most basic ideas of consumer market theory, tracing back to Vilfredo Pareto and Eugen Slutsky in the first decade of the last century. This subject once was one of the most important analyses.

Our households, composed of different income classes, should no longer be considered homogeneous. This observation has been already achieved by Hildenbrand [3]) and others. It is a well-known fact that the Pareto-Slutsky equation cannot confirm the demand law, which means the correspondence between prices falling and demands non-decreasing. That is to say, the individualistic principle fails in making the demand law. Hildenbrand's aim was to elucidate a certain sufficient condition under which the demand law holds. He has solved the demand law by introducing heterogeneous households of different income levels. His contribution actually is regarded as a great contribution to economic science in verifying that the demand law just works only if a macroscopic order is to be incorporated. A further extension of this idea to Japanese consumer data has been given in [1].[1]

This study newly focuses on the correlative factors on consumer demand fluctuations. We will mainly have two routes of the effects to households demands: income effect and the other effects generated by some correlative links either due to either the necessities of life or some macroscopic business fluctuation. The former is traditionally observed as a microscopic behavior caused by an income effect due to price change. But this effect, as we stated in the above, has been not confirmed without introducing a certain macroscopic factor of heterogeneous consumers, e.g., an assumption of greater variances of expense items as a class income is higher. This was solved theoretically as well as empirically by Hildenbrand [3]. He then employed a so-called Family Expenditure Survey to examine the empirical evidences in some major Western countries. In this paper, we will also do it in Japanese economy to monitor the effects of macroscopic correlation among consumption categories of different income classes or heterogeneous agents, and then identify the other non-random effects due to business fluctuations, if any.[2]

[1]See Appendix 1.

[2]In this example, middle class property is a state property, if the income range of the middle class is pertinently defined. The ingredients of various types in the middle class of income are then ignored. Such a simplification is called a metonymy from types to state properties, according to [3]. Exchangeable agents are virtually used, instead of types in a precise sense. Types are replaced with state variables. A state variable then becomes a surrogate variable for type ([2], p. 133).

The sources of our idea in this paper depends on Iyetomis idea to distinguish the principal modes at the industrial levels (to construct macroscopic effective demands) from random modes by the use of the principle of the random matrix. That is to say, the so-called effective macroscopic demand generated by industries is subject to a series of business cycles. Iyetomi et al. [4, 5] has shown that there is implemented a structural (nonrandom) cycle in the Japanese economy. The paper has judged an existence of business cycle by applying the random matrix theory (RMT) to the empirical data of the Indices of Industrial Production (IIP) of Japan.

Following Iyetomi's idea, we may identify a nonrandom co-movement in the consumption of heterogeneous households. Such a principal factor, if it exists, suggests that its mode components may form a driving force for consumption in a certain direction. Thus our first step is to find a statistically significant principal eigenvalue explicitly distinguished from a random distribution of eigenvalues. On the other hand, the remaining configuration except for nonrandom modes may then be regarded as randomly distributed modes over the different income classes. The properties of the latter modes may be unspecified in our analysis. They seemingly look statistically random. But we cannot know from our present analytical tool whether there is still embedded any true correlation or not.

In our new study, we will employ the Monthly Expenditure per Household (MEH). It is important to mention that both data have a common subset at background of categories in the part of Consumer Goods. Production is normally connected with household consumption. If some macroeconomic variables should have any correlative relation, consumption could naturally be affected by the macroeconomic fluctuations. Some variables like Industrial-Production Index, Unemployment, and NIKKEI Stock Average Index may be regarded as surrogate variables of income. By using such a surrogate variable, we can then argue the effect of income fluctuation to the household consumption on the average common to the different income classes. A more detailed discussion will come up again later.

10.2 An Application of the Random Matrix Theory (RMT) to Detect Some Causal Relationships

So we illustrate more details about our plan. According to our traditional idea, the demand law is distorted by income effect. However, the demand law may be equally likely affected by other forces like nonrandom modes reflecting natural or social necessities, and possibly business fluctuations. If it should be the case, it could be conceived that another force as distinct from individualistic inner preference is working. We thus have another problem on the demand law.

However, Slutsky [7] believed that such a force does not exist. He rather believed that the summation of pure random shocks might generate cyclical fluctuations. But, according to Iyetomi et al. [4, 5], his prediction was wrong. Iyetomi has proven this reasoning by applying the random matrix theory to the data to estimate some causal relationship running over the concerned data, that is to say, the Dominant Factors identified by the random matrix theory. Iyetomi et al. [4, 5] applied to the data of the industry classification with three properties to derive the effective demand structure of a national economy. So we will have a new work on consumer data to be challenged.

We will apply the latter idea to the household consumer demand with income class properties. In such way, first of all, we aim at detecting whether our household demand data at hand really is randomly distributed or not in view of random matrix theory. We then use Yearly Income Group (All Household) of 5 income rank classifications as our usual data in , Monthly Receipts and Disbursements per Household, for convenience. While the item of expenses, i.e., 10 categories, is listed as follows:

1. Food
2. Housing
3. Fuel, light, & water charges
4. Furniture & household utensils
5. Clothing & footwear
6. Medical care
7. Transportation & communication
8. Education
9. Culture & recreations
10. Other consumption expenditure

In brief, in the Report on the Family Income and Expenditure, we have usually 10 categories with 5 ranks of income classes. On the other hand, in the Indices of Industrial Production (IIP) of Japan, we can use normally 21 categories of goods with 3 ranks of production (value added), shipments, and inventory.

Thus we will work on the hyper-plane of the above statistics characterized by 5 income ranks and 10 spending categories. If we should find any major principal mode by a spectra distribution derived from our empirical data, and it also be separated explicitly from the derived random distribution by the random matrix theory, it could be verified that we found a common factor over the different income classes. This must be the first work. We will then identify the effects of macroeconomic variables on the spending categories over the different income classes.

Actually the members belonging to an income class always is being replaced. A member who earned a higher income may be fallen into a lower income class. The converse process also holds. Even taking into account such an effect of canceling out, the members of an income class is actually moving. But we assume that the shift from lower classes cancel out the shift from upper classes, due to a limitation of statistical data. It then means that types are replaced with state variables. Hildenbrand [3] called this assumption the so-called metonymy on transitions of sample members among the income classes.

Now we thus follow a preceding study of Iyetomi et al. [4, 5] on the industrial data. The variable employed here is put $S_{\alpha,g}(t_j)$, where α denotes an income class, labeled by from 1 at the bottom to 5 at the top, g denotes 10 items of expenses of households, and t_j is expressed by $t_j = j\Delta t$. Here $\delta t = 1$ implies a month, and j runs from 1 to N ($N = 146$ for Jan 2001 to Feb 2012). It then holds for the logarithmic growth rate:

$$r_{\alpha,g}(t_j) := \log_{10} \frac{S_{\alpha,g}(t_{j+1})}{S_{\alpha,g}(t_j)}. \tag{10.1}$$

We furthermore normalize them to define as follows:

$$w_{\alpha,g} := \frac{\sigma_{\alpha,g}(t_j) - \langle r_{\alpha,g} \rangle_t}{\sigma_{\alpha,g}}. \tag{10.2}$$

Here $\langle r_{\alpha,g} \rangle_t$ is the average all over t_j, and $\sigma_{\alpha,g}$ is the standard deviation of $r_{\alpha,g}$ over time. Such an operation lead to bringing the property that the average is set zero, and 1 the standard deviation 1. This is a set up for the application to random matrix theory.

Next, we apply the Fourier decomposition to $w_{\alpha,g}(t_j)$ in the following manner:

$$w_{\alpha,g} = \frac{1}{\sqrt{N'}} \sum_{k=0}^{N'-1} \tilde{\omega}_{\alpha,g}(\omega_k) e^{i\omega_k t_j}. \tag{10.3}$$

Here the frequency is defined as

$$\omega_k := \frac{2\pi k}{N' \Delta t}. \tag{10.4}$$

Taking notice that the data figures are all real-valued, we can set as follows:

$$\tilde{\omega}^*_{\alpha,g}(\omega_k) = \tilde{\omega}_{\alpha,g}(\omega_{N'-k}). \tag{10.5}$$

Taking into account these operations, it then follows:

$$\begin{aligned} &\tilde{\omega}_{\alpha,g}(0) = 0, \\ &\sum_{k=0}^{N'-1} \left|\tilde{\omega}_{\alpha,g}(\omega_k)\right|^2 = N'. \end{aligned} \tag{10.6}$$

Thus we can define the averaged power spectrum $p(\omega_k)$:

$$p(\omega_k) = \frac{1}{M} \sum_{\alpha=1}^{5} \sum_{g=1}^{10} \left|\tilde{\omega}_{\alpha,g}(\omega_k)\right|^2. \tag{10.7}$$

In our case, $M = 50$. It also follows:

$$\sum_{k=0}^{N'-1} p(\omega_k) = N'. \tag{10.8}$$

We define the correlation matrix as C. This is composed of elements $w_{\alpha,g}, w_{\beta,l}$ whose diagonal one is set 1, due to the definition of $w_{\alpha,g}$:

$$C_{\alpha,g,\beta,l} := \langle w_{\alpha,g} w_{\beta,l} \rangle_t. \tag{10.9}$$

Here α and β run from 1 to 5, g and l from 1 to 10. The eigenvectors $V(n)$ and eigenvalues $\lambda(n)$ are associated with this coefficient matrix C. It then is expressed as follows:

$$C = \sum_{n=1}^{M} \lambda^{(n)} V^{(n)} V^{(n)^{Transpose}}. \tag{10.10}$$

If the data should not be all random in the concerned space, we could usually find such a configuration of eigenvalues distribution as a few isolated large eigenvalues explicitly separated distribution on the range of smaller values. The largest group of eigenvalues and associated eigenvectors then corresponds to the dominant factors, which explains co-movements of the fluctuations of different goods. Here we call the largest eigenvector the first eigenvector. This vector can be interpreted as the first major component (Principal Components) that accounts for the most influential combination of correlated factors (different goods) over the adopted space of M-dimension. Similarly, the eigenvector for the second largest eigenvalue (the second eigenvector) is the second principal component that accounts for as much of the correlation as possible in the subspace orthogonal to the first eigenvector, and so on [4]. In other words, the second component can state another influence independent of the first influence.

10.3 The Results of Our Statistical Verifications Obtained

10.3.1 Their Obtained Distributions of Eigenvalues

As we have already noted, we employed the Report on the Family Income and Expenditure by the Statistics Bureau, Cabinet Office, Japan. By calculating eigenvalues by solving the equations formulated in the above section, we have obtained the following distributions: Fig. 10.1 (The Eigenvalue distribution of household consumption in Japan for Jan 2000 to Feb 2012).

Here the data has been seasonally adjusted as the default setting of X-12-ARIMA. We also produce the two largest modes of correlated components over the different income classes. See Fig. 10.2 and Table 10.1 as for the detailed components of the largest eigenvector (the mode 1). Also see Fig. 10.3 and Table 10.2 as for the detailed components of the second largest eigenvector (the mode 2).

10.3.2 The Comparison Among the Alternative Seasonal Adjustments

When we deal with consumer data in general, we always come to grip with seasonal adjustments. So our test must also be engaged into seasonal adjustments. In

Fig. 10.1 The eigenvalue distribution of household consumption

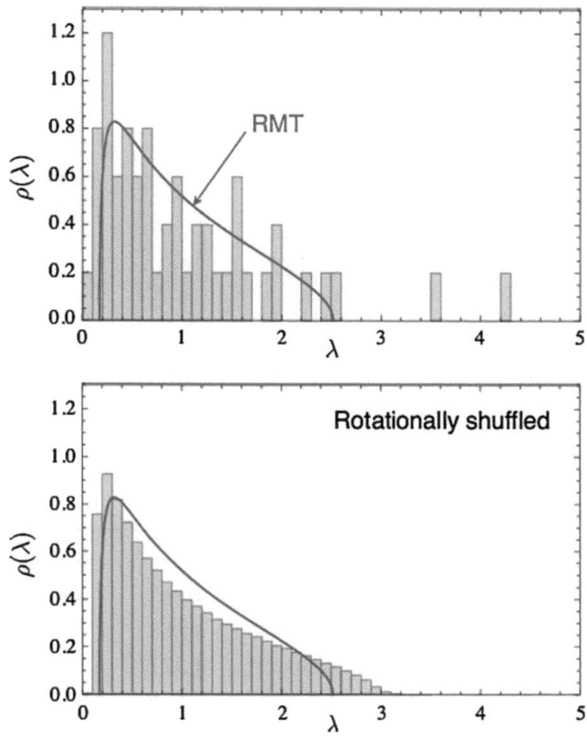

this study, two alternative methods of adjustments are applied. First one is X-12-ARIMA given by the Statistics Bureau, Cabinet Office, Japan, which is providing us with our sample data. The other one is DECOMP, which is developed by the Institute of Statistical Mathematics, Japan. The ideas of two alternatives actually depend on considerably different ideas. The illustration for DECOMP will be argued later.[3] In our application, surprisingly, the derived results are not unchanged as for the principal components identification in our context of random matrix application. Hence, *a fortiori*, we can articulate the same finding for both ways of verification that the largest eigenvalue is dominated by FOOD, and the second by FUEL, LIGHT, & WATER. In other words, the first component is constructed by the explanatory category dominated by FOOD, while the second one by the category dominated by FUEL, LIGHT & WATER.

We thus summarize the obtained results to diagram forms by producing the following diagrams: Fig. 10.4.

[3] See Appendix 2.

Fig. 10.2 The eigenvalue distribution of the mode 1

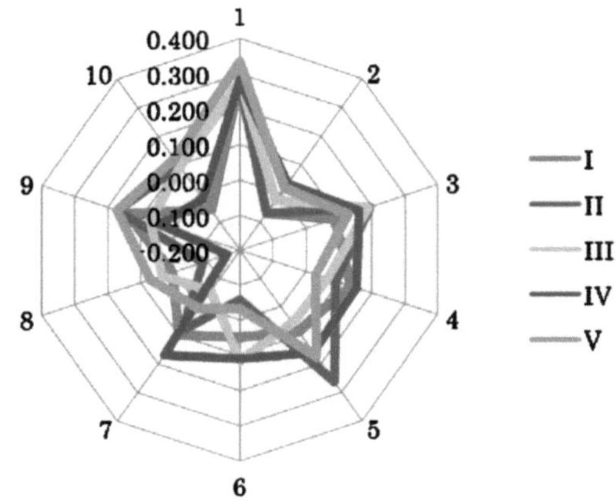

Table 10.1 Mode 1 (eigenvalue $\lambda_1 = 4.24$)

	I	II	III	IV	V
1 Food	0.288	0.275	0.281	0.283	0.335
2 Housing	−0.069	−0.070	−0.024	0.034	0.018
3 Fuel, light & water charges	0.088	0.193	0.189	0.161	0.128
4 Furniture & household utensils	0.137	0.091	0.067	0.153	0.030
5 Clothing & footwear	0.086	0.264	0.058	0.161	0.176
6 Medical care	0.049	−0.059	0.115	0.109	−0.032
7 Transportation & communication	0.098	0.083	−0.056	0.169	0.005
8 Education	0.005	−0.088	0.035	−0.160	0.071
9 Culture & recreation	0.141	0.171	0.071	0.043	0.171
10 Other consumption expenditures	−0.056	0.096	0.094	−0.019	0.122

[a]The data is used from Jan 2000 to Feb 2012

10.4 Some Implications Derived from Our Statistical Tests

10.4.1 Main Findings

Now we can derive several implications from our statistical observations. In our approach described in the first section, at first, we have successfully found two statistically significant principal modes distinguished from a random distribution of eigenvalue. In this case, we have found the modes 1 and 2 associated with the eigenvalues $\lambda_1 > \lambda_2$.

The first principal mode is represented by FOOD. It then turns out that the mode 1 may contribute by 8.5 percent to the whole variations of the consumption. The second principal one is represented by FEEL, LIGHT & WATER. It then turns out that the mode 2 may contribute by 7.1 percent to the whole variations of the consumption. The result becomes similar, irrespective of the kinds of the seasonal adjustment

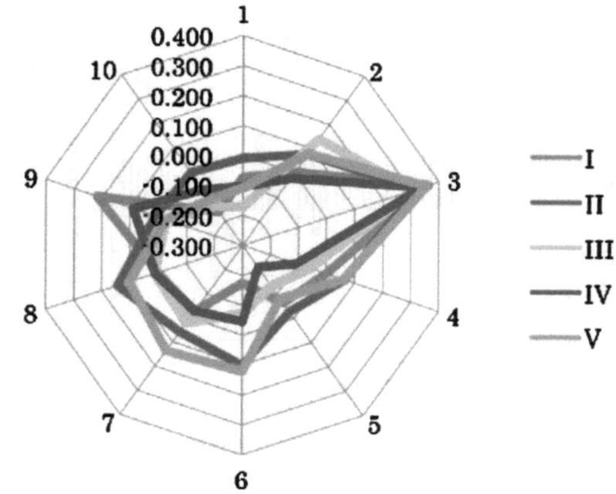

Fig. 10.3 The eigenvalue distribution of the mode 2

Table 10.2 Mode 2 (eigenvalues $\lambda_2 = 3.54$)

	I	II	III	IV	V
1 Food	−0.073	−0.009	−0.171	−0.104	−0.108
2 Housing	0.002	0.076	0.139	−0.023	0.079
3 Fuel, light & water charges	0.362	0.327	0.310	0.342	0.355
4 Furniture & household utensils	0.000	0.065	−0.058	−0.106	0.061
5 Clothing & footwear	−0.064	−0.027	−0.133	−0.210	−0.081
6 Medical care	−0.175	0.102	−0.073	−0.046	0.121
7 Transportation & communication	−0.025	0.059	0.021	−0.028	0.137
8 Education	0.001	0.143	0.012	0.010	0.101
9 Culture & recreation	0.217	0.010	−0.028	0.089	−0.026
10 Other consumption expenditures	−0.162	−0.002	−0.141	−0.072	−0.111

[a]The data is used from Jan 2000 to Feb 2012

method adopted. We thus conclude that the FOOD dominant mode and the FUEL dominant mode are the driving modes for the whole consumption.

Next, we can then notice the different signs of the components each modes. In particular, the signs between the FOOD component and the FUEL component in the mode 1 are all positive, while the signs among the same components in the mode 2 are mutually opposite, as seen from Fig. 10.1.

The result on the mode 1 seems natural, if we interpret the FOOD mode in the context of Engels coefficient connections. If we cook FOOD items, we will also spend FUEL items. In the mode 2, however, it seems difficult to give an instinctive interpretation. But we may notice that the CLOTHING has all negative signs all over the different income classes. In the mode 2, then, the FUEL consumption is opposite to the FOOD as well as CLOTHING. We thus suggest that the mode 2 might be close to something like a consumption activity induced by the use of personal computer. In the recent community we cannot neglect such activity at all. In the event, whatever

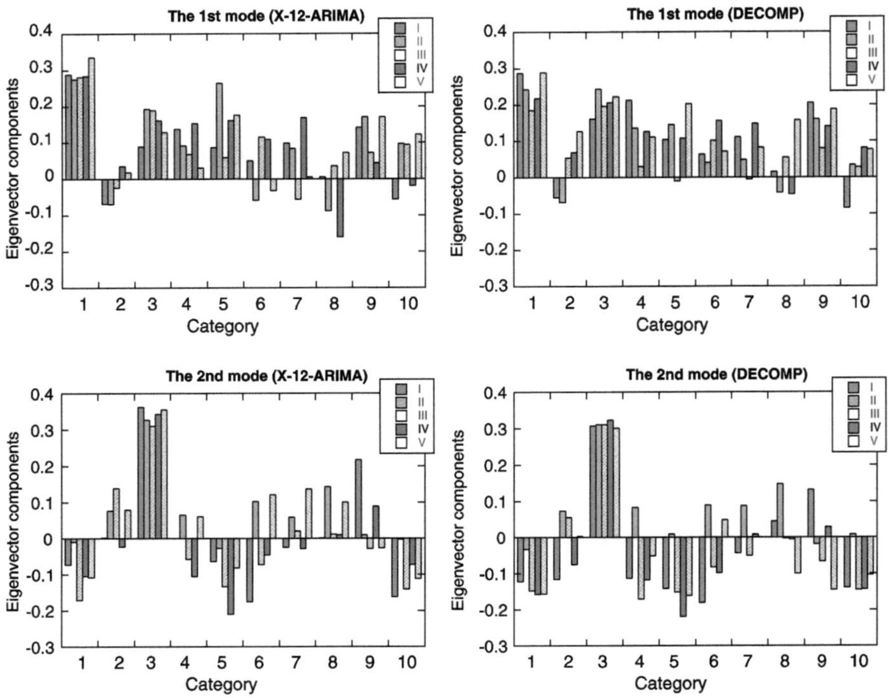

Fig. 10.4 The comparison between X-12-ARIMA and DECOMP

interpretation is to be added, we have shown that at least about 15 percent of the whole variation of consumption is generated by an organized co-movement in a particular way, whether natural or cultural. As we noted, however, we cannot know whether the remaining random distribution really is purely random or not, in view of the present random matrix theory.

In order to detect any correlation induced by hidden factors, however, we can furthermore investigate a possible correlative relation. We look for the mode 3 ($\lambda_1 > \lambda_2 > \lambda_3$). While the mode accompanying the eigenvalues below 0.1 may be regarded as completely random. Inspecting more precisely, we notice the size of the standard deviation, i.e., the size of fluctuations of the concerned mode. If the mode 3 accompanies a bigger size of the standard deviation in comparison to the random variable, the effects due to another factor could be discernible. At the end of the first section, we have already suggested to add several macroeconomic variables to the given data set. We can add to Industrial-Production Index, Unemployment, and NIKKEI Stock Average Index. Instead of these three variables, we can alternatively add to three random variables. However, a tentative principal analysis in the expanded data did not show any big difference from the previous analyses. But another calculation indicated that a mode around the added variables showed some big change of the standard deviation size. This may imply that it does seem to be

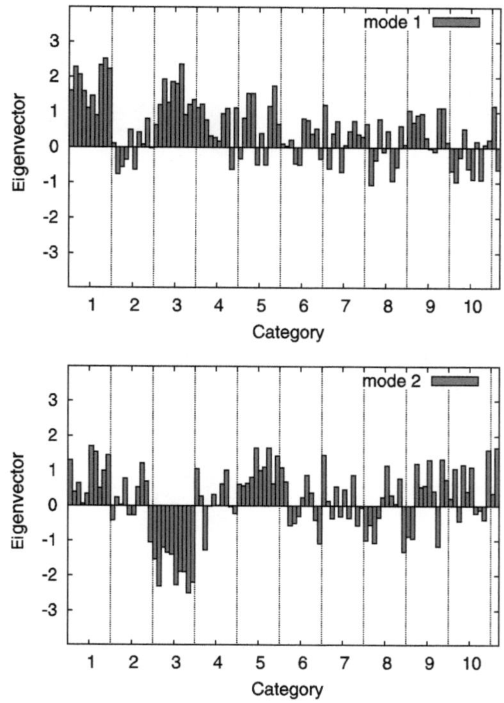

Fig. 10.5 Contributions due to the extra working hours/temperatures

any visible correlation with business fluctuation. This verification will be our further challenging task.

10.4.2 Further Findings

Finally, we note our another attempt to detect another hidden factor affecting consumer demands. We at present chose two factors such as *Extra Working Hours* and *TEMPERATURE*. In particular, we may regard the extra working hours as a factor being sensitive to business fluctuations. After calculations, we may briefly show our additionally obtained results:

1. After we moreover added the time series of extra working hours to the original 10 data, we have calculated the eigenvector distributions. The extra working hours as an additional 11th element marked a positive high value, in particular, in mode 1. We also discovered a similar result in mode 2.
2. We have the *anomaly* time series of monthly average temperature in Tokyo Metropolitan City. As we similarly added the anomaly time series of temperatures to the original 10 data, it also turns out that we still have an unchanged pattern both for mode 1 and mode 2. The components of mode 2, in particular, is magnified by temperature.

See Fig. 10.5. Here, on the horizontal axis, the column before the last one corresponds to the component from *Extra Working Hours*, the last column to the component *Temperatures*.[4]

The first principal eigenvector as a whole retains the positive correlation over all components. Once the time series of Extra Working Hours, we inspected that FOOD , FUEL and Extra Working Hours are contributing much to the first principal eigenvector. Hence we may regard the Extra Working Hours as an important parameter for business fluctuations. Thus mode 1 may depend on business fluctuations. In mode 2, we can discern that FUEL and CLOTHING are closely correlated with TEMPERATURE. Hence the consumption pattern of mode 2 may be most sensitive to TEMPERATURE.

10.5 Concluding Remarks

The observation of the consumer demand should not be limited to a special facet where price sensitive behaviors could be dominant as if prices look like mainly independent variables. Income effects must not be merely auxiliary. More specifically, we can suppose the consumer demand to be a composition of different forces: prices, demand and different income classes. We can thus depict an image of the real facets of interacting prices demands with different income classes as the following diagram. See Fig. 10.6. If the case should be one-commodity world, we could summarize how prices and demands interact such as Fig. 10.7.

As we demonstrated in the abstract of this article, the consumer theory was microscopically too restrictive to overlook many important scenes of the whole consumption activities. In some sense, such a narrow interest may be dangerous because other decisive factors contributing to the consumption activities may be missed. In this article, we have detected a particular scene where some natural or social correlative relations i.e., some dominant forces, may work in the consumption activities over the different income classes. By introducing the different income classes, we can just analyze a new facet of interactive correlations among the heterogeneous consumers. Here we can find any correlative relation, irrespective of price variations. Such a way of thinking may lead us observing another hidden force of the consumption activities. It is considerably interesting to notice that our consumer behaviors are closely connected with some socially combinatorial pattern formations. Our present study will contribute to this interest in a near future.

[4] We have employed the database in terms of year-over year basis.

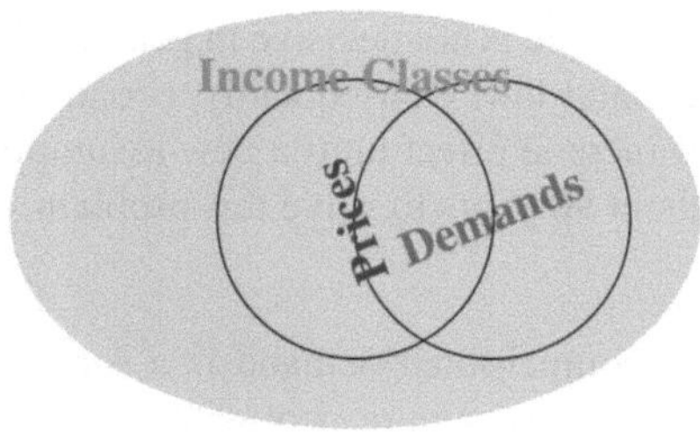

Fig. 10.6 The real facets of prices, demands, and income classes

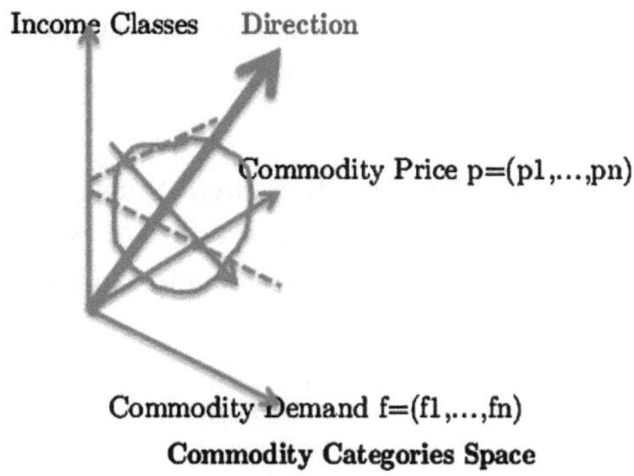

Fig. 10.7 A composition of different forces

Appendix 1

We denote a demand at price p by $f(p)$. We then have a next formula between two distinct price vector p^1 and p^2: ?

$$(p^1 - p^2)(f(p^1) - f(p^2)) \leq 0 \tag{10.11}$$

i.e.,

$$\mathrm{d}p\mathrm{d}f \leq 0. \tag{10.12}$$

This is the demand law. The law does not state any analytical confirmation with respect of price when an income variation x derived by price changes is taken account into. The tentative formulation is called the Pareto-Slutsky equation:

$$\frac{\partial f_j}{\partial p_k} = \frac{\partial h_j}{\partial p_k} - \frac{\partial f_j}{\partial x} f_k. \tag{10.13}$$

In other words,

- demand change = substitution effects + income effects

Here j and k are indices of good. f_j is a demand of good k. h_j is a compensated demand of good j. x is an income level. Demand for good depend on prices of goods p_k including itself (p_j) and also depend on income. Income x is to be measured in terms of goods. So income may be changeable depending on price variations Δp. A change of income naturally induces a change of demand. But the sign of a change of income is not decisive in general. This is a complicated factor for establishing the demand law. Consequently, economists never were successful to confirm the sign of income effect until a new assumption was invented by Hildenbrand [3]). It took about 90 years to solve this problem.

Compensated demand is a sophisticated idea. This demand h always has a negative sign respect with a price rise. Actually, h is supposed to be a special form only reactive to price variation but inactive to income level to guarantee the same level of satisfaction by supplementing a new injection of income if short, or reducing if long.

We introduce into the Pareto-Slutsky equation a very small footnote-sized perturbation of Δp and $h : \Delta p$ and Δh. A variation of $\mathrm{d}p\mathrm{d}f$ caused by Δp and Δh i.e., $\Delta p \Delta f$ may be approximately estimated as $\Delta p \frac{\partial f_j}{\partial p_k} \Delta p$ In other words, it holds:

$$\Delta p \frac{\partial f_j}{\partial p_k \partial p_k} = \Delta p \frac{\partial h_j}{\partial p_k \partial p_k} - \Delta p \frac{\partial f_j}{\partial x_k} \Delta p f. \tag{10.14}$$

We cannot find any reason that the first item is always equal to the second item. If the second item should be negative, the total variation caused by Δp_k could be positive, leading the contrary to the demand law. Thus, the Pareto-Slutsky equation, as it is, is not successful to guarantee a definite analytical relation in general. A so-called *Giffen effect* could not be removed.

Appendix 2

Most of the items of expenditure have strong seasonal dependence as easily expected. The excepted items are Housing, Medical and Transport. Removal of seasonal components out of the original data is thus a critical procedure to elucidate possible correlations embedded in expenditure of Japanese consumers.

Here we adopt two seasonal adjustment methods. One of them is the X-12-ARIMA,[5] developed and used by the U.S. Census Bureau. It is also a standard seasonal adjustment method for the Statistic Bureau in Japan. The X-12-ARIMA program, having a long history in the development, is full of experimental knowledge with a number of degrees of freedom left for users to optimize the procedure. However, we allowed the program to determine a best regARIMA model by itself.

To assess the reliability of such an automatic seasonal adjustment, we applied another program called DECOMP [6] to the same data. The DECOMP based on state-space-modeling is free from the moving average procedure that plays an important role in X-12-ARIMA. As indicated in its name, the DECOMP decomposes

[5]http://www.census.gov/srd/www/x12a/.

a given time series data into trend, seasonal and irregular components in a transparent way. In return, there is no much room for us to play with the program for optimization of the procedure. The parameter set for the program that we used is

- Log Transformed: Yes
- Seasonal frequency: 12
- Trend order: 1
- AR order: 0
- Trading Day Effects: Yes

As will be shown later, the X-12-ARIMA and DECOMP bring about no fundamentally different results for the principal component analysis. The RMT (random matrix theory) tells us that there are two statistically meaningful principal components for both of the seasonally adjusted data. And the characteristic features of the principal components so obtained are essentially the same between the two alternative data.

References

1. Aruka Y (2001) Family expenditure data in Japan and the law of demand: macroscopic microeconomic view. In: Takayasu H (ed) Empirical science of financial fluctuations. Springer, Tokyo, pp 294–303. Reprinted in Aruka Y (2011) Complexities of production and interacting human behaviour. Physica-Verlag, Heidelberg, pp 129–139
2. Aruka Y, Akiyama E (2009) Non-self-averaging of a two-person game with only positive spillover: a new formulation of Avatamsaka's dilemma. J Econ Interact Coord 4(2):131–165
3. Hildenbrand W (1994) Market demand. Princeton University Press, Princeton
4. Iyetomi H et al (2011) What causes business cycles? Analysis of the Japanese industrial production data. J Jpn Int Econ 25:246272
5. Iyetomi H et al (2011) Fluctuation-dissipation theory of input-output interindustrial correlations. Phys Rev E 83:016103
6. Kitagawa G, Gersch W (1984) A smoothness priors-state space approach to the modeling of time series with trend and seasonality. J Am Stat Assoc 79(386):378–389. http://ssnt.ism.ac.jp/inets2/title.html
7. Slutsky EE (1937) The summation of random causes as the source of cyclical processes. Econometrica 5:105–146

Chapter 11
Uncovering the Network Structure of the World Currency Market: Cross-Correlations in the Fluctuations of Daily Exchange Rates

Sitabhra Sinha and Uday Kovur

Abstract The cross-correlations between the exchange rate fluctuations of 74 currencies over the period 1995–2012 are analyzed in this paper. The eigenvalue distribution of the cross-correlation matrix exhibits a bulk which approximately matches the bounds predicted from random matrices constructed using mutually uncorrelated time-series. However, a few large eigenvalues deviating from the bulk contain important information about the global market mode as well as important clusters of strongly interacting currencies. We reconstruct the network structure of the world currency market by using two different graph representation techniques, after filtering out the effects of global or market-wide signals on the one hand and random effects on the other. The two networks reveal complementary insights about the major motive forces of the global economy, including the identification of a group of potentially fast growing economies whose development trajectory may affect the global economy in the future as profoundly as the rise of India and China has affected it in the past decades.

11.1 Introduction

At whatever scale one studies economic phenomena, we can find complex systems, comprising relatively large number of mutually interacting elements often connected to each other in non-trivial topologies, at work. The components can be individual traders, firms, banks, markets or countries, but however complicated the behavior of the individual agents in the system, an even richer collective behavior is manifested at the scale of the entire group of interacting agents. Explaining the emergence of such systems-level phenomena which may be qualitatively different from the properties exhibited by the individual components is one of the key goals of many physicists working on socio-economic questions, an enterprise that is of-

S. Sinha (✉)
The Institute of Mathematical Sciences, CIT Campus, Taramani, Chennai 600113, India
e-mail: sitabhra@imsc.res.in

U. Kovur
Department of Mathematics, Birla Institute of Technology & Science, Pilani 333031, India

F. Abergel et al. (eds.), *Econophysics of Agent-Based models*,
New Economic Windows, DOI 10.1007/978-3-319-00023-7_11,

ten referred to as *econophysics* [1]. An important step in this direction will be to identify features of economic systems that are *universal*, in the sense of occurring at many different scales, suggesting that their existence is not contingent upon the particular conditions prevailing in a specific situation. This will help econophysicists to focus on phenomena that are not just the outcome of a series of historical accidents and which can therefore be potentially explained by generalizable mechanisms.

Market dynamics has been identified by many physicists as a particular area of economics that has the potential for yielding several such universal features. In particular, one can mention the identification of scale-invariant distributions in price fluctuations, the trading volume and number of trades [2, 3] in equities markets (but see also Ref. [4]). However, in order to get an understanding of how qualitatively new features emerge at the level of the collective dynamics of the entire market, one needs to understand the nature and structure of interactions between the agents. While several studies on the networks underlying equities markets (e.g., Ref. [5]) have been done, we need to compare between markets of different kinds in order to distinguish those features that are particular to specific systems and those which are universal. With this aim, we undertake a detailed investigation of the world currency market in this article. While several previous studies have looked at the cross-correlations between the foreign exchange rates of different currencies (e.g., see Refs. [7–9]), our results reveal several novel insights and unexpected features of the network of interactions between the currencies that we reconstruct from the cross-correlations data. The period of the preceding sixteen years we have chosen for our study has seen remarkable transformations in the world economy with the emergence of new economic powerhouses such as China and India, but it has also shown how our world is vulnerable to massive system-spanning crises (such as that of 2007–2008). The study of networks in the global currency market provides an important perspective with which to view the positive as well negative impacts of globalization. It has been argued that globalization is neither a completely new phenomenon in world history nor are its effects always beneficial to the economy [10]. We hope that by investigating the collective dynamics of the international trade in currencies in order to identify the major motive forces of the world economy, one can potentially understand the long-term trends and prospects of globalization.

11.2 The World Currency Market

The foreign exchange (FX) market, representing the entire global decentralized trading of various currencies, is the largest financial market in the world with an average daily trading volume estimated in 2010 to be 4×10^{12} US Dollars [11]. A typical trade in the FX market consists of a pair of agents exchanging a certain amount of a particular currency for a mutually agreed amount of another currency. The ratio of the amounts of the two currencies changing hands specify the corresponding *exchange rate* for the pair of currencies concerned. Thus the exchange rates determine the value of a currency with respect to another (the numeraire). The modern FX market characterized by a large number of currencies having floating exchange rates which continuously fluctuate over time date from the 1970s. The varying rates reflect the changing demand and supply for the currencies, and are thought to be directly influenced by the trade deficit/surplus of the corresponding countries [12] as well as macroeconomic variables such as changes in growth of the gross domestic product, interest rates, etc. However, international events can often trigger large perturbations in the FX market and it is possible that sudden changes in the exchange rates of a certain group of currencies can spread over time, eventually affecting a much larger number of currencies. Our article aims at uncovering the network of interactions between the different currencies of the FX market along which perturbations can propagate in the world currency market.

Description of the data set. We have considered the daily exchange rate of currencies in terms of US Dollars (i.e., the base currency) publicly available from the website of the financial services provider company, Oanda Corporation [13]. We have chosen the US Dollar as the numeraire as it is currently the primary reserve currency of the world and is most widely used in international transactions. The daily rates are computed as the average of all exchange rates (taken as the midpoint of the bid and ask rates) quoted during a 24-hour period prior to the day of posting the rate. For cross-correlation analysis, we have focused on the price data of $N = 74$ currencies from October 23, 1995 to April 30, 2012, which corresponds to $T = 6034$ working days. The choice of currencies was governed by our decision to only include those which either follow a free float or a managed float exchange rate regime. We have thus avoided currencies such as the Chinese yuan whose rate of exchange is pegged against another currency so that the value of currency does not vary appreciably in time (resulting in trivial cross-correlations). We have also excluded countries having a dollarized economy such as Panama, Ecuador Vietnam or Zimbabwe, that use a foreign currency—in majority of cases, the US Dollar—instead of or alongside the domestic currency, as this introduces strong artifacts in the cross-correlations. The period of observation was chosen so as to maximize the volume of available data. Using the MSCI Market Classification Framework [14] we have divided the countries to which the currencies belong into three categories: developed, emerging and frontier markets. This classification is based on a number of criteria including market accessibility, size and liquidity of the market and the sustainability of economic development. While many of the OECD countries belong to the developed category, the rapidly growing economies of Asia, Africa and Latin America (such as the BRICS group comprising Brazil, Russia, India, China and South Africa) are in the emerging category with the frontier markets category being populated by the remainder. The individual currencies, along with the above economic classification of the corresponding countries and the geographical regions to which they belong are given in Table 11.1.

Table 11.1 The list of 74 currencies analyzed in this article arranged according to type of market and grouped by geographical region

i	Currency code	Currency name	Type of market	Geographical region
1	CAD	Canadian Dollar	Developed	Americas
2	DKK	Danish Krone	Developed	Europe and Middle-East
3	EUR	Euro	Developed	Europe and Middle-East
4	ILS	Israeli New Shekel	Developed	Europe and Middle East
5	ISK	Iceland Krona	Developed	Europe and Middle-East
6	NOK	Norwegian Kroner	Developed	Europe and Middle-East
7	SEK	Swedish Krona	Developed	Europe and Middle-East
8	CHF	Swiss Franc	Developed	Europe and Middle-East
9	GBP	Great Britain Pound	Developed	Europe and Middle-East
10	AUD	Australian Dollar	Developed	Asia-Pacific
11	HKD	Hong Kong Dollar	Developed	Asia-Pacific
12	JPY	Japanese Yen	Developed	Asia-Pacific
13	NZD	New Zealand Dollar	Developed	Asia-Pacific
14	SGD	Singapore Dollar	Developed	Asia-Pacific
15	BOB	Bolivian Boliviano	Emerging	Americas
16	BRL	Brazilian Real	Emerging	Americas
17	CLP	Chilean Peso	Emerging	Americas
18	COP	Colombian Peso	Emerging	Americas
19	DOP	Dominican Republic Peso	Emerging	Americas
20	MXN	Mexican Peso	Emerging	Americas
21	PEN	Peruvian Nuevo Sol	Emerging	Americas
22	VEB	Venezuelan Bolivar	Emerging	Americas
23	ALL	Albanian Lek	Emerging	Europe, Middle-East and Africa
24	DZD	Algerian Dinar	Emerging	Europe, Middle-East and Africa
25	CVE	Cape Verde Escudo	Emerging	Europe, Middle-East and Africa
26	CZK	Czech Koruna	Emerging	Europe, Middle-East and Africa
27	EGP	Egyptian Pound	Emerging	Europe, Middle-East and Africa
28	ETB	Ethiopian Birr	Emerging	Europe, Middle-East and Africa
29	HUF	Hungarian Forint	Emerging	Europe, Middle-East and Africa
30	MUR	Mauritius Rupee	Emerging	Europe, Middle-East and Africa
31	MAD	Moroccan Dirham	Emerging	Europe, Middle-East and Africa
32	PLN	Polish Zloty	Emerging	Europe, Middle-East and Africa
33	RUB	Russian Rouble	Emerging	Europe, Middle-East and Africa
34	ZAR	South African Rand	Emerging	Europe, Middle-East and Africa
35	TZS	Tanzanian Shilling	Emerging	Europe, Middle-East and Africa
36	TRY	Turkish Lira	Emerging	Europe, Middle-East and Africa
37	INR	Indian Rupee	Emerging	Asia
38	IDR	Indonesian Rupiah	Emerging	Asia

Table 11.1 (Continued)

i	Currency code	Currency name	Type of market	Geographical region
39	KRW	South Korean Won	Emerging	Asia
40	PHP	Philippine Peso	Emerging	Asia
41	PGK	Papua New Guinea Kina	Emerging	Asia
42	TWD	Taiwan Dollar	Emerging	Asia
43	THB	Thai Baht	Emerging	Asia
44	GTQ	Guatemalan Quetzal	Frontier	Americas
45	HNL	Honduran Lempira	Frontier	Americas
46	JMD	Jamaican Dollar	Frontier	Americas
47	PYG	Paraguay Guarani	Frontier	Americas
48	TTD	Trinidad Tobago Dollar	Frontier	Americas
49	HRK	Croatian Kuna	Frontier	Europe and CIS
50	KZT	Kazakhstan Tenge	Frontier	Europe and CIS
51	LVL	Latvian Lats	Frontier	Europe and CIS
52	BWP	Botswana Pula	Frontier	Middle-East and Africa
53	KMF	Comoros Franc	Frontier	Middle-East and Africa
54	GMD	Gambian Dalasi	Frontier	Middle-East and Africa
55	GHC	Ghanaian Cedi	Frontier	Middle-East and Africa
56	GNF	Guinea Franc	Frontier	Middle-East and Africa
57	KES	Kenyan Shilling	Frontier	Middle-East and Africa
58	KWD	Kuwaiti Dinar	Frontier	Middle-East and Africa
59	MWK	Malawi Kwacha	Frontier	Middle-East and Africa
60	MRO	Mauritanian Ouguiya	Frontier	Middle-East and Africa
61	MZM	Mozambique Metical	Frontier	Middle-East and Africa
62	NGN	Nigerian Naira	Frontier	Middle-East and Africa
63	STD	Sao Tome and Principe Dobra	Frontier	Middle-East and Africa
64	SYP	Syrian Pound	Frontier	Middle-East and Africa
65	ZMK	Zambian Kwacha	Frontier	Middle-East and Africa
66	JOD	Jordanian Dinar	Frontier	Middle-East and Africa
67	BND	Brunei Dollar	Frontier	Asia
68	BDT	Bangladeshi Taka	Frontier	Asia
69	KHR	Cambodian Riel	Frontier	Asia
70	FJD	Fiji Dollar	Frontier	Asia
71	PKR	Pakistan Rupee	Frontier	Asia
72	WST	Samoan Tala	Frontier	Asia
73	LKP	Lao Kip	Frontier	Asia
74	LKR	Sri Lankan Rupee	Frontier	Asia

11.3 The Return Cross-Correlation Matrix

To quantify the degree of correlation between the exchange rate movements for different currencies, we first measure the fluctuations using the logarithmic return so that the result is independent of the scale of measurement. If $P_i(t)$ is the exchange rate of the i-th currency at time t (in terms of USD), then the logarithmic return is defined as

$$R_i(t, \Delta t) \equiv \ln P_i(t + \Delta t) - \ln P_i(t). \tag{11.1}$$

For daily return, $\Delta t = 1$ day. By dividing the time-series of returns thus obtained with their standard deviation (which is a measure of the volatility of the currency exchange rate), $\sigma_i = \sqrt{\langle R_i^2 \rangle - \langle R_i \rangle^2}$, we obtain the normalized return, $r_i(t, \Delta t) \equiv R_i/\sigma_i$. We observed that the cumulative distribution of the returns displayed power-law scaling in the tails, i.e., $P(r_i > x) \sim x^{-\alpha}$ where α is the corresponding exponent value. Using maximum likelihood estimation, the exponents for the different currencies were obtained and they were found to be distributed over a narrow range of values with a peak around $\alpha \simeq 3$. This indicates that the so-called *inverse-cubic law* distribution of returns, reported in many studies of stock price fluctuations [15–18], also holds for currency exchange rate movements [19, 20]. This further strengthens the *universality* of this empirical fact about the nature of market fluctuations and supports the validity of explaining this feature using very general models which do not consider details of particular markets or economies (see, e.g., Ref. [21]).

After obtaining the return time series for all N currencies over the period of T days, we calculate the cross-correlation matrix $\mathbf{C}$ whose individual elements $C_{ij} = \langle r_i r_j \rangle$, represent the correlation between returns for a pair of currencies i and j. If the fluctuations of the different currencies are uncorrelated, the resulting random correlation matrix (referred to as a Wishart matrix) has eigenvalues distributed according to [22]:

$$P(\lambda) = \frac{Q}{2\pi} \frac{\sqrt{(\lambda_{max} - \lambda)(\lambda - \lambda_{min})}}{\lambda}, \tag{11.2}$$

with $N \to \infty$, $T \to \infty$ such that $Q = T/N \geq 1$. The bounds of the distribution are given by $\lambda_{max} = [1 + (1/\sqrt{Q})]^2$ and $\lambda_{min} = [1 - (1/\sqrt{Q})]^2$. For the data we have analyzed, $Q = 81.54$, which implies that in the absence of any correlation the spectral distribution should be bounded between $\lambda_{max} = 1.23$ and $\lambda_{min} = 0.79$. We observe from Fig. 11.1 that the bulk of the empirical eigenvalue distribution indeed falls below the upper bound given by λ_{max}, although a significant fraction of the eigenvalues are smaller than what we expect from the lower bound λ_{min}. Also, a small number ($\simeq 8$) of the largest eigenvalues are seen to deviate from the bulk of the distribution predicted by random matrix theory, and we focus our analysis on these modes to obtain an understanding of the interaction structure of the world currency market.

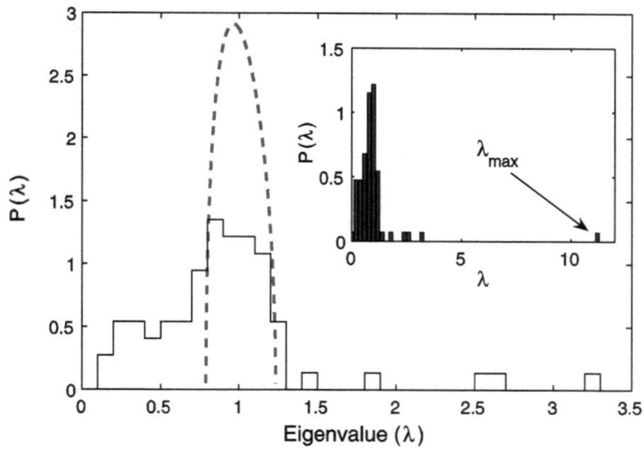

Fig. 11.1 The probability density function of the eigenvalues of the cross-correlation matrix **C** for fluctuations in the exchange rate in terms of US Dollars of 74 currencies for the period Oct 1995–April 2012. For comparison the theoretical distribution predicted by Eq. (11.2) is shown using *broken curves*. We explicitly verified that the theoretical distribution fits very well the spectral distribution of surrogate correlation matrices generated by randomly shuffling the returns for the different currencies. The *inset* shows the largest eigenvalue corresponding to the global mode of market dynamics

The random nature of the eigenvalues occurring in the bulk of the distribution is also indicated by the distribution of the corresponding eigenvector components. Note that, these components are normalized for each eigenvalue λ_j such that, $\sum_{i=1}^{N}[u_{ji}]^2 = N$, where u_{ji} is the i-th component of the jth eigenvector. For random matrices generated from uncorrelated time series, the distribution of the eigenvector components follows the Porter-Thomas distribution,

$$P(u) = \frac{1}{\sqrt{2\pi}} \exp\left[-\frac{u^2}{2}\right]. \tag{11.3}$$

We have explicitly verified this form for the corresponding distribution of the random surrogate matrices obtained by shuffling the empirical return time series so that all correlations between the different currencies are destroyed. As seen from Fig. 11.2, it also approximately fits the distributions of the eigenvector components for the eigenvalues belonging to the bulk of the empirical spectral distribution. However, the eigenvectors of the largest eigenvalues (e.g., the largest eigenvalue λ_{max}, as shown in the inset) deviate quite significantly, indicating its non-random nature.

The largest eigenvalue λ_0 for the cross-correlation matrix is about 9 times larger than the upper bound of the random spectral distribution. While this is similar to the situation for cross-correlations of stock movements in financial markets (e.g., see Refs. [5, 6]), the corresponding eigenvector does not show a relatively uniform composition unlike the case in equities markets where almost all stocks contribute to this mode with all elements having the same sign. Instead, there is large variation in the relative contributions of the different components to the largest eigenmode, with those of four currencies (VEB, PYG, NGN, BND) having a different sign than the

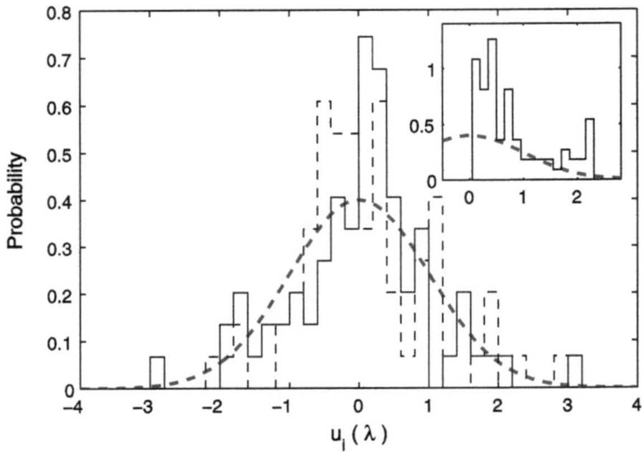

Fig. 11.2 The probability distribution of the eigenvector components corresponding to two eigenvalues belonging to the bulk of the spectral distribution predicted by random matrix theory and (*inset*) that corresponding to the largest eigenvalue. In both cases, the corresponding distribution obtained from the surrogate correlation matrices obtained by randomly shuffling the returns is shown using a broken curve for comparison

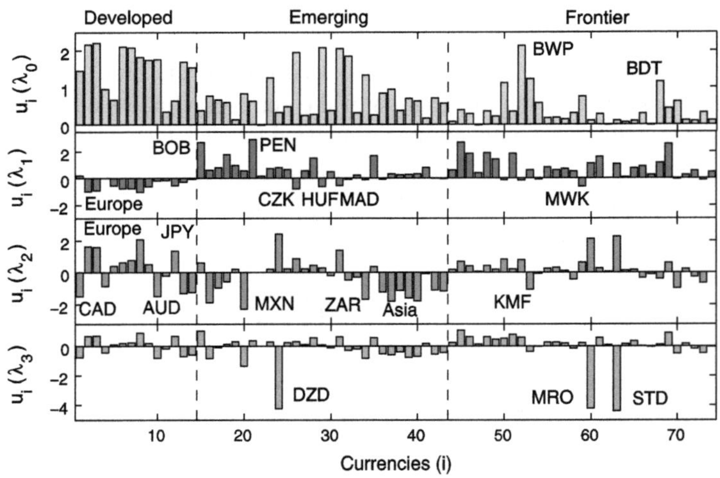

Fig. 11.3 The eigenvector components $u_i(\lambda)$ for the four largest eigenvalues of the correlation matrix **C**. The currencies are arranged according to the market classification of the corresponding country (developed, emerging or frontier) separated by *broken lines*. Some of the prominent components for each eigenvector (discussed in the text) are individually identified by the respective currency codes

rest—although with an extremely low magnitude (Fig. 11.3, top). This eigenmode represents the global component of the time-series of currency fluctuations which is common to all currencies. Thus, the strength of the relative contribution of a currency to the leading eigenvector can be construed as the extent to which the currency is in sync with the overall movement of the world currency market reflecting the collective response of the world economy to information shocks (which may include major perturbations such as the worldwide financial crisis of 2007–2008). Note that, this suggests that the relative strengths of the components in the leading eigenvector may be used as a measure of the role the corresponding currency plays in the world market (and to an extent, that the country plays in the international economy). Seen from this point of view, it is perhaps not surprising that most of the currencies belonging to countries in the developed markets category contribute significantly to this mode which reflects their dominance in the world economy. We also see that the countries in the emerging markets category can be very different from each other in terms of their role in the global mode, with components corresponding to the East European economies such as Czech Republic, Hungary and Poland having some of the largest contributions. Turning to the frontier markets category, while the contributions of most of these currencies have very low magnitude, a few countries (most notably Botswana but also Bangladesh, Kazakhstan and Comoros) stand out for the relatively high strength of the corresponding eigenvector component. The strong contribution from these countries could be either because of their impressive economic performance (e.g., Botswana has maintained one of the world's highest economic growth rates from the time of its independence in 1966 [23]) or possibly due to remittances in foreign currencies from expatriates working abroad having a large contribution to the national economy (as in the case of Bangladesh). As newly developing economies are potentially highly profitable but risky targets for foreign investment, it may be of interest to explore the possibility of using this measure to identify frontier markets having strong interaction with the world market which may make them relatively safer to invest in. On the other hand, from the point of view of portfolio diversification for reducing risk, one may use such a measure to identify economies whose fluctuations have the least in common with the global mode.

Of even more interest for understanding the topological structure of interactions in the world currency market are the intermediate eigenvalues in between the largest eigenvalue λ_0 and the bulk predicted by random matrix theory. For equities markets, it has been shown that in many cases the eigenvectors corresponding to these eigenvalues are localized, i.e., a relatively small number of stocks, usually having similar market capitalization or belonging to the same business sector, contribute significantly to these modes [6, 24, 25]. Figure 11.3 shows that the different currencies contribute to the different eigenvectors corresponding to the three largest intermediate eigenvalues very unequally. For example, from the eigenvector corresponding to λ_1, the second largest eigenvalue, we observe that many Latin American currencies such as those of Bolivia and Peru, have a dominant contribution in this mode with the contribution of European currencies (and a few non-European ones, such as those of Morocco and Malawi, whose economy is closely connected to that of Europe) being not only different but actually having the opposite sign. The third eigenvector shows that contributions from European and Japanese currencies have a different sign from that of established as well as rapidly developing economies of America, Asia-Pacific and Africa (such as Canada, Mexico, South Africa, Australia, New Zealand, Israel, Singapore and India). The fourth eigenvector has significant contributions from only three currencies, those of Algeria, Mauritania and Sao

Tome & Principe. This may reflect existing economic linkages between these countries that has resulted in such strong coupling in the movements of their currency exchange rates with respect to the US Dollar.

Despite the above insights, a direct inspection of eigenvector composition for the intermediate eigenvalues does not very often yield a straightforward interpretation of the group of currencies dominantly contributing to a particular mode. This is because apart from information about interactions between currencies, the cross-correlations are also affected strongly by the global mode corresponding to the overall market movement. In addition, there are a large number of modes belonging to the random bulk which correspond to idiosyncratic fluctuations. Both the global and random modes can mask significant intra-group correlations. Thus, in order to identify the topological structure of interactions between the currencies we need to remove the global mode corresponding to the largest eigenvalue and also filter out the effect of random noise (contributed by the eigenvalues belonging to the bulk of the spectral distribution). For this we use the filtering method proposed in Ref. [26] based on the expansion of a matrix in terms of its eigenvalues λ_i and the corresponding eigenvectors $\mathbf{u}_i$: $\mathbf{C} = \sum_i \lambda_i \mathbf{u}_i \mathbf{u}_i^T$. This allows the correlation matrix to be decomposed into three parts, corresponding to the global, group and random components:

$$\mathbf{C} = \mathbf{C}_{global} + \mathbf{C}_{group} + \mathbf{C}_{random} = \lambda_0 \mathbf{u}_0^T \mathbf{u}_0 + \sum_{i=1}^{N_g} \lambda_i \mathbf{u}_i^T \mathbf{u}_i + \sum_{i=N_g+1}^{N-1} \lambda_i \mathbf{u}_i^T \mathbf{u}_i, \quad (11.4)$$

where, the eigenvalues have been arranged in descending order (the largest labelled 0) and N_g is the number of intermediate eigenvalues. From the empirical data it may not be obvious what is the value of N_g, as the bulk may differ from the predictions of random matrix theory because of underlying structure induced correlations. For this reason, we use visual inspection to choose $N_g = 6$, and verify that small changes in this value do not alter the results. Our results are robust with respect to small variations in the estimation of N_g because the error involved is only due to the eigenvalues closest to the bulk that have the smallest contribution to $\mathbf{C}_{group}$. Figure 11.4 shows the result of the decomposition of the entire cross-correlation matrix (the distribution of whose elements is shown in the inset) into the three components. In contrast to the case of stock-stock correlations in financial markets (e.g., Ref. [6]), in the currency market the group correlation matrix elements C_{ij}^{group} show a significantly reduced tail and is completely enveloped by the distribution of the global correlation matrix elements C_{ij}^{global}. This indicates that there is a relatively small fraction of strongly interacting currencies, implying that the segregation into groups may be weak in this market.

In order to graphically present the interaction structure of the stocks using the information in the group correlation matrix $\mathbf{C}_{group}$, we first use a method suggested by Mantegna [27] to transform the correlation between currencies into distances to produce a connected network in which co-moving currencies are clustered together. The distance d_{ij} between two currencies i and j are calculated from the

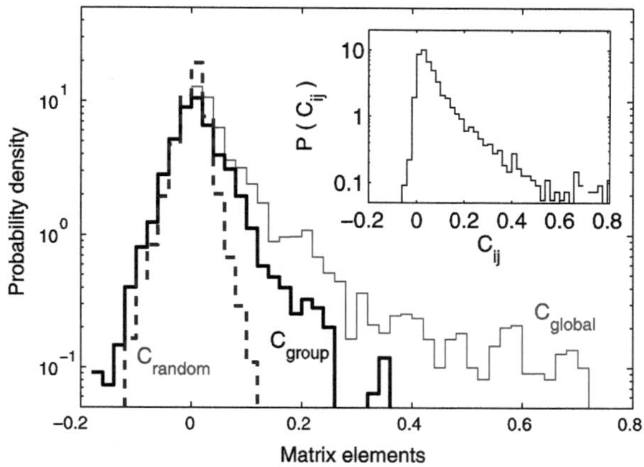

Fig. 11.4 The probability distribution of the matrix elements following decomposition of the correlation matrix **C** into global ($\mathbf{C}_{global}$), group ($\mathbf{C}_{group}$) and random effects ($\mathbf{C}_{effects}$) with $N_g = 7$. The distribution of the components C_{ij} of the original cross-correlation matrix **C** is shown in the inset for comparison

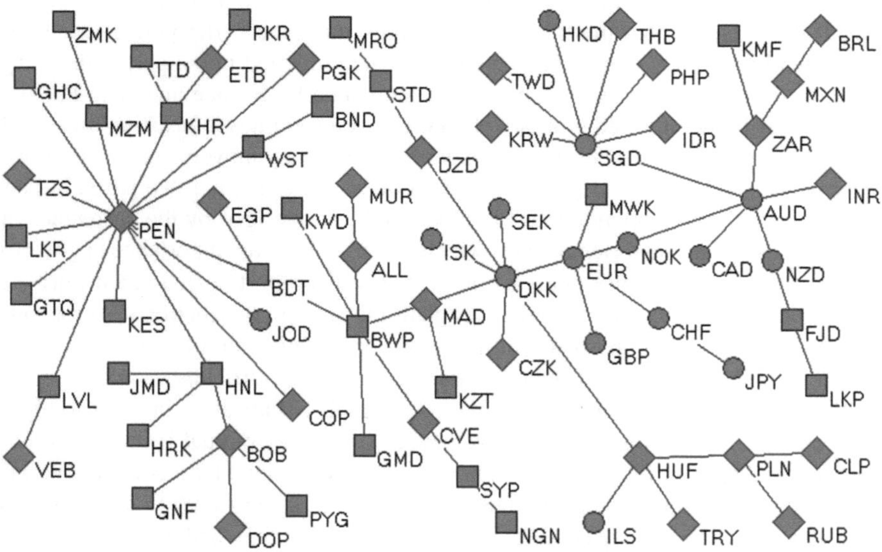

Fig. 11.5 The minimum spanning tree connecting the 74 currencies considered here. The node shapes indicate the type of the underlying economy of the country to which the currency belongs (*circles* indicate developed, *diamonds* indicates emerging and *squares* indicate frontier markets). The figure has been drawn using the Pajek software

cross-correlation matrix **C**, according to $d_{ij} = \sqrt{2(1 - C_{ij})}$. These are used to construct a minimum spanning tree, which connects all the N nodes of a network with $N - 1$ edges such that the total sum of the distance between every pair of nodes, $\sum_{i,j} d_{ij}$, is minimum. As seen in Fig. 11.5, for the currency market this method reveals clusters of currencies belonging to countries having similar economic profile and/or belonging to the same geographical region. In particular, note the cluster centered around the hub node (i.e., a node having significantly more connections than the average) corresponding to SGD which consists exclusively of currencies belonging to developed or emerging economies of the Asia-Pacific region such as those of Hong Kong, Taiwan, Thailand, Indonesia etc. On the other hand, the currencies clustered around the hub AUD are related by the geo-economic status of the corresponding countries of being major non-European players in the world economy (e.g., Canada, Mexico, Brazil, South Africa and India). It should be noted that the hubs of these two clusters (SGD and AUD) are directly linked to each other and are in turn connected to the cluster of European currencies (comprising two hubs corresponding to the Euro and the Danish currency) suggesting a close interplay in the currency movements of all the important countries driving international economic dynamics. Possibly more intriguing is the occurrence of a much bigger cluster (containing a third of all the currencies considered) arranged around the largest hub in the network which corresponds to the Peruvian currency. This cluster comprises a wide assortment of currencies belonging to countries spread geographically around the world but which share an economic resemblance in that most of them are in a relative state of underdevelopment compared to the economies considered earlier. It thus appears that the tree network representing the underlying interactions in the world currency market can be approximately divided into a part comprising developed or rapidly growing economies (dominated by Europe and Asia-Pacific) and another part composed of relatively underdeveloped ones (consisting mostly of Latin American and African countries), with the currency movements of these two groups being relatively independent of each other. Note that the two parts, in particular, the hubs corresponding to PEN and DKK, are bridged by the currencies of Morocco, Botswana and Bangladesh, which therefore have an importance in governing the collective dynamics of the world economy disproportionate to their intrinsic economic status. This can potentially explain the strong contribution of these currencies to the leading eigenvector of the cross-correlation matrix that represents the global eigenmode which has been discussed earlier in this article.

We have also used an alternative method of graph visualization in order to highlight any existing groups of currencies having significant mutual interactions. For the case of stocks in financial markets, the modules obtained by this technique often represent strongly performing business sectors in the economy [5, 6]. It is thus plausible that the currency communities identified using this method will represent important groupings driving the world economy. The binary-valued adjacency matrix $\mathbf{A}$ of the network is generated from $\mathbf{C}_{group}$ by using a threshold c_{th} such that $A_{ij} = 1$ if $C_{ij}^{sector} > c_{th}$, $A_{ij} = 0$ otherwise. An appropriate choice of the threshold makes apparent any clustering in the network that is implied by the existence of a tail in the C_{ij}^{group} distribution. Figure 11.6 shows the resultant network for the best choice of $c_{th} = c^*$ $(= 0.133)$ in terms of creating the largest clusters of interacting currencies (isolated nodes have not been shown). The five clusters differ considerably in size, with two of them corresponding to strongly interacting currency triads (with the DZD-MRO-STD triad being the currencies having the dominant contribution to the fourth largest eigenmode identified earlier in Fig. 11.3). The next largest cluster, having nine currencies, consists of rapidly emerging economies outside Europe—including Brazil, India and South Africa of the BRICS group as well

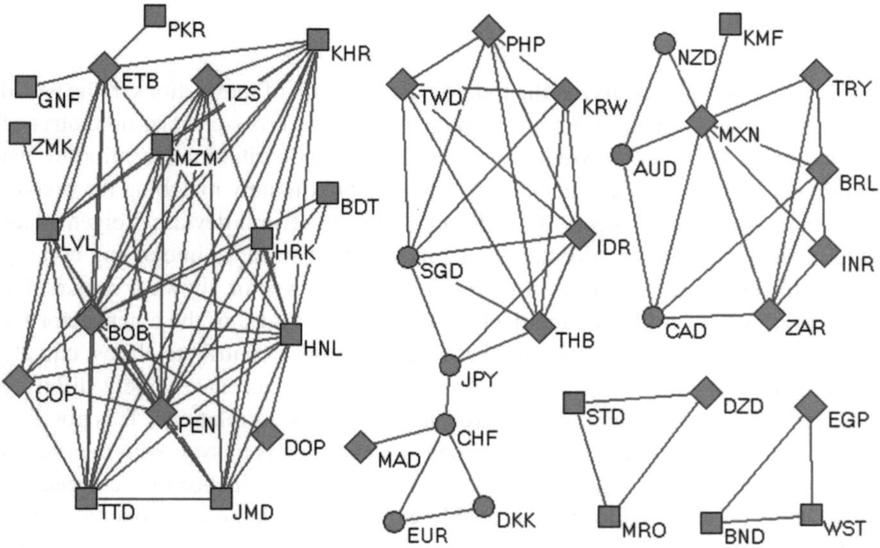

Fig. 11.6 The network of interactions among currencies generated from the group correlation matrix $\mathbf{C}_{group}$ with threshold $c^* = 0.133$. The node shapes indicate the type of the underlying economy of the country to which the currency belongs (*circles* indicate developed, *diamonds* indicates emerging and *squares* indicate frontier markets). The cluster at the center consists mostly of countries belonging to the Asia-Pacific region including several members of the ASEAN group, although it is also connected via the Japanese Yen to a smaller sub-group of European currencies. The cluster at top right consists of three of the BRICS countries as well as several economies outside Europe which are important in the global economy (such as Australia, Canada, Mexico and Turkey). The cluster at the left comprises mostly Latin American and African currencies—although note the presence of Bangladesh and Brunei. The two small clusters at the bottom connect triads of currencies. The figure has been drawn using the Pajek software

as Turkey and Mexico from the "Next Eleven" (N-11) group identified in Ref. [28] as countries having the potential of becoming some of the largest economies in the world in the coming years—and a few non-European developed economies such as Australia and Canada. The even larger cluster comprising eleven currencies is dominated by the countries of Asia-Pacific such as Taiwan and Singapore as well as the N-11 countries Indonesia, Korea and Philippines, which have either developed or fast growing economies; however, through the Japanese Yen, these currencies are also connected to a smaller sub-cluster of European currencies which contains the Euro apart from the Swiss and Danish currencies (note also the presence of the currency of Morocco, a north African country but one that has strong economic ties with Europe). The largest cluster has seventeen densely inter-connected currencies which are geographically spread around the world, although half of them are from Latin America or the Caribbean. Possibly this cluster reflects a new wave of fast growing economies (e.g., it includes two N-11 countries, Bangladesh and Pakistan) whose development trajectory may affect the global economy in the future as profoundly as the rise of India and China has affected it in the past decades.

11.4 Conclusions

In this article we have analyzed the topological structure of interactions in the world currency market by using the spectral properties of the cross-correlation matrix of exchange rate fluctuations. We see that the eigenvalue distribution is similar to that seen in equities markets and consists of a bulk approximately matching the predictions of random matrix theory. In addition, there are several deviating eigenvalues which contain important information about groups of strongly interacting components. However, the composition of the leading eigenvector shows a remarkable distinction in that, unlike the relatively homogeneous nature of the eigenvector for cross-correlations in the equities market where all stocks contribute almost equally to the market or global mode, the different currencies can have widely differing contributions to the global mode for exchange rate cross-correlations. This possibly reflects the extent to which the fluctuations of a currency is in sync with the overall market movement and may also be used to measure the influence of a currency in the world economy. While, as is probably expected, the large components of this mode mostly belong to currencies of the developed economies of western Europe as well as the rapidly growing economies of the Asia-Pacific region, there are unexpectedly strong contributions from currencies outside this group—such as those of Botswana, Bangladesh and Kazakhstan. This indicates that these economies may be playing an important role in directing the collective dynamics of the international currency market that is not exclusively dependent on their intrinsic economic strength, but rather the position they occupy in the network of interactions among the currencies. This is confirmed by the reconstructed network of interactions among the currencies as a minimum spanning tree. This network shows a segregation between clusters dominated by developed or rapidly growing economies on the one hand, and relatively underdeveloped economies on the other. While these two parts can show dynamics relatively independent of each other, a few currencies—those of Morocco, Botswana and Bangladesh—act as a bridge between them. Thus the role of these currencies as vital connecting nodes of the world currency market possibly give them a much more important position than would be expected otherwise. We have also used an alternative graph representation technique to identify several groups of strongly interacting currencies. Some of the smaller clusters may be reflecting possible economic or other relations between the corresponding countries. However, the largest cluster comprises a densely interconnected set of currencies belonging to countries that are geographically spread apart. We speculate that these could well belong to the next wave of fast emerging economies that will drive the economic growth of the world in the future. This is significant from the point of view of applications, as such economies are potentially lucrative targets for foreign investment and are eagerly sought after by portfolio fund managers. Methods of identifying early the next fast growth economies assume critical importance in such a situation. Our analysis of cross-correlations of exchange rate fluctuations suggests that prominent clusters in the reconstructed networks of interactions in the world currency market may potentially provide us with such methods.

Acknowledgements We would like to thank R K Pan who helped in developing the software used for the analysis reported here and S Sridhar for stimulating discussions. Part of the work was supported by the Department of Atomic Energy, Government of India through the IMSc Complex Systems Project (XII Plan).

References

1. Sinha S, Chatterjee A, Chakraborti A, Chakrabarti BK (2011) Econophysics: an introduction. Wiley-VCH, Weinheim
2. Gabaix X, Gopikrishnan P, Plerou V, Stanley HE (2003) A theory of power-law distributions in financial market fluctuations. Nature (London) 423:267–270
3. Farmer JD, Smith DE, Shubik M (2005) Is economics the next physical science? Phys Today 58(9):37–42
4. Vijayaraghavan VS, Sinha S (2011) Are the trading volume and the number of trades distributions universal? In: Abergel F, Chakrabarti BK, Chakraborti A, Mitra M (eds) Econophysics of order-driven markets. Springer, Milan, pp 17–30
5. Sinha S, Pan RK (2007) Uncovering the internal structure of the Indian financial market: large cross-correlation behavior in the NSE. In: Chatterjee A, Chakrabarti BK (eds) Econophysics of markets and business networks. Springer, Milan, pp 3–19
6. Pan RK, Sinha S (2007) Collective behavior of stock price movements in an emerging market. Phys Rev E 76:046116
7. Ausloos M, Ivanova K (2001) Correlations between reconstructed EUR exchange rates versus CHF, DKK, GBP, JPY and USD. Int J Mod Phys C 12:169–195
8. Mizuno T, Takayasu H, Takayasu M (2006) Correlation network among currencies. Physica A 364:336–342
9. Drozdz S, Gorski AZ, Kwapien J (2007) World currency exchange rate cross-correlations. Eur Phys J B 58:499–502
10. Jennings J (2011) Globalizations and the ancient world. Cambridge University Press, Cambridge
11. Bank for International Settlements (2010) Foreign exchange and derivatives market activity in April 2010. Triennial Central Bank Survey
12. Sarkar A, Barat P (2006) Fluctuation dynamics of exchange rates on Indian financial market. In: Chatterjee A, Chakrabarti BK (eds) Econophysics of stock and other markets. Springer, Milan, pp 67–76
13. http://www.oanda.com/currency/historical-rates/
14. http://www.msci.com/products/indices/market_classification.html
15. Lux T (1996) The stable Paretian hypothesis and the frequency of large returns: an examination of major German stocks. Appl Financ Econ 6:463–475
16. Gopikrishnan P, Meyer M, Amaral LAN, Stanley HE (1998) Inverse cubic law for the distribution of stock price variations. Eur Phys J B 3:139–140
17. Pan RK, Sinha S (2007) Self-organization of price fluctuation distribution in evolving markets. Europhys Lett 77:58004
18. Pan RK, Sinha S (2008) Inverse-cubic law of index fluctuation distribution in Indian markets. Physica A 387:2055–2065
19. Koedijk KG, Schafgans MMA, de Vries CG (1990) The tail index of exchange rate returns. J Int Econ 29:93–108
20. Muller UA, Dacorogna MM, Olsen RB, Pictet OV, Schwarz M, Morgenegg C (1990) Statistical study of foreign exchange rates, empirical evidence of a price change scaling law, and intraday analysis. J Bank Finance 14:1189–1208
21. Vikram SV, Sinha S (2011) Emergence of universal scaling in financial markets from mean-field dynamics. Phys Rev E 83:016101

22. Sengupta AM, Mitra PP (1999) Distribution of singular values for some random matrices. Phys Rev E 60:3389–3392
23. http://www.cia.gov/library/publications/the-world-factbook/geos/bc.html
24. Gopikrishnan P, Rosenow B, Plerou V, Stanley HE (2001) Quantifying and interpreting collective behavior in financial markets. Phys Rev E 64:035106
25. Plerou V, Gopikrishnan P, Rosenow B, Amaral LAN, Guhr T, Stanley HE (2002) Random matrix approach to cross correlations in financial data. Phys Rev E 65:066126
26. Kim D-H, Jeong H (2005) Systematic analysis of group identification in stock markets. Phys Rev E 72:046133
27. Mantegna RN (1999) Hierarchical structure in financial markets. Eur Phys J B 11:193–197
28. O'Neill J, Wilson D, Purushothaman R, Stupnytska (2005) How solid are the BRICs. GS global economics paper 134

Chapter 12
Systemic Risk in Japanese Credit Network

Hideaki Aoyama

Abstract In this work, we study a comprehensive Japanese credit network of banks and firms with links representing the lending/borrowing relationships between the banks and the firms. We examine these relationships in order to identify key nodes in regard to the risk levels that they impose on the bank-firm system. By assigning some level of distress to a bank and letting the distress propagate to the firms and banks according to relative node exposures to the distressed node, we find final states of the distress distribution. We then define DebtRank as the asset-weighted average of distress distribution and identify the level of threat that the bank poses to the entire system.

12.1 Introduction

In any economic system, systemic risk, stability of the system as a whole, as well as evaluation of key nodes, is one of the topmost important issue, especially in view of the economic crisis, from recent Lehman crisis to other crises that were repeated over and over and affected huge number of people worldwide.

In approaching issue, it is important to note that economy as a whole is made of vast number of economic agents who interact in a (large) number of ways. As ways they choose partners with whom they interact are affected by a large number of factors, including each agent's individual situations, even outside the economy, the network is far from regular. In other words, we are faced with complex network of heterogeneous interacting economics agents, necessitating simulation approach with suitable measures for risk.

There are basically two ways to approach to this problem, which are fundamentally different from each other:

- Realistic, but complex approach: One may set up some numbers of equations for dynamic interactions between agents, whose variables are entries in financial statements, like saving, borrowings, sales, profit and what not.

H. Aoyama (✉)
Kyoto University, Kyoto 606-8502, Japan
e-mail: hideaki.aoyama@scphys.kyoto-u.ac.jp

F. Abergel et al. (eds.), *Econophysics of Agent-Based models*,
New Economic Windows, DOI 10.1007/978-3-319-00023-7_12,

- Abstract, but simple approach: One may define some abstract quantities that reflects the most important characterization of each agent and define interactions among them.

Both approaches have advantages and short comings: The former may be close to the reality and one can may look into various aspects of the system, like the tax system, on how they affect the system and what not, while as one tries to construct more realistic models the numbers of the variables and the numbers of the equations increase. The parameters in the equations may rise accordingly and one may have hard time in determining their values and identifying which parameter values are significant cause of any particular economic phenomena. On the other hand, in the latter approach, the simple fact that it is abstract may raise discussion on how it can be important and what is really analyzed in comparison with reality. The merit of the this approach, however, is that its structure can be made quite simple, being tailored to the particular phenomena (the type of the systemic risk in that particular economic network).

In this paper, we take the latter approach, in particular, DebtRank approach, which was originally proposed in [1] in the banking system in U.S. We first extend this approach, using the framework we proposed in [2] so that it applies to Japanese bipartite network of banks and firms linked by lending/borrowing relationship, and then carry out the calculation of the DebtRank using annual data from 1980 to 2010.

12.2 The Network and Distress Propagation

The network linkage data we use is provided by Nikkei Inc. and contains amount of the long-term and short-term yearly lending by banks to the firms from 1980 to 2011. The banks and firms are identified by respective Nikkei codes and the names. All the yearly BS, PL and CS of both banks and firms are given. The number of firms and banks are plotted in Figs. 12.1 and 12.2.

Some network properties, including Minimum Spanning Tree (MST) of the banks, were studied in [3], whose main conclusion was that there are two kinds of major branches, (1) branch made of city (mega) banks, and (2) branches that are made of banks in the same region, *i.e.*, a branch made of banks in Kansai region, a branch for banks in Tohoku region, etc. This would have indication as to the spread of distress in this bi-partite network.

In order to evaluate key nodes of this network, we first assign a maximum distress to a bank and study the spread of the distress to the whole network. We then quantify the distress the initial bank causes to the whole network, which is called DebtRank. Let us explain the actual calculation procedure.

The degree of distress is quantified (in an abstract manner) as variables h_β assigned to the bank β and h_f assigned to the bank f, each ranging from 0 to 1, 1 being the maximum distress (bankruptcy) and 0 corresponding to a 'healthy' state. Thus, in order to study the importance of a bank β_0, we initially assign $h_{\beta_0} = 1$ and

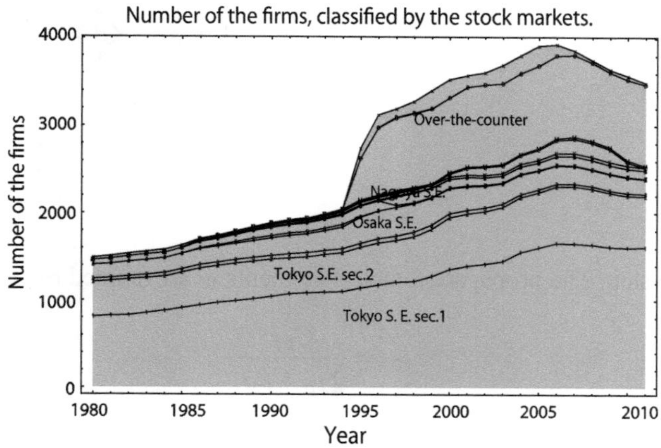

Fig. 12.1 Number of firms

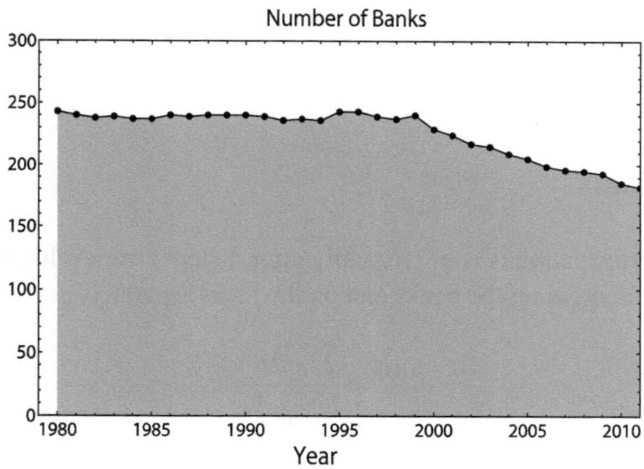

Fig. 12.2 Number of banks

all the rest of h_β and h_f to zero. We then let propagate h from the banks (β) to the firm f as,

$$h_f \to h_f + \sum_\beta w_{f\beta} h_\beta, \tag{12.1}$$

and from the firms f to the bank β as,

$$h_\beta \to h_\beta + \sum_f w_{\beta f} h_f \tag{12.2}$$

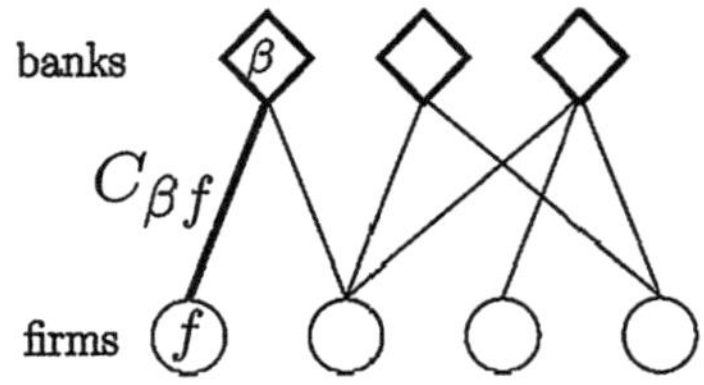

Fig. 12.3 Distress propagation from the bank β to the firm f

at each time step. The propagation matrix elements w are defined by the following,

$$w_{f\beta} := \frac{C_{\beta f}}{\sum_{\beta'} C_{\beta' f}}, \tag{12.3}$$

$$w_{\beta f} := \frac{C_{\beta f}}{\sum_{f'} C_{\beta f'}}, \tag{12.4}$$

where we denote the lending from the bank β to the firm f by $C_{\beta f}$, as in Fig. 12.3. We also note that we do not allow multiple visit around loops: once distress propagates from the bank β to the firm f and then back to the bank β, it no longer travels back the firm f, and similarly from the firm to the bank and back to the firm.

12.3 DebtRank

After all the propagation is over (typically after 4 steps), we evaluate the weighted average of the distress on the banks and on the firms separately:

$$d_b^{(\beta_0)} = \frac{\sum_\beta A_\beta h_\beta}{\sum_\beta A_\beta}, \tag{12.5}$$

$$d_f^{(\beta_0)} = \frac{\sum_f A_f h_f}{\sum_f A_f}, \tag{12.6}$$

where A_β is the total asset of the bank β and A_f that of the firm f. We remind the readers that β_0 is the bank that was assigned $h = 1$ initially.

Since the bank β_0 has $h = 1$ and the propagation matrix elements w are in general small compared to 1 (resulting from the fact that most banks lend to many firms and most firms borrow from several banks), the resulting weighted average $d_b^{(\beta_0)}$ is large compared to $d_f^{(\beta_0)}$. Therefore if we directly sum-up those two, the distress on firms are much less counted in compared to the distress on banks. Therefore here we choose to define 'normalized' sum

$$d(\beta_0) = \frac{d_b^{(\beta_0)}}{\mathrm{E}[d_b^{(\beta_0)}]} + \frac{d_f^{(\beta_0)}}{\mathrm{E}[d_f^{(\beta_0)}]}, \tag{12.7}$$

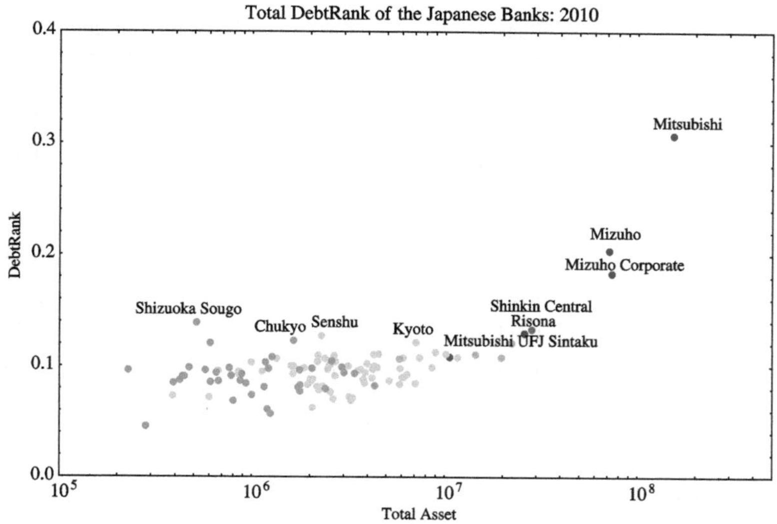

Fig. 12.4 The total asset and the total DebtRank of the banks

where E[·] is the average of the respective quantity. We call this **DebtRank** of the bank β_0.

Figure 12.4 gives the scatter plot of the bank's asset A_β and the total DebtRank $d(\beta_0)$ and Table 12.1 give the listing of 10 banks with high DebtRanks. The mega banks, such as Mitsubishi, have large assets and as a result of the weighted average (12.5) they achieve high total DebtRank, which is natural in the sense that they play major role in Japanese economy.

On the other hand, it is surprising that below the level of the those mega banks, the scatter plot flattens, with small regional banks such as Shizuoka achieving high DebtRank. This demonstrates clearly their importance in the local economy. One may speculate that they have strong ties with the local firms, which results in large propagation matrix and thus large DebtRank. One may claim that in spite of large DebtRank, these regional banks are not that important. But it is not true: Since we take weighted average in Eqs. (12.5) and (12.6), if those local firms with strong ties to those regional banks are small, they do not contribute much to the DebtRank. The fact that the DebtRanks of those regional banks are large means they have strong ties to the large and important firms in the same region.

This we think is a strong finding: DebtRank provides us with a tool, a measure, with which we can identify key notes that are sometimes hidden from surface. With DebtRank we can uncover the important banks other than (trivially important) mega banks.

Table 12.1 Top 10 high-DebtRank banks in 2010. Bank category is 1: City Bank, 2: Regional Bank, 3: 2nd Regional Banks (which are somewhat close to Saving and Loan in U.S.), and 7: Credit Association

Bank	Total Asset (10^9 yen)	Category	DebtRank
Mitsubishi	153,924	1	4.70
Mizuho	71,537	1	3.38
Mizuho Corporate	73,599	1	3.09
Shizuoka Sougo	515	3	2.62
Senshu	2,292	2	2.53
Shinkin Chuou	28,400	7	2.52
Risona	26,050	1	2.47
Chukyo	1,646	3	2.45
Kyoto	7,104	2	2.38
Matsue Sougo	331	3	2.34

12.4 Summary

In this talk we have presented new extension of the DebtRank concept to the bipartite network made of banks and firms. We defined the DebtRank as the normalized sum of the distress distribution on the two layers of the network. From this we clearly see not only that mega-banks, having wide influence over all the Japanese economy, has large influence on economy, but also that there are several regional banks with large influence, possibly through strong financial relationship with local important firms. This conclusion implies the need for appropriate governmental support on not only the mega banks, but also influential regional banks who are identified by the high DebtRank.

Acknowledgements Part of this work is the result of collaboration with Yoshi Fujiwara and Stefano Battison, whom the author would like to thank.

The author would like to thank Y. Ikeda, H. Iyetomi, W. Souma, I. Vodenska, and H. Yoshikawa for discussions at various stages of this work.

Part of this work is supported in part by the Program for Promoting Methodological Innovation in Humanities and Social Sciences by Cross-Disciplinary Fusing of the Japan Society for the Promotion of Science, and the European Community Seventh Framework Programme (FP7/2007-2013) under Socio-economic Sciences and Humanities, grant agreement no. 255987 (FOC-II).

Appendix

Here we show the plot of the DebtRank vs. Total Asset in Fig. 12.5 (as in Fig. 12.4) for the year 1980–2010. Going through these figures, we find that

1. Basic structures, big city banks dominating the high DebtRank region (toward the upper left) and smaller (mostly regional) banks in the left-bottom region.
2. Existence of regional banks with high DebtRanks almost all the years. Most notables are 2002 where Matsue Sougo bank achieves DebtRank equal to 3.10, followed by other years after 2000.

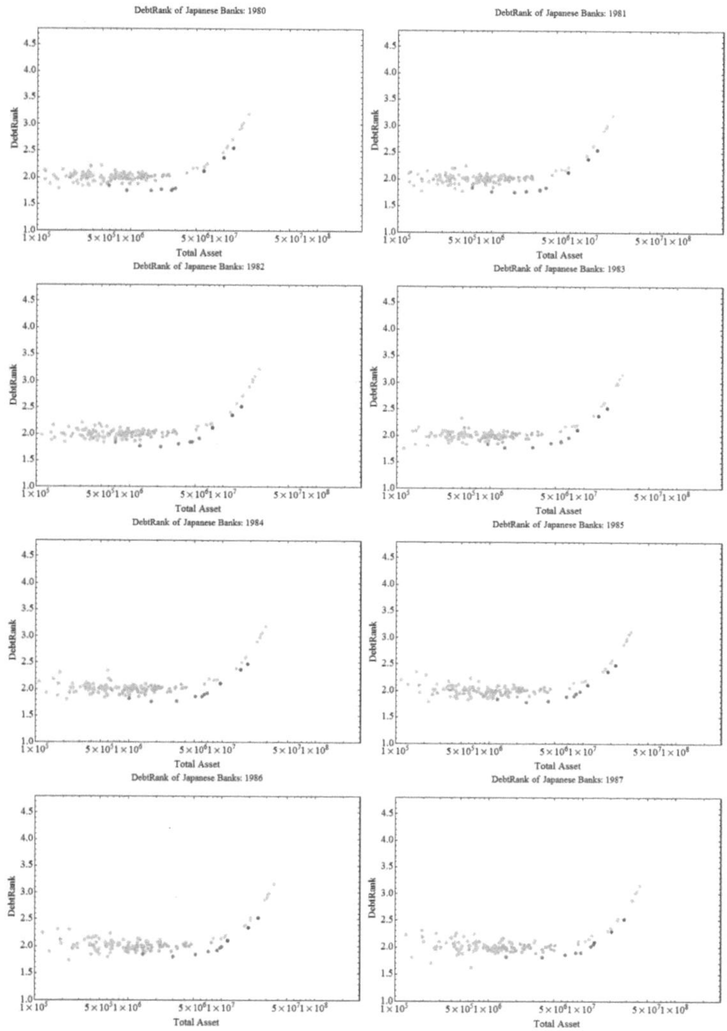

Fig. 12.5 DebtRank vs. Total Asset from 1980–2010

3. Spread of the left-bottom flat region: this tail-part fattens notably from 2000 and on, which coincides with period right after financial crisis in Japan. The real cause, whether it is due to shrinkage of credits and some number of mergers of banks, is not clear at this moment, but is worth looked into in-depth.

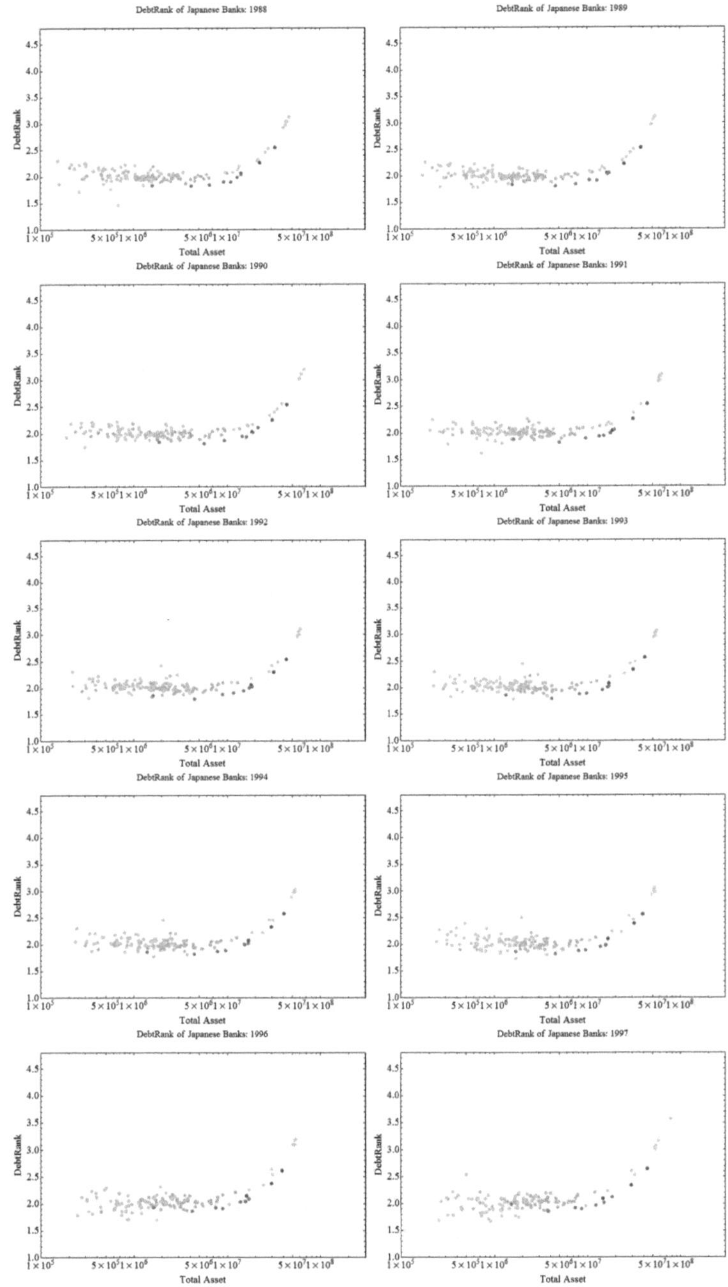

Fig. 12.5 (Continued)

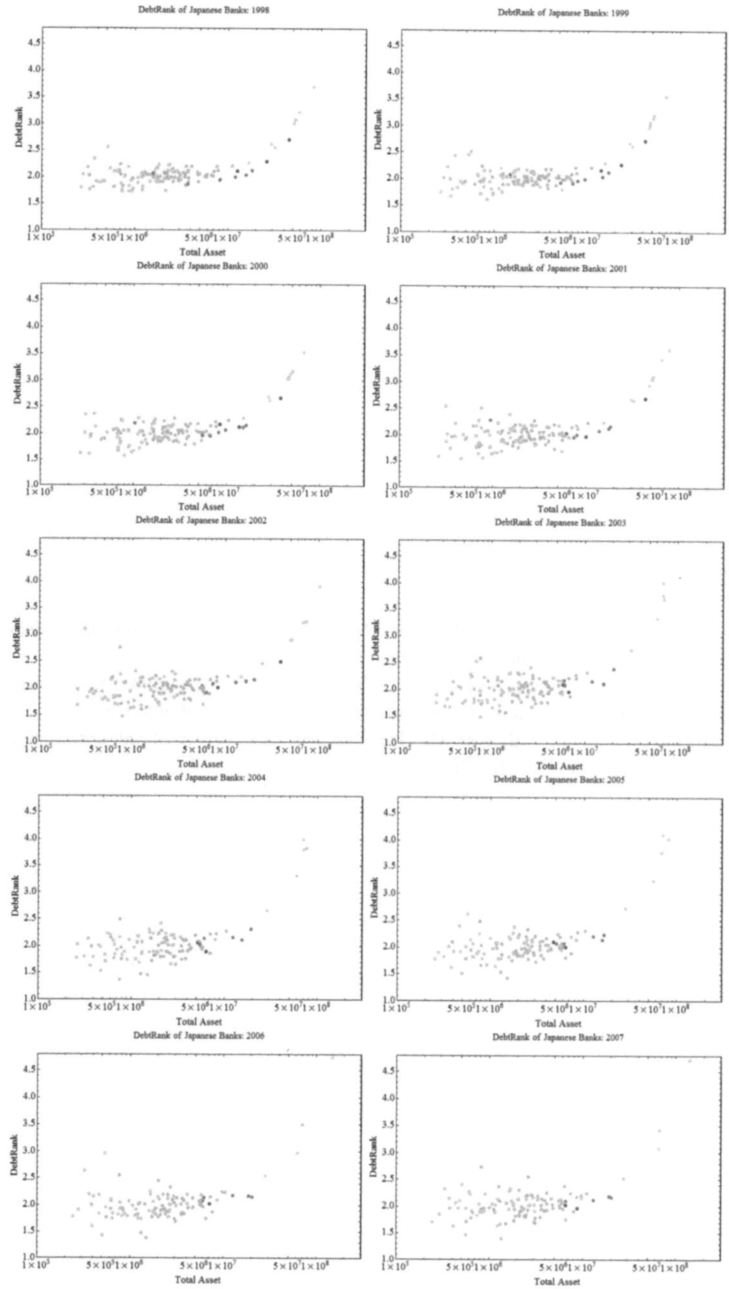

Fig. 12.5 (Continued)

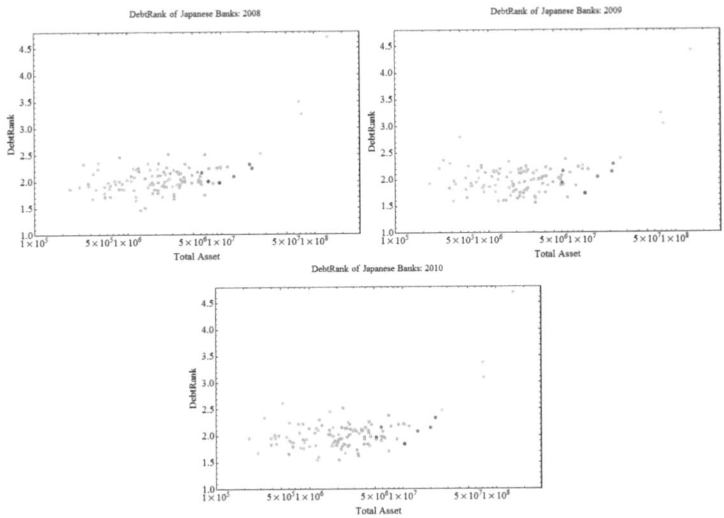

Fig. 12.5 (Continued)

References

1. Battiston S, Puliga M, Kaushik R, Tasca P, Caldarelli G (2012) Debtrank: too central to fail? Financial networks, the fed and systemic risk. Sci Rep 2:541
2. Fujiwara Y, Aoyama H, Ikeda Y, Iyetomi H, Souma W (2009) Structure and temporal change of the credit network between banks and large firms in Japan. Economics 3:7
3. De Masi G, Fujiwara Y, Gallegati M, Greenwald B, Stiglitz JE (2009) An analysis of the Japanese credit network. Evol Inst Econ Rev 7:209

Chapter 13
Pricing of Goods with Bandwagon Properties: The Curse of Coordination

Mirta B. Gordon, Jean-Pierre Nadal, Denis Phan, and Viktoriya Semeshenko

Abstract In this article, we briefly review the models of social interactions concerning the pricing of goods with Bandwagon properties.

13.1 Introduction

Social interactions play an important role on the collective outcomes. The decision of leaving a neighborhood, to attend a seminar or a crowded bar, to participate to collective actions such as strikes and riots, are particular examples. In market situations like the subscription to a telephone network or the choice of a computer operating system, the willingness to pay generally depends not only on the individual preferences but also on the choice made by others. Long after the pioneering works of T.C. Schelling [6] and of M. Granovetter [5], there has been a growing economic literature that recognizes the influence of social interactions on consumers. When the utility of a good increases with the number of buyers, there generically exist multiple equilibria for some range of prices. The Pareto-optimal equilibrium

M.B. Gordon · V. Semeshenko
Laboratoire d'Informatique de Grenoble (LIG), CNRS, Grenoble, France

J.-P. Nadal (✉)
Centre d'Analyse et de Mathématique Sociales (CAMS, UMR 8557 CNRS–EHESS), Ecole des Hautes Etudes en Sciences Sociales, Paris, France
e-mail: nadal@lps.ens.fr

J.-P. Nadal
Laboratoire de Physique Statistique (LPS, UMR 8550 CNRS–ENS–UPMC–Université Paris Diderot), Ecole Normale Supérieure, Paris, France

D. Phan
Groupe d'Etude des Méthodes de l'Analyse Sociologique de la Sorbonne (GEMASS, UMR 8598 CNRS), Université Paris Sorbonne-Paris IV, Paris, France

V. Semeshenko
Departamento de Economía, Facultad de Ciencias Económicas, Universidad de Buenos Aires, Buenos Aires, Argentina

F. Abergel et al. (eds.), *Econophysics of Agent-Based models*, New Economic Windows, DOI 10.1007/978-3-319-00023-7_13,

corresponds to the high demand solution, but this equilibrium may be not achieved due to a lack of coordination.

In contrast, the analysis of the supply has deserved much less attention. A particular insightful paper is Becker's note [1] attributing to social interactions the fact that popular restaurants do not increase their prices despite a persistent excess demand. In this paper we present a recent work where we go one step further, addressing the pricing issue in a monopoly market for goods with bandwagon effects in its generality.

13.2 Demand with Multiple Equilibria

We consider a family of models of interacting heterogeneous agents, which are variants of the "Dying seminar" model of Schelling [6], and can also be seen as directly related to statistical physics models, the so called Random Field Ising Models [7]. Different versions have been studied in social and economics contexts (see e.g. [2] for a review).

For the models we consider, a full analytical study can be made. Let us first consider the Demand side. We consider an homogeneous not-divisible good proposed at a posted price p, and a large population of customers heterogeneous in their idiosyncratic willingness to pay (IWP). In the absence of social influence, each customer buys a unit of the good if the price is below its IWP. In the presence of social influence ('bandwagon effect'), the reservation price of the agent (the maximum price at which the agent buys) becomes its IWP increased by a quantity proportional to the fraction of buyers in the population. The proportionality coefficient j measures the strength of the social influence. The model can be studied for an arbitrary distribution of the IWPs in the population, in term of the control parameters which are the mean value h of the IWP distribution, the posted price p and the social influence strength j (after an appropriate rescaling so that the variance of the IWP distribution is normalized to unity), and the (qualitative) results are shown to only depend on the number of maxima of the IWP distribution (see [4] for details). Figure 13.1 illustrates the main collective outcome for a monomodal IWP distribution. It shows the fraction of buyers η as function of the posted price p. For a social strength below some threshold value j_B, or large enough mean willingness to pay h, the demand has a standard behaviour, that is the fraction of buyers is a continuously decreasing function for the price. Above the critical value j_B, there exists a large range of p and $h < 0$ values where the demand is multi-valued. There is a low η-branch, and a large η-branch, with a range of price values $[p_L, p_U]$ where the two solutions coexist. If the customers find themselves on the large η branch, one says that the customers 'coordinate', whereas in the other case it is said that there is failure of coordination.

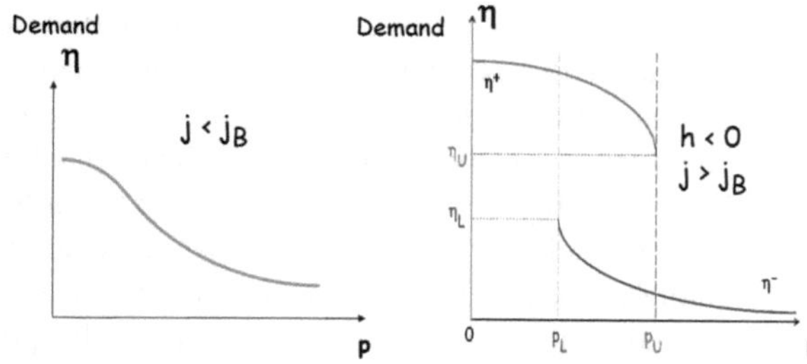

Fig. 13.1 Schematic representation of the demand curve (fraction of buyers vs. price) for different values of the (normalized) social strength j: (**a**) classical behaviour at low social influence or high mean willingness to pay; (**b**) multiply valued function at large social influence $j > j_B$ and low enough mean willingness to pay

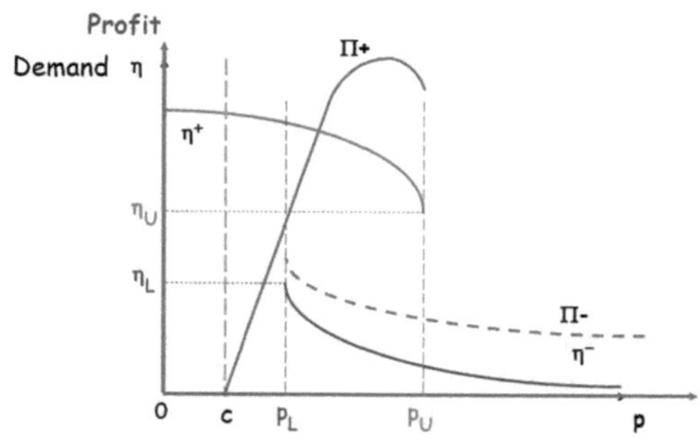

Fig. 13.2 Schematic representation of the interplay between supply and demand. As a function of price, the figure shows the demand together with the associated profit for the seller

13.3 The Curse of Coordination

Consider now the supply side, with a single seller, a monopolist. The seller wants to post a price which will maximize its profit, $\Pi = N\eta(p - c)$, where N is the number of customers and c is the cost for producing one unit of good. Again, one can study the generic properties of the seller's optimization problem, for an arbitrary smooth distribution of the IWPs assuming here that this distribution has a single maximum) [3].

As expected, when the demand is multivalued, there exist two possible prices (an optimal and a suboptimal one) as pointed out by Becker. However, surprisingly, we also find (1) that there is a range of parameters where the profit presents two relative maxima (corresponding to different prices) in a region where the demand is a standard monotonic function of the price, and (2) that there is a very large range of parameters where the seller is facing systemic risk when increasing the supply to meet the large (optimal) demand equilibrium: this strategy may not give the expected payoff unless the customers coordinate themselves. The later case is illustrated on Figure 13.2. In addition, as can be seen on the figure, the optimal equilibrium is found to correspond to a price very close to the critical value at which the large demand disappears: a small decrease of this critical value—which can result from a small change in the distribution of consumers preferences—, will thus induce a sudden large drop in consumer demand. The model suggests possible strategies that might be worth to implement in those cases. If the demand is found to be small, that is on the low η branch, under the assumption that a smooth change of parameter provoques a smooth change of demand whenever possible, a possible strategy for the seller is to lower the price below the value p_L; then the demand jumps onto the high-η branch, and the seller can smoothly increase its price towards larger profits. Alternatively, the risk adverse seller might just want to avoid the domain of multiple demand solutions, and thus post a price just below the value p_L, so that a reasonably large demand is maintained as being the unique equilibrium at this posted price.

Acknowledgements We are grateful to David Martimort for helpful suggestions and Annick Vignes for useful comments. This work was part of the project ELICCIR supported by the joint programme "Complex Systems in Human and Social Sciences" of the French Ministry of Research and of the CNRS, and of the project DyXi supported by the Programme SYSCOMM of the French National Research Agency, the ANR (grant ANR-08-SYSC-008). M.B.G., J.-P.N. and D.P. are CNRS members.

References

1. Becker GS (1991) A note on restaurant pricing and other examples of social influences on price. J Polit Econ 99:1109–1116
2. Bouchaud J-P (2013) Crises and collective socio-economic phenomena: cartoon models and challenges. J Stat Phys 151(3):567–606
3. Gordon MB, Nadal J-P, Phan D, Semeshenko V (2013) Entanglement between demand and supply in markets with bandwagon goods. J Stat Phys 151(3):494–522
4. Gordon MB, Nadal J-P, Phan D, Semeshenko V (2009) Discrete choices under social influence: generic properties. Math Models Methods Appl Sci 19(1):1441–1481
5. Granovetter M (1978) Threshold models of collective behavior. Am J Sociol 83(6):1360–1380
6. Schelling TS (1978) Micromotives and macrobehavior. Norton, New York
7. Sethna JP (2006) Statistical mechanics: entropy, order parameters and complexity. Oxford University Press, Oxford

Part III
Discussions and Commentary on Econophysics

Chapter 14
Evolution of Econophysics

Kishore C. Dash

Abstract Econophysics is a transdisciplinary research field, in which laws, theories, methods of physics are applied to economics. The term "econophysics" was coined by H. Eugene Stanley in 1995 in Kolkata. Before the term 'Econophysics' was coined many people from different branches of science had worked and applied their knowledge in the field of economics leading to evolution of econophysics. We can divide the evolution of econophysics into three different parts, such as (a) Pre-Classical era—the period before a systematic Physics started with Newton and much before the development of social science, (b) Classical era—from Newton dealing with the jump and bridging the gap(?) between the science and social science and (c) Modern era—'Institutionalised econophysics' after the name 'Econophysics' was coined. I have attempted to bring out the work of philosophers and scientists from different fields and from different ages, which has lead to the present day 'Econophysics'. Besides, contribution of different institutions, role of journals and books for development of 'Econophysics' has also been discussed. Opinion of people interested in the subject has also been expressed in the form of there replies to some specific questions.

14.1 Introduction

Formal education started in different parts of the world from ancient times. The writing systems developed around 3500 BCE The earliest Sumerian versions of the epic date from as early as the Third Dynasty of Ur (2150–2000 BC) (Dalley 1989:41–42) [1]. Some of the earliest written records show that formal education, in which basic communication skills, language, trading customs, agricultural and religious practices were taught, began in Egypt some times between 3000 and 500 BCE. In ancient India, during the Vedic period from about 1500 BC to 600 BC, most education was based on the Veda (hymns, formulas, and incantations, recited or chanted by priests of a pre-Hindu tradition) and later Hindu texts and scriptures.

K.C. Dash (✉)
Neelashaila Mahavidyalaya, Rourkela, India
e-mail: kesidash@gmail.com

F. Abergel et al. (eds.), *Econophysics of Agent-Based models*,
New Economic Windows, DOI 10.1007/978-3-319-00023-7_14,

Vedic education included: proper pronunciation and recitation of the Veda, the rules of sacrifice, grammar and derivation, composition, versification and meter, understanding of secrets of nature, reasoning including logic, the sciences, and the skills necessary for an occupation [2]. Some medical knowledge existed and was taught. In China, during the Zhou Dynasty (1045 BC to 256 BC), there were five national schools in the capital city, Pi Yong (an imperial school, located in a central location) and four other schools for the aristocrats and nobility, including Shang Xiang. The schools mainly taught the Six Arts: rites, music, archery, charioteering, calligraphy, and mathematics. Hippocrates (c. 460–370 BCE), Socrates (c. 470–399 BCE), and Aristotle (c. 384–322 BCE) all speculated about what drives human will, motivation, and learning. According to modern educational theorist Howard Gardner, "Greek philosophers may have been the first to raise questions about the nature of matter, living entities, knowledge, will, truth, beauty, and goodness. In recent centuries, however, philosophy has steadily been yielding ground, enthusiastically or reluctantly, to empirical science" (Gardner, 2000, p. 1).

India was the first country, where formal education started, before it had started at any other place in the world. Sanskrit is probably the first language from which many other languages are generated. According to a quote by American Historian Will Durant (1885–1981)

> "India was the motherland of our race and Sanskrit the mother of Europe's languages. India was the mother of our philosophy, of much of our mathematics, of ideals embodied in Christianity... of self-government and democracy. In many ways, Mother India is the mother of us all." [3–5]

According to the Forbe (July 1987), Sanskrit was considered the most suitable language for speech recognition in computer. During the early years of software revolution when western scientists were figuring out a way of constructing protocol for writing code, they realized that the principles had already been laid down in the Sanskrit language by Panini some 2500 years ago.

14.1.1 Ancient Universities

The world's first University was established in Takshasila (1000 BC to 500 AD). It became well known in 700 BC [6]. The University was one of its kind. Each teacher was an institution and enjoyed full autonomy in his work. Takshasila became the center of higher education because several teachers who were recognized as authorities in their subjects resided there. At that time the education began at home and the students got their secondary education in the Ashrams. Taksashila thus was the intellectual capital of Bharat (India). Nalanda University (425–1205 AD) was another university, the Harvard of its times was one of the greatest achievements in the field of education. More than 10,500 students studied over 60 subjects at the Nalanda University which included Brahminical and Buddhist, sacred and secular, philosophical and practical. Nalanda attracted students not only from parts of India but also from far off lands. The University was founded by Sakraditya.

14.1.2 Economics (Arthashastra)

Arthashastra, whose meaning is '*Science of Economics*' is an extraordinary detailed manual on statecraft by one of classical India's greatest minds—Kautilya, Vishnugupta or Chanakya (350–283 BC) and is read in Europe even today. Chanakya is touted as the "Pioneer Economist of India". Chanakya was the adviser and Prime Minister of Emperor Chandragupta. Chanakya was a professor at the University of Takshila (located in present day Pakistan) and was an expert in commerce, warfare, economics, etc. Artha, literally wealth, is one of the four supreme aims prescribed by Hindu tradition. It is used in the sense of (a) material well being, (b) Livlihood, (c) Economically productive activity particularly in agriculture, cattle rearing and trade, (d) Wealth of Nations.

Therefore, Arthashastra is the science of economics, including starting productive enterprises, taxation, revenue collection, budget and accounts. Arthashastra contains 15 books which cover numerous topics on economics, administration, government etc. It is written mainly in prose but also incorporates 380 shlokas. The first five books deal with internal administration and the last eight on a state's relations with its neighbors [7, 8]. 'Arthashastra'—Science of Economics & Government existed even before Kautilya. Unfortunately, all the earlier works are lost and Kautilya's Arthashastra is the earliest text that has come down to us. Another useful manual in ancient times was '*Oeconomicus*' written by Xenophon (431–355 BC). Aristotle had also written on many subjects including economics in his book Politics (c.a. 350 BC). In Politics, Book II, Part V, he argued that, 'Property should be in a certain sense common, but, as a general rule, private; for, when everyone has a distinct interest, men will not complain of one another, and they will make more progress, because everyone will be attending to his own business... And further, there is the greatest pleasure in doing a kindness or service to friends or guests or companions, which can only be rendered when a man has private property. These advantages are lost by excessive unification of the state'. In Politics Book I, Aristotle discusses the general nature of households and market exchanges [8].

14.1.3 Econophysics

Econophysics is a trans-disciplinary research field, in which laws, theories, methods of physics are applied to economics. The term 'Econophysics' was coined by H. Eugene Stanley in 1995, during statistical physics, Kolkata II, conference. In this conference, there were many papers on stock and other markets written by physicists. The first meeting on 'Econophysics' was organized in 1998 in Budapest by János Kertész and Imre Kondor. Among the formal courses on Econophysics, that offered by the Physics Department of the Leiden University, from where the first Nobel-laureate in economics Jan Tinbergen came, is particularly noteworthy [9].

However before the term 'Econophysics' was coined many people from different branches of science had worked and applied their knowledge in the field of economics leading to evolution of econophysics. We can divide the evolution of econophysics into three different parts, such as

(a) **Pre Classical era**
(b) **Classical era**
(c) **Modern era**

(a) Pre-Classical era It can be considered as the period when there was no boundary between studies of different subjects. A philosopher was free to think and work in any field. There was no boundary, no specialization of the fields as on to-day. There were no sharply defined fields. Essentially, it is the period before a systematic Physics started with Newton and much before the development of social science (which started towards third quarter of the eighteenth century).

(b) Classical era It can be considered to exist since Newton. There were many branches of science after Newton (although mathematical science existed before physics). Later on Social sciences came into existence with different fields like economics, political science, sociology etc. In this period, people were shifting from their main stream to other branches. For example physical scientists started shifting from their field to social sciences and vice-versa. It was just like two sides of a river and there was no bridge. In this section, I shall deal with the jump and bridging the gap (?) between the science and social science.

(c) Modern era Finally I bring an account of Modern era of econophysics i.e., 'Institutionalized econophysics' after the name 'Econophysics' was coined by Prof. E.H. Stanley, in Kolkata, India, in 1995. In this era, there were bridges on the river in the form of conferences, publication of articles in journals of physical sciences, publication of books etc. reducing the gap between physical sciences and social sciences. It is probably going on a way towards a grand unification of different forms of sciences.

14.1.3.1 Pre-Classical Era

Thomas Aquinas (1215–1274) According to Thomas Aquinas, an Italian theologists and writer on economic problems, Just Price, a concept enunciated by him, is just sufficient to cover the costs of production, including the maintenance of a worker and his family. He considers that raising prices in response to high demand was a type of theft.

Duns Scotus (1265–1308) He is a philosopher from Scotland and had taught in Oxford, Paris and Cologne. He criticized the concept of just price and defended merchants. According to him, if people did not benefit from a transaction, they would not trade. He argued that by transporting goods and making them available to the public, merchants are doing a useful social role.

Nicole Oresme (1320–1382) was a great philosopher before Copernicus, who has done a lot of work in almost all fields in fourteenth century. He wrote influential works on economics, mathematics, physics, astronomy, philosophy, and theology; was Bishop of Lisieux, a translator, a counsellor of most original thinkers of 14th century. With his Treatise on the origin, nature, law, and alterations of money, one of the earliest manuscripts devoted to an economic matter, Oresme brings an interesting insight on the medieval conception of money.

Nicolaus Copernicus (1473–1543) During 1516–1521, Copernicus resided at Olsztyn Castle as economic administrator of Warmia, including Olsztyn (Allenstein) and Pieniężno (Mehlsack). While there, he wrote a manuscript, Locationes mansorum desertorum (Locations of Deserted Fiefs), with a view to populating those fiefs with industrious farmers and so bolstering the economy of Warmia. He participated in discussions in the East Prussian diet about coinage reform in the Prussian countries; a question that concerned the diet was who had the right to mint coin. In 1526 Copernicus wrote a study on the value of money, Monetae cudendae ratio. In it he formulated an early iteration of the theory, now called Gresham's Law, that "bad" (debased) coinage drives "good" (un-debased) coinage out of circulation-70 years before Thomas Gresham. He also formulated a version of quantity theory of money. Copernicus' recommendations on monetary reform were widely read by leaders of both Prussia and Poland in their attempts to stabilize currency. [Wikipedia, April 2012]

14.1.3.2 Classical Era

There was no other science before physics started to develop, during the time of Galileo Galelei (1564–1642). The only science that was somewhat mature was mathematics, which is an analytical science (based on logic) and not synthetic (based on observations/ experiments carried out in controlled environments or laboratories). Developments of mathematics were probably a need for astronomical studies. Astronomical studies had a deep impact in the development of physics, giving rise to the birth of classical physics, almost single-handedly by Sir Isaac Newton (1643–1727). Mathematics remained at the core of physics since then. The rest of main stream sciences, like chemistry, biology etc. all tried to get inspiration from, utilize, and compare with physics since then. In principle, development in social sciences started much later. However, people with science background had a jump to social science during this era as I shall discuss their contribution to the field of economics, remaining in their core field.

Isaac Newton (1642–1727) His explanation of gravity and his investigations into the properties of light had a profound impact and he is rightly regarded as one of the greatest scientists of his or any other generation. At the same time he is considered as the originator of Gold Standard. He has spent a precious thirty years of his life contributing to economics, by reforming the coinage of England. He was also the master of the mint, who overvalued Gold so that Gold became a standard in Great Britain. He was warden of the Royal Mint in 1696. In 1699 Isaac Newton became Master of the Royal Mint; a post which he occupied until his death in 1727. He not only standardized Britain's coinage, but he also profited from it nicely. As Master of the Mint he was paid six hundred pounds a year. Newton received additional payment for each bag of silver or gold coins produced by the Mint, adding up to one thousand pounds a year to his salary.

Edmond Halley (1656–1742) Edmond Halley was first to predict the return of the comet named after him. In 1691 Halley suggested that such a transit of Venus would be ideal situation to make measurements from all locations of the Earth. More importantly he laid the actuarial foundations of life assurance. Underwriters play a big part in the insurance industry. They're the ones who calculate the risk, based on statistics, and decide what the premiums will be. In 1693, the astronomer Edmond Halley created a basis for underwriting life insurance by developing the first mortality table. He combined the statistical laws of mortality and the principle of compound interest. However, this table used the same rate for all ages. In 1756, Joseph Dodson corrected this error and made it possible to scale the premium rate to age.

James Dodson (1705–1757) James Dodson was a British mathematician, actuary and innovator in the insurance industry. Actuarial science is the discipline that applies mathematical and statistical methods to assess risk in the insurance and finance industries. Actuarial science includes probability, mathematics, statistics, finance, economics, financial economics, and computer programming. Historically, actuarial science used deterministic models in the construction of tables and premiums. Dodson had formed a new society on a plan of assurance that would be more "equitable". He built on the statistical mortality tables developed by Edmund Halley in 1693. His great works include his work on 'The Anti-Logarithmic Canon' (published in 1742) and 'The Mathematical Miscellany'. Dodson published 'The Calculator ... adapted to Science, Business, and Pleasure', which is a large collection of small tables, with some seven-figure logarithms in 1747. The same year he started the publication of 'The Mathematical Miscellany,' whose Vol. iii were published in 1755. It is devoted to problems relating to annuities, reversions, insurances, leases on lives, etc. [based on Wikipedia, May, 2012]

Daniel Bernoulli (1687–1759) The expected utility hypothesis is a theory of utility in which "betting preferences" of people with regard to uncertain outcomes (gambles) are represented by a function of the payouts (whether in money or other goods), the probabilities of occurrence, risk aversion, and the different utility of the same payout to people with different assets or personal preferences. This theory has proved useful to explain some popular choices that seem to contradict the expected value criterion (which takes into account only the sizes of the payouts and the probabilities of occurrence), such as occurring in the contexts of gambling and insurance. Daniel Bernoulli initiated this theory in 1738. Nicolas Bernoulli described the

St. Petersburg paradox (involving infinite expected values) in 1713, prompting two Swiss mathematicians to develop expected utility theory as a solution. The theory can also more accurately describe more realistic scenarios (where expected values are finite) than expected value alone. In 1738, Nicolas' cousin Daniel Bernoulli published the canonical 18th century description of this solution in Specimen theoriae novae de mensura sortis or Exposition of a New Theory on the Measurement of Risk. Bernoulli's paper was the first formalization of marginal utility, which has broad application in economics in addition to expected utility theory. He used this concept to formalize the idea that the same amount of additional money was less useful to an already-wealthy person than it would be to a poor person.

Francois Quesnay (1694–1774) Quesnay was a country surgeon, who applied his ideas of blood circulation to economic circulation. Descriptions of Quesnay's economic theory are normally based on the texts which are read from the point of view of today's mainstream neoclassical theory. According to Galen blood has a one-way flow from the heart to the organs where it is consumed. Quesnay based his argument on the systemic circulation of blood rediscovered by William Harvey (1578–1657) in 1628, which became conclusive only when Malpighi in 1661 discovered the capillaries. So Quesnay's argument supposed that blood was recycled, something incomprehensible within the system of Galen. But there is an interesting analogy in economic theory: As for Galen, arterial blood from the heart and venous blood from the liver is consumed by all organs, for Harvey blood is recycled, so in neoclassical economics commodities flow one-way to be destroyed by producing personal utility and in classical economics at least the output of "productive" labour is input to the next economic circle.

Pierre-Simon Laplace (1749–1827) Laplace was a mathematical physicist quite famous for his 'Laplace's demon' in 1814. According to him "We may regard the present state of the universe as the effect of its past and the cause of its future". Laplace stressed out, in 1812, that events that might seem random and unpredictable can in fact be predictable. In his 'Essai Philosophique Surles Probabilities' he pointed out that events that might seem random & unpredictable such as the number of letters in the Paris dead letter office can be quite predictable and can be shown to obey simple power laws. Adolphe Quetelet further amplified the Laplace's ideas by studying the existence of patterns in data sets ranging from economic to social problems.

Lambert Adolphe Jacques Quetelet (1796–1874) Quetelet was a man of both natural and social science. He was one among the first person in introducing statistical methods to the social sciences. Quetelet was the enunciator of the word 'social physics', named so because he had applied probability and statistics to 'social science'. Social phenomena has a lot of complexity. He thought of measuring the variables it is associated with. He had tried to and aimed at understanding social phenomena as crime rates, marriage rates or suicide rates using statistical laws and wanted to explain the values of these variables. It is interesting to note that he describes his concept of the "average man" (l'homme moyen) who is characterized by the mean values of measured variables that follow a normal distribution in his most influential book *Sur l'homme et le développement de ses facultés, ou Essai de physique sociale*, published in 1835. It has also an English translation and is titled as *Treatise on Man*, literally its meaning is "On Man and the Development of his Faculties, or Essays on Social Physics".

Isidore Auguste Marie François Xavier Comte (1798–1857) Comte had studied in École Polytechnique in Paris and then at the medical school at Montpellier. August Comte was quite a visionary personality, who had predicted biophysics, Geophysics, Sociophysics and may be econophysics. He had developed a systematic classification of all sciences, including inorganic physics (astronomy, earth science and chemistry) and organic physics (biology and, for the first time, physique sociale, later renamed sociologie). In this way he had almost unified all branches of science. Comte re-invented "sociologie," (sociology) and introduced the term as a neologism, in 1838 (which has introduced by Emmanuel Joseph Sieyès in 1780). In his work 'The Course in Positive Philosophy' and 'A General View of Positivism he first described the epistemological perspective of positivism'. There were five volumes, out of which the first three were devoted to physical sciences already in existence (mathematics, astronomy, physics, chemistry, biology), and the latter two emphasized the coming of social science. In this regard, Comte is regarded as the first philosopher of science. He was also the first to distinguish natural philosophy from science. According to Comte, the physical sciences had necessarily to arrive first, before humanity could adequately channel its efforts into the most challenging and complex "Queen Science" of human society (sociology) itself.

Antoine Augustin Cournot (1801–1877) Cournot was a French philosopher and mathematician. He earned a doctoral degree in mathematics, with mechanics as his main thesis supplemented by astronomy. Cournot was mainly a mathematician, but did have some influence over economics. His theories on monopolies and duopolies are still famous. In 1838 the book "Researches on the Mathematical Principles of the Theory of Wealth" was published, in which he used the application of the formulas and symbols of mathematics in economic analysis. Today many economists believe this book to be the point of departure for modern economic analysis. Cournot introduced the ideas of functions and probability into economic analysis. He derived the first formula for the rule of supply and demand as a function of price and in fact was the first to draw supply and demand curves on a graph, anticipating the work of Alfred Marshall by roughly thirty years. Cournot is credited with the "one monopoly profit" theorem, which says that a monopolist can extract only one premium for being a monopolist, and getting into complementary markets does not pay. An exception occurs when the monopolist's market is price-regulated (Baxter's Law). Today, Cournot's work is recognized in econometrics. In the field of economics he is best known for his work in the field of oligopoly theory-Cournot competition which is named after him.

Léon Walras (1834–1910) Walras was studying in the Paris School of Mines, but grew tired of engineering. He also tried careers as a bank manager, journalist, romantic novelist and a clerk at a railway company before turning to economics. He was a French mathematical economist. He discovered the marginal theory of value (independently of William Stanley Jevons and Carl Menger) and pioneered the development of general equilibrium theory. Much like the Fabians, Walras called for the nationalization of land, believing that land's value would always increase and that rents from that land would be sufficient to support the nation without taxes. In 1874 and 1877 Walras published Elements of Pure Economics, a work that led him to be considered the father of the general equilibrium theory. The problem that Walras set out to solve was one presented by Cournot, that even though it could be demonstrated that prices would equate supply and demand to clear individual markets, it was unclear that an equilibrium existed for all markets simultaneously.

Jules Regnault (1834–1894) was a French economist who first suggested a modern theory of stock price changes in *Calcul des Chances et Philosophie de la Bourse* (1863) and used a random walk model. He is also one of the first authors who tried to create a "stock exchange science" based on statistical and probabilistic analysis. His hypotheses were used by Louis Bachelier.

Clausius (1865–?) An alternative way to study stock market volatility is by applying concepts of physics which significant literature has already proven to be helpful in describing financial and economic phenomena. One measure that can be applied to describe the nonlinear dynamics of volatility is the concept of entropy. This concept was originally introduced in 1865 by Clausius to explain the tendency of temperature, pressure, density and chemical gradients to flatten out and gradually disappear over time. Based on this, Clausius developed the Second Law of Thermodynamics which postulates that the entropy of an isolated system tends to increase continuously until it reaches its equilibrium state.

William Stanley Jevons (1835–?) was a British economist and logician. But his interest in natural science was by no means exhausted throughout his life & he continued to write occasional papers on scientific subjects, and his intimate knowledge of the physical sciences greatly contributed to the success of his chief logical work, *The Principles of Science*. Jevons arrived quite early in his career at the doctrines that constituted his most characteristic and original contributions to economics and logic. The theory of utility became the keynote of his general theory of political economy. The degree of utility of a commodity is some continuous mathematical function of the quantity of the commodity available, together with the doctrine that economics is essentially a mathematical science, took more definite form in a paper on "A General Mathematical Theory of Political Economy", written for the British Association in 1862. Jevons first received general recognition for writing on practical economic question. *A Serious Fall in the Value of Gold* (1863) and *The Coal Question* (1865) placed him in the front rank as a writer on applied economics and statistics; and he would be remembered as one of the leading economists of the 19th century even had his *Theory of Political Economy* never been written.

Josiah Willard Gibbs (1839–1903) Gibbs is a great scientist who had a lot of contributions to the field of physics, chemistry, and mathematics. He had been awarded the first American doctorate in engineering by Yale University. Gibbs is considered one of the founders of statistical mechanics along with James Clerk Maxwell and Ludwig Boltzmann. He coined the term "statistical mechanics", and introduced the term phase space and used it to define the micro-canonical, canonical, and grand canonical ensembles, thus obtaining a more general formulation of the statistical properties of many-particle systems than what had been achieved previously by Maxwell and Boltzmann. Many scientist, economists and Nobel prize winner were influenced by Gibbs. Indirectly Gibbs had influence on mathematical economics and on general equilibrium theory. The thesis of Irving Fisher, who received the first Ph.D. in economics from Yale in 1891 was supervised by Gibbs. Nobel Laureate Paul Samuelson, the second Nobel prize winner in economics in 1970 explicitly acknowledged the influence of the classical thermodynamic methods of Gibbs and described Gibbs as "Yale's great physicist."

Alfred Marshall (1842–1924) Alfred Marshall was one of the most influential economists of his time. His book, Principles of Economics (1890), was the dominant economic textbook in England for many years. It brings the ideas of supply and demand, marginal utility, and costs of production into a coherent whole. He is known as one of the founders of economics. Marshal, after experiencing a mental crisis abandoned Physics and took a broad approach to social science in which economics plays an important but limited role. Marshall envisioned dramatic social change involving the elimination of poverty and a sharp reduction of inequality. He saw the duty of economics was to improve material conditions. Alfred Marshall was the first to develop the standard supply and demand graph demonstrating a number of fundamentals regarding supply and demand including the supply and demand curves, market equilibrium, the relationship between quantity and price in regards to supply and demand, law of marginal utility, law diminishing returns, and the ideas of consumer and producer surpluses.

Vilfredo Pareto (1848–1923) Vilfredo Federico Damaso Pareto was an Italian engineer, sociologist, economist, political scientist and philosopher. He made several important contributions to economics, particularly in the study of income distribution and in the analysis of individuals' choices. He introduced the concept of Pareto efficiency and helped develop the field of microeconomics. He also was the first to discover that income follows a Pareto distribution, which is a power law probability distribution. The Pareto principle was named after him and built on observations of his such as that 80 % of the land in Italy was owned by 20 % of the population. He also contributed to the fields of sociology and mathematics. His books look more like modern economics than most other texts of that day: tables of statistics from across the world and ages, rows of integral signs and equations, intricate charts and graphs.

Louis Bachelier (1870–1946) He is credited with being the first person to model the stochastic process now called Brownian motion, which was part of his Ph.D. thesis *The Theory of Speculation*, (published 1900). His thesis, which discussed the use of Brownian motion to evaluate stock options, is historically the first paper to use advanced mathematics in the study of finance. Thus, Bachelier is considered a pioneer in the study of financial mathematics and stochastic processes. Also notable is that Bachelier's work on random walks was more mathematical and predated Einstein's celebrated study of Brownian motion by five years.

George Udny Yule (1871–1951) was a British statistician. In the 1920's Yule wrote three influential papers on time series analysis, "On the time-correlation problem" (1921), a critique of the variate difference method, "Why Do We Sometimes Get Nonsense Correlations between Time-series?" (1926), an investigation of a form of spurious correlation, and "On a Method of Investigating Periodicities in Disturbed Series, with Special Reference to Wolfer's Sunspot Numbers" (1927), which used an autoregressive model to model the sunspot time series instead of the established periodogram method of Schuster. In 1925 Yule published the paper "A Mathematical Theory of Evolution, based on the Conclusions of Dr. J. C. Willis, F.R.S.", where he proposes a stochastic process that leads to a distribution with a power-law tail—in this case, the distribution of species and genera. This was later called the Yule process, but is now better known as preferential attachment. Herbert A. Simon dubbed the resulting distribution the Yule distribution in his honour.

Albert Einstein (1879–1955) His paper on Brownian motion was on the motion of small particles suspended in a stationary liquid. This paper showed that Brownian movement can be construed as firm evidence that molecules exist. Now-a-days Brownian motion is relevant in analysis of stock market as it has desirable mathematical characteristics, where statistics can be estimated and probabilities can be calculated. M.F.M. Osborne showed that the logarithms of common-stock prices, and the value of money, can be regarded as an ensemble of decisions in statistical equilibrium, and that this ensemble of logarithms of prices, each varying with time, has a close analogy with the ensemble of coordinates of a large number of molecules. R.N. Mantegna showed that the daily variations of the price index are distributed on a 'Lévy' stable probability distribution, and that the spectral density of the price index is close to one expected for a Brownian motion. William Smith also applied Einstein's theory using the method of regulated Brownian motion to analyze the effects of price stabilization schemes on investment when demand is uncertain.

Meghnad Saha (1893–1956) was an astrophysicist noted for his development in 1920 of the thermal ionization equation. Meghnad Saha (1893–1956), the founder of the Saha Institute of Nuclear Physics, in Kolkata, and collaborators had already discussed at length in their text book in the 1950s, the possibility of using a Maxwell-Boltzmann velocity distribution (a gamma distribution) in an ideal gas to represent the income distribution in societies: "suppose in a country, the assessing department is required to find out the average income per head of the population. They will proceed somewhat in the similar way . . .(the income distribution) curve will have this shape because the number of absolute beggars is very small, and the number of millionaires is also small, while the majority of the population have average income." ("Distribution of velocities" in A Treatise on Heat, M.N. Saha and B.N. Srivastava, Indian Press, Allahabad, 1950; pp. 132–134).

Jan Tinbergen (1903–1994) was a Dutch economist. He was awarded the first Bank of Sweden Prize in Economic Sciences in Memory of Alfred Nobel in 1969, which he shared with Ragnar Frisch for having developed and applied dynamic models for the analysis of economic processes. Tinbergen studied mathematics and physics at the University of Leiden under Paul Ehrenfest. During 1929 he earned his Ph.D. degree at this university with his thesis entitled "Minimum problemen in de natuurkunde en de economie" (Minimisation problems in Physics and Economics). Tinbergen became known for his 'Tinbergen Norm', which is the principle that, if the difference between the least and greatest income in a company exceeds a rate of 1:5, that will not help the company and may be counterproductive. In his work on macroeconomic modeling and economic policy making, Tinbergen classified some economic quantities as targets and others as instruments. Targets are those macroeconomic variables the policy maker wishes to influence, whereas instruments are the variables that the policy maker can control directly. Tinbergen emphasized that achieving the desired values of a certain number of targets requires the policy maker to control an equal number of instruments. Tinbergen's classification remains influential today, underlying the theory of monetary policy used by central banks. Many central banks today regard the inflation rate as their target; the policy instrument they use to control inflation is the short-term interest rate.

Ettore Majorana (1906–1938) Majorana earned his undergraduate degree in engineering and completed his physics doctorate, both at the University of Rome La Sapienza. He wrote the paper (probably during his first disappearance) on *The value of statistical laws in Physics and the Social Sciences* which was found among his papers by his brother Luciano, and was published after his disappearance by Giovanni Gentile junior. It is essentially a paper giving analogy between physics and social science. According to him 'The deterministic conception of nature holds in its very being a real motive of weakness because irremediably contradicts the most evident data of our conscience. G. Sorel tried to compose this dysfunction by distinguishing between artificial nature and natural nature (this later being a-causal), although in this way he denied the unity of Science. On the other hand, the formal analogy between the statistical laws of Physics and those of Social Sciences supports the opinion that also human actions were submitted to a rigid determinism. It is important then, that the principles of Quantum Mechanics have lead to a recognition (as well as a certain absence of objectivity in the description of phenomena) of the statistical character of the ultimate laws of elemental processes. This conclusion has made substantial the analogy between Physics and Social Sciences, and has produced between them an identity of value and method' [10].

Tjalling Charles Koopmans (1910–1985) was the joint winner, with Leonid Kantorovich, of the 1975 Nobel Memorial Prize in Economic Sciences for his contributions to the field of resource allocation, specifically the theory of optimal use of resources. He began his university education at the Utrecht University specializing in mathematics in 1927. Later, in 1930, he switched to theoretical physics. In 1933, he met Jan Tinbergen, and moved to Amsterdam to study mathematical economics under him. In addition to mathematical economics, Koopmans extended his explorations to econometrics and statistics. His early works on the Hartree-Fock theory are associated with the Koopmans' theorem, which is very well known in quantum chemistry. His article of deriving the distribution of the serial correlation coefficient was recognized by John von Neumann.

Paul Anthony Samuelson (1915–2009) was an American economist, and the first American to win the Nobel Memorial Prize in Economic Sciences. Samuelson was one of the first economists to generalize and apply mathematical methods developed for the study of thermodynamics to economics. As a graduate student at Harvard, he was the sole protégé of the polymath Edwin Bidwell Wilson, who had himself been a student of Yale physicist Willard Gibbs. Gibbs, the founder of chemical thermodynamics, was also mentor to American economist Irving Fisher and he influenced them both in their ideas on the equilibrium of economic systems. Samuelson also published one of the first papers on nonlinear dynamics in economic analysis.

Ilya Prigogine (1917–2003) Prigogine studied chemistry at the Free University of Brussels. Prigogine is best known for his definition of dissipative structures and their role in thermodynamic systems far from equilibrium, a discovery that won him the Nobel Prize in Chemistry in 1977. Dissipative structure theory led to pioneering research in self-organizing systems, as well as philosophical inquiries into the formation of complexity on biological entities and the quest for a creative and irreversible role of time in the natural sciences. The 'invisible hand' mechanism of the market to evolve towards the 'most efficient' (beneficial to all participating agents) predates by ages the demonstration of 'self-organization' mechanism in physics or chemistry of many-body systems, where each constituent cell follows very local (space and time) dynamic rules yet the collective system evolves towards a globally organized pattern. His work is seen by many as a bridge between natural sciences and social sciences.

Kenneth J. Arrow (1921–) He earned a Bachelor's degree from the City College of New York in 1940 in mathematics, where he was a member of Sigma Phi Epsilon. At Columbia University, he received a Master's degree in 1941. From 1946 to 1949 he spent his time partly as a graduate student at Columbia and partly as a research associate at the Cowles Commission for Research in Economics at the University of Chicago. During that time he also held the rank of Assistant Professor in Economics at the University of Chicago. In 1951 he earned his Ph.D. from Columbia. His first major work, forming his doctoral dissertation at Columbia University was *Social Choice and Individual Values* (1951), which brought economics into contact with political theory. This gave rise to social choice theory with the introduction of his "Possibility Theorem". In the 1950s, Arrow and Gerard Debreu developed the Arrow-Debreu model of general equilibria. In 1971 Arrow with Frank Hahn co-authored *General Competitive Analysis* (1971), which reasserted a theory of general equilibrium of prices through the economy.

John Forbes Nash, Jr. (1928–) is an American mathematician whose works in game theory, differential geometry, and partial differential equations have provided insight into the forces that govern chance and events inside complex systems in daily life. His theories are used in market economics, computing, evolutionary biology, artificial intelligence, accounting, politics and military theory. Serving as a Senior Research Mathematician at Princeton University during the latter part of his life, he shared the 1994 Nobel Memorial Prize in Economic Sciences with game theorists Reinhard Selten and John Harsanyi. Nash has developed work on the role of money in society. Within the framing theorem that people can be so controlled and motivated by money that they may not be able to reason rationally about it, he has criticized interest groups that promote quasi-doctrines based on Keynesian economics that permit manipulative short-term inflation and debt tactics that ultimately undermine currencies.

Benoît Mandelbrot (1924–2010) Mandelbrot worked on a wide range of mathematical problems, including mathematical physics and quantitative finance, but is best known as the father of fractal geometry. He coined the term fractal and described the Mandelbrot set. Mandelbrot also wrote books and gave lectures aimed at the general public. From 1949 to 1958 Mandelbrot was a staff member at the Centre National de la Recherche Scientifique. From 1951 onward, Mandelbrot worked on problems and published papers not only in mathematics but in applied fields such as information theory, economics, and fluid dynamics. He became convinced that two key themes, fat tails and self-similar structure ran through a situation of problems encountered in those fields. Mandelbrot found that price changes in financial markets did not follow a Gaussian distribution, but rather Lévy stable distributions having theoretically infinite variance. He found, for example, that cotton prices followed a Lévy stable distribution with parameter α equal to 1.7 rather than 2 as in a Gaussian distribution. "Stable" distributions have the property that the sum of many instances of a random variable follows the same distribution but with a larger scale parameter.

M.F.M. Osborne (1917–2003) According to Professor Joseph L. McCauley (University of Huoston), Osborne first introduced the lognormal stock pricing model in 1958 and should be honored as the first econophysicist. According to Steve Hue, Prof. of Physics at the University of Oregaon, Osborne's book "The Stock Market and Finance from a Physicist's Viewpoint [Paperback]" is quite interesting as he explores market micro-structure, market making, supply-demand (bid-ask) in detail, going far beyond the usual idealizations made by economists. M.F.M. Osborne showed that the logarithms of common-stock prices, and the value of money, can be regarded as an ensemble of decisions in statistical equilibrium, and that this ensemble of logarithms of prices, each varying with time, has a close analogy with the ensemble of coordinates of a large number of molecules. Using a probability distribution function and the prices of the same random stock choice at random times, he was able to derive a steady state distribution function, which is precisely the probability distribution for a particle in Brownian motion. A similar distribution holds for the value of money, measured approximately by stock market indices. Thus, it was shown in his paper that prices in the market did vary in a similar fashion to molecules in Brownian motion.

Wolfgang Weidlich (1931–) Prof. Weidlich obtained Diploma in physics, and Ph.D. in Physics from free University, Berlin in 1955 and 1957 respectively. Besides main stream physics he has a number of publications, books and book reviews in interdisciplinary field basically in socio dynamics. In his book 'Sociodynamics'—A Systematic Approach to Mathematical Modeling in the Social Sciences, he has developed systematic approach to initiating and evaluating mathematical models for a broad class of collective dynamic social processes in different sectors of society. He has characterized and compared the hierarchies of complex structures in nature and society and their methods of description. Then, universally applicable methods originating in statistical physics, non-linear dynamics and synergetics are combined with concepts of social science to construct sociodynamics, a general strategy for designing mathematical models for the quantitative description of a wide range of collective dynamical phenomena within society. The central equation of sociodynamics that is the master equation for the probability distribution of socio configurations, has been derived by him and the general properties of its solutions has been treated. The evolution equations for quasi-mean values and variance have been derived from this equation.

14.1.3.3 Modern Era

From the previous section we observe that people from the fields of Physics, Mathematics, Physiology, and Engineering have jumped from their branch, in which they are formally trained, to Economics. Similarly some of the social scientists have appreciated the works done by the natural scientist in the field of social science and even worked under them leading to their Ph.D. degree. Again some physical scientists have been awarded Nobel Memorial Prize in economics. So also some social scientists have been awarded Nobel Prize in Economics working on the principles of natural sciences. These are the developments which laid a foundation stone for the new branch, 'Econophysics'. Even the word 'Phynance' has come up which is a combination of Physics and Finance. In this section I attempt to bring an account of Modern era of econophysics i.e., 'institutionalised econophysics' after the name 'Econophysics' was coined by Prof. E.H. Stanley, in Kolkata, India, in 1995. In this era there were bridges on the river in the form of conferences, publication of articles in journals of physical sciences, publication of books etc., which have been instrumental for reducing the gap between physical sciences and social sciences. I will deal with the bridging of natural science with social science extensively in this section. It is probably going on a way towards a grand unification of different forms of sciences.

(In this section name of the Scientists/ Professors have been arranged in ascending alphabetic order of their family name.)

Frederic Abergel Frederic is now working as the Director of the laboratory of Mathematics Applied to Systems at Ecole Centrale Paris. He was a Consultant for BNP Paribas, Equity Derivatives from 1996–2001, Senior quant, Equity Derivatives, Calyon from 2001–2004, Head of Paris Analytics, Barclays Capital from 2004–2005, Head of Equity and Commodity Analytics and Systems, NatIxis, from 2005–2007. He specialises in financial modelling, electronic markets, trading, algorithmics and systems, risk management and management. He has a large number of publications and has 675 numbers of citations. His papers on Econophysics: Empirical facts and agent-based models and Agent-based models & Econophysics review (Quantitative Finance, 2011) have received many citations.

John Angle John Angle was initially employed as an assistant professor (University of Arizona). Most of his career since has been as a statistical consultant to the Economic Research Service (ERS) of the U.S. Department of Agriculture in Washington, DC. He took early retirement from ERS to found The Inequality Process Institute LLC (TIPI), an organization that does both pure and applied research on personal income and wealth. At TIPI, he works full time on research related to the Inequality Process, a particle system model of personal income and wealth distribution, their dynamics, and the dynamics of income and wealth at the micro-level. In 1990 he published a paper discussing the similarity of the Inequality Process to the stochastic particle system model of the kinetic theory of gases. Prof. John Angle is a mathematical sociologist interested in income and wealth phenomena. He has an outsider's interest in the natural sciences, particularly statistical physics, and envy the excitement of discovery among physicists.

Masanao Aoki Aoki is an emeritus professor of Economics at ULCA. His research interests are as follows. New approach to macroeconomic modeling by means of jump Markov processes by specifying transition rates appropriately in the backward Chapman-Kolmogorov (master equation); solutions of master equations to obtain aggregate dynamic equations, and fluctuations by solving the associated Fokker-Planck equations. Modeling and analysis of multi-agent models to investigate such things as herding behavior and return dynamics, i.e., power-laws in share or stock markets; Modeling and analysis of multiple country models by state space time series technique; aggregation of economy with heterogeneous agents by neural network methods; adaptive learning algorithms. He has many books, reviews and publications.

Hideaki Aoyama Hideaki Aoyama, is one of Japan's leading theoretical physicists. Though his interests span many areas, current focus is on econophysics, the new field that hopes to move economics closer to being an exact science.

He is currently Professor of Physics in the Graduate School of Science, Kyoto University. Some of his books are Pareto Firms (2007), Econophysics (2008), all in Japanese, and a new study forthcoming from the Cambridge University Press, Econophysics and Corporations, Statistical Life and Death in Complex Business Networks. According to Aoyama the prospect for the econophysics is very high in Japan: Econophysics is expected to make huge impact on economics, with its new ideas and approaches suitable for economic phenomena, imported from various areas of physics. Prof. Aoyama with others has run four domestic econophysics conferences at Yukawa Institute for Theoretical Physics at Kyoto University in 2003, 2005, 2007 and 2009. Very recently they have analyzed business cycles in Japan using the indices of industrial production (IIP), an economic indicator which measures the current conditions of production activities over the nation on a monthly basis.

W. Brian Arthur is an economist credited with influencing and describing the modern theory of increasing returns. He is an authority on economics in relation to complexity theory, technology and financial markets. Presently, he is an external faculty at the Santa Fe Institute, and a Visiting Researcher at the Intelligent Systems Lab at PARC. He received his B.Sc. in Electrical Engineering at Queens University Belfast (1966), an M.A. in Operational Research (1967), at Lancaster University, Lancaster, England, and an M.A. in Mathematics at the University of Michigan (1969). Arthur received his Ph.D. in Operations Research (1973) and an M.A. in Economics (1973) from the University of California, Berkeley. Arthur is noted for his seminal works "studying the impacts of positive feedback or increasing returns in economies, and how these increasing returns magnify small, random occurrences in the market place." These principles are especially significant in technology-specific industries. Arthur is one of the early economic researchers in the emerging complexity field.

Yuji Aruka Yuji Aruka is a Professor of Economics at Chuo University, Tokyo, Japan. He is also a visiting professor to many Universities. He has many publications in physics journals besides economics journals. His academic Interests and Teachings are: Macroscopic microeconomics and heterogeneous interacting agents under uncertainty, Social games, Econophysics and their interdisciplinary fields. He is the Editor of Journal of Economic Interaction and Coordination since 6/2006. He was the Chairman, IWSEP (International Workshop-on Socio- and Econo-Physics) 2003, Co-Chairman, INSC08 (3rd International Nonlinear Science Conference) at Chuo University, 2008 and Co-Chairman, CS09 (The 9th Asia-Pacific Complex System Conference) at Chuo University, 2009. He is also a referee of journals like: Structural Change and Economic Dynamics, Nonlinear Dynamics, Psychology and Life Sciences (NDPLS), Advances in Complexity etc.

Marcel Ausloos is a Professor of Physics at University of Liège, Belgium, where his group works on applications of Physics in Economy and Sociology. Besides field like magnetism, superconductivity, optics, transport properties, phase transitions, fractals, evolution, growth, he has worked on econophysics. His recent papers on econophysics are quite interesting like "Econophysics of a religious cult: the Antoinists in Belgium [1920–2000]," "Benford's law and Theil transform of financial data, "Has the world economy reached its globalization limit?" etc. He has also many articles published in Physica A, Elsevier etc.

Belal E. Baaquie is working as a Professor of Physics in National University of Singapore since 1984. His research interest includes Quantum Field Theory and financial modeling based on techniques of quantum theory. Prof. Baaquie is an authority and very much special in quantum finance. He has written a number of books in this field like 'Interest Rates and Coupon Bonds in Quantum Finance', 'Quantum Finance: Path Integrals and Hamiltonians for Options and Interest Rates'. In these books he has analyzed interest rates and coupon bonds using quantum finance. In the second one he shows how to approach problems related to financial markets with mathematical techniques that are traditionally used in quantum field theory. He has written many articles on general topics such as finance, education, history, economics and philosophy. He has developed many modules for teaching science to a broad based non-specialist audience. He has published a number of papers on Physics and Finance.

Fulvio Baldovin Fulvio is a professor at University of Padova in the Dept. of Physics. His research interests are Statistical Physics, Chemical Physics & Material Physics and Condensed Matter Physics. He has more than 40 publications with 120 numbers of citations. His publications on econophysics are mainly based on high frequency financial market dynamics, Noise-induced dynamical phase transitions in long-range systems, Thermodynamics and dynamics of systems with long-range interactions etc. His papers have been published in refereed journals like Physical Review E, Physica A etc.

Jean-Philippe Bouchaud Jean-Philippe Bouchaud graduated from the Ecole Normale Supérieure in Paris in 1985, where he also obtained his Ph.D. in physics. He became interested in economics and theoretical finance in 1991. His work in finance includes extreme risk models, agent based simulations, market microstructure and price formation. He has been very critical about the standard concepts and models used in economics and in the financial industry (market efficiency, Black-Scholes models, etc.). He is a Pioneer of Econophysics. His work covers the physics of disordered and glassy systems, granular materials, the statistics of price formation, stock market fluctuations and the modeling of financial risks.

Bikas K. Chakrabarti is a Profesor of Physics in Saha Institute of Nuclear Physics and Visiting Professor of Economics in Indian Statistical Institute, Kolkata. The research activity of Chakrabarti is mainly focused on statistical physics, condensed matter physics, computational physics, and their application to social sciences; Econophysics and Sociophysics. Among other awards, he received the Shanti Swarup Bhatnagar prize in 1997. He is the editorial Board member of Physics journals including "European Physical Journal" (European Physical Society) and

Economics journals including “Journal of Economic Interaction & Coordination” (Springer). He is the co-author of many books like ‘Econophysics: An Introduction (Wiley, 2010)’, ‘Econophysics of Income & Wealth Distributions (Cambridge Univ Press, 2013)’. He has initiated the Econophys-Kolkata series of Conferences and co-edited their Proc Volumes (seven so far), all published in New Economic Windows Series of Springer. He has published many papers in Physics, Economics and inter-disciplinary journals, including 4 reviews in esteemed Reviews of Modern Physics. A random saving gas model by the “Kolkata School” (published during 1995–2005), led by Prof. Chakrabarti could capture both the initial Gamma/lognormal distribution for the income distribution of poor and middle income groups and also pareto tail for distribution for the riches.

Anirban Chakraborti Chakraborti is an associate professor at École Centrale Paris. He is the first Ph.D. in econophysics from India (under supervision of Prof. Bikas K. Chakrabarti, SINP, Kolkata). He has contributed to several interesting and important areas, such as, Simulations of agent-based market models and their relation to different theories in physics such as the kinetic theory of gases, percolation theory, and theory of self-organization. With co-workers, he has introduced a self-organizing model where agents trade with a single commodity with the money they possess, and studied the role of money in the economic market. His research on statistical physics focuses not only to its application to problems in economic systems (Econophysics) but also to combinatorial optimization such as the Traveling Salesman problem and study of “complex systems”. He has also co-authored and co-edited at least six books so far and has contributed his papers in five books. He has many reviews and invited articles. He has a number of publications in refereed journals. He was awarded the Young Scientist Medal of the Indian National Science Academy (2009) for his pioneering studies on statistical models related to Econophysics.

Satya R. Chakravarty is a Professor of Economics at the Indian Statistical Institute, Kolkata. Professor Chakravarty worked as a visiting Professor in many Universities. Professor Chakravarty’s main areas of interest are Welfare Economics, Public Economics, Mathematical Finance, Industrial Organization and Game theory. He has over 80 publications in prestigious journals and edited volumes. He has authored six books and has co-edited two books on Quantitative Economics and Econophysics. He is a member of the editorial board of “The Journal of Economic Inequality” and a member of the advisory board for the book series “Economic Studies in Inequality, Social Exclusion and Well-Being”, Springer-Verlag. Professor Chakravarty is a Co-Editor of the Economics e-Journal, Kiel Institute for the World Economy. He has acted as a referee for many journals. He received the Mahalanobis Memorial Award of the Indian Econometric Society in 1994. He has acted as an external adviser to the World Bank, Washington, D.C., and as an adviser to the National Council of Social Policy Evaluation, Mexico.

Arnab Chatterjee Arnab Chatterjee was awarded Ph.D. from Jadavpur University on 'Statistical Physics of Two Model Dynamical Systems: Magnets and Trading Markets' under supervision of Prof. Bikas K. Chakrabarti, SINP, Kolkata. He is working as a Postdoctoral Researcher at BECS, Aalto University, Espoo, Finland (since Feb. 2012). His main area of research is basically applications of Statistical Physics to Condensed Matter and Social Sciences. He has worked on Dynamic transition in Ising Systems, Statistical physics of socio-economic systems (Econophysics and Sociophysics), kinetic exchange models with quenched and annealed disorder etc. He has a number of publications in the field of 'Econophysics' in refereed journals. Besides, along with Bikas Chakrabarti and others, he has edited and written a number of books, reviews on Econophysics.

Carl Chiarella is a Professor of Quantitative Finance at the University of Technology, Sydney. He completed his Ph.D. in applied mathematics at the University of New South Wales in 1969 for a thesis on nuclear reactor theory. He completed the M. Com (Hons) in economics at the University of New South Wales and took out a Ph.D. in economics in 1987 from the same University for a thesis in economic dynamics. Carl has held visiting appointments at a number of Universities. He has authored more than 150 research articles in international and national journals and edited volumes and the author/co-author of 5 books. Carl is a Co-Editor of the Journal of Economic Dynamics and Control and Associate Editor of Quantitative Finance, Studies in Nonlinear Dynamics and Econometrics and European Journal of Finance.

Morrel Cohen Cohen did his Ph.D. from University of California, Berkeley in Physics in 1952. He became Professor of Physics, James Frank Institute, University of Chicago from 1960–1972. From 1968–1972 he became Professor of Theoretical Biology at University of Chicago and from 1972–1981 he became Louis Block Professor of Physics and Theoretical Biology in University of Chicago. Later on from 1981 to 2000 he became a scientist (Senior and Emeritus) at Exxon Research and Engineering Company. Presently he is a distinguish scientist at Rutgers University. He is engaged in economics studies since 1975. One of his recent paper 'Econophysical visualization of Adam Smith's invisible hand' with Iddo I. Eliazar published in Physica A 392 (2013), 813–823 they extend that general connection beyond entropy to a concept analogous to that of the free energy of statistical thermodynamics and this extension allows them to introduce into general non-physical contexts the concept of a deterministic or systematic "intrinsic force" which is analogous to a physical force. This intrinsic force acts on a quantity, a measurable microscopic property of a complex system, shaping its macroscopic probability distribution function (PDF). They have shown how the intrinsic force can be extracted quantitatively up to a scale factor from the macroscopic PDF when the latter is known, thus solving the top-down inversion problem.

Rama Cont Rama Cont is the Director of Research, CNRS, Laboratoire de Probabilités et Modeles Aléatoires, Université Paris VI–VII. He did his Ph.D. in theoretical physics from University of Paris XI (Orsay) in 1998. His research interests are computational finance, stochastic modeling of financial markets, Lévy processes and applications, interest rate and credit risk modeling, modeling of social networks, ill posed inverse problems. He has received Europlace Institute Research Grant for project on "Measuring systemic risk", 2010, Grand Prix Louis Bachelier (French Academy of Sciences, SMAI and Natixis Foundation), 2010, Best paper in Mathematical Finance 2006, Europlace Institute of Finance etc. He has written different books on Quantitative Finance. He has published many papers in refereed journals on Stochastic analysis, systemic risk, complex networks, limit order markets etc.

Nivedita Deo Nibedita Deo did her M.Sc. from Delhi University and Ph.D. from Purdue University, USA. Besides statistical physics, her research interest includes Physics and Society: Econophysics, Applications of Statistical Physics to Economics and Finance. She has published many papers on econophysics and is also supervising Ph.D. students.

Deepak Dhar is a distinguished Professor, Department of Theoretical Physics, Tata Institute of Fundamental Research Homi Bhabha Road, Mumbai, India. He has received Young Scientist Award in 1983, S.S. Bhatnagar award in Physical Sciences 1991, J.R. Schrieffer Prize in Condensed Matter Physics 1993 to name a few. He was an advisory Editor, Physica A (till 2004) Member of the Editorial Board, J. Stat. Phys. (1993–1996, 1999–2002, 2005–), Phys. Rev. E, (2008–), Pramana (2008), J. Phys. A, (2010–) and Member, IUPAP Commision on Statistical Physics (1992–1995). Prof. Dhar is one among the scientists who have contributed to econophysics and his econophysics works are mainly based on minority games.

J. Doyne Farmer Doyne is an External Professor at the Santa Fe Institute. He has broad interests in complex systems, and has done research in dynamical systems theory, time series analysis and theoretical biology. At present his main interest is in developing quantitative theories for social evolution, in particular for financial markets (which provide an accurate record of decision making in a complex environment) and the evolution of technologies (whose performance through time provides a quantitative record of one component of progress). During the eighties he worked at Los Alamos National Laboratory, where he was an Oppenheimer Fellow, founding the Complex Systems Group in the theoretical division. He began his career as part of the U.C. Santa Cruz Dynamical Systems Collective, a group of physics graduate students who did early research in what later came to be called "chaos theory". He received his Ph.D. in Physics there in 1981. Doyne Farmer then took up a post-doctoral appointment at the centre for non-linear studies at the Los Alamos National Laboratory. In 1991 Farmer gave up his position at Los Alamos to start Prediction Company, with Norman Packard and Jim McGill. The purpose of this company was to create automated trading systems for a variety of commodity and securities markets, making predictions of trends using principles of physics, particularly Chaos Theory.

Yoshi Fujiwara Fujiwara is a professor at Hyogo University in the graduate school of simulation studies. His interests are Theoretical Physics, Behavioral Economics, Macroeconomics, Complex Systems, Complex Networks. He has 40 publications in journals like Adv. Operations research, Physical Review E, Philosophy & Methodology of Economics eJournal. With other co-authors, he has written a book 'Econophysics and Companies: Statistical life and Death in Complex Business Networks'.

Xavier Gabaix Xavier Gabaix is a French economist, currently a Professor of Finance at the New York University Stern School of Business. He has been listed among the top 8 young economists in the world by The Economist (Wikipedia). He has graduated in Mathematics and did his Ph.D. in Economics. His research interest includes Asset Pricing, Behavioral economics, Origins and consequences of scaling behavior etc. He has received a number of prizes and awards like Lagrange Prize for research on complex systems, 2012 (CRT Foundation), Rising Star in Finance Award, 2012 , Fischer Black Prize, Best Young French Economist Prize, 2011, Bernacer Prize for best European economist under 40 working in macroeconomics/finance, Young Scientist Award for Socio- and econophysics, 2006. He has more than 1900 citations as on December 2011. Besides economics journals, he has many publications in physics journals also.

Mauro Gallegati is a Professor of Economics at the Polytechnic University of Marche, Ancona. He has been visiting Professor in several Universities and research institutes, including Cambridge, Stanford, MIT, Columbia, Santa Fe Institute etc. Prof. Gallegati is on the editorial board of several economics journals. His research includes business fluctuations, nonlinear dynamics, models of financial fragility and heterogeneous interacting agents. Mauro Gallegati is well known from his widely cited work with Joseph E. Stiglitz, developing theory of asymmetric information and heterogeneous agents and their applications. He has publications in the top journals on economic, economic history and history of economic analysis, nonlinear mathematics, applied economics, complexity and econophysics. The research group lead by Prof. Gallegati studies agent-based models of economic phenomena, with a special focus on the performance of heterogeneous, interacting agents, generating aggregate fluctuations, coordination failures and emerging phenomena in general. The group has extensive knowledge in simulation design and analysis of complex economic phenomena.

Diego Garlaschelli Garlaschelli has become the first Professor of Econophysics from April, 2011 in Lorentz Institute for Theoretical Physics, Leiden, Institute of Physics, University of Leiden. He did his Ph.D. in Physics, from University of Siena (Italy), on "Statistical Physics Approach to the Topology and Dynamics of Complex Networks". He is the referee of many international journals like Nature, Physical Review Letters, Physical Review E, New Journal of Physics, EuroPhysics Letters, European Physical Journal B, Physica A, Journal of Physics A, Advances in Complex Systems, BioMed Central Systems Biology, BioMed Central Bioinformatics,

Ecological Modelling, Journal of Economic Behavior and Organization, Social Networks etc. He has been awarded for best talk presented by young researchers at the "Second International Conference on Frontier Science 'a Nonlinear World: the Real World'", Pavia (Italy), Collegio Cairoli inSeptember 2003 and at the "First Bonzenfreies Colloquium on Market Dynamics and Quantitative Economics", Alessandria (Italy), in September 2004.

Dirk Helbing Dirk Helbing has formal education in physics and mathematics at the Gottingen. But he was awarded Ph.D. from Stuttgart University, on modeling social processes by means of game-theoretical approaches, stochastic methods, and complex systems theory. Although he is a physicist by training he is a professor of sociology. He is the head of the ETH Zurich Competence Center "Coping with Crises in Complex Socio-Economic Systems" and the "Physics of Socio-Economic Systems" Division of the Deutsche Physikalische Gesellschaft (German Physical Society). His education in physics, traffic dynamics and optimization helped him to model and simulate social problems and he is known for the social force model, for applying to self-organizing phenomena in pedestrian crowds. Helbing is working on many social models like evolutionary game theory, optimization of urban and freeway traffic, socio-inspired technology and techno-social systems, disaster spreading and crisis management etc. He is the Principal Investigator on a project named Future ICT Knowledge Accelerator and Crisis Relief System, a computing system working on big datasets, conceived as sort of a crystal ball of the world.

Janusz A. Holyst Janusz is a Full Professor at Faculty of Physics, Warsaw University of Technology where he leads a Center of Excellence of Complex Systems Research. His current research field includes models of emotions in cyber communities, economic and social networks, collective bankruptcies, collective opinion formation, non-equilibrium statistical physics, cellular automata, self- organized criticality and phase transitions. He is one of the pioneers in applying physical methods to economical and social systems and is the Co-Founder and Chairman of the Section Physics in Economy and Social Sciences of Polish Physical Society. His list of publications includes over120 papers (www.if.pw.edu.pl/jholyst) in peer reviewed journals that have been cited over 800 times. Prof. Holyst has organized or co-organized 15 international interdisciplinary workshops or conferences on complex systems. He acts as an Associate Editor of European Journal of Physics B, Guest Editor of Physcia A, Acta Physica Polonica, Acta Physica Polonica B and as a referee many other physical journals. Prof. Holyst has worked as an adviser on modelling of marketing and economic processes for American Company Bunge.

Roberto Iglesias is a Professor of the Ph.D. programs in Instituto de Física and Faculdade de Ciências Econômicas, UFRGS, Porto Alegre, Brasil and Head of the Applied Theoretical Physics Group, Instituto de Fisica—UFRGS. His research activities include Statistical Physics applied to Economics and Social Sciences, game theory applied to economic systems and in opinion dynamics, Magnetism, Kondo effect and Strongly Correlated Electron Systems. They develop models of economic and social organizations explicitly including the interactions between individuals and its capacities of learning and changing their behavior, analyzing the emergency of collective social behaviors. He has many publications on physics applied to economics and social sciences, published in Physics A, Revista Mexicana de Física, The European Physical Journal B, The European Physical Journal—Special Topics, J. Stat Mech. etc. since 2003.

Giulia Iori Iori did her Ph.D. in physics from Università di Roma la Sapienza Roma, Italy in 1993 and is now working as a Professor of Economics at City University, London. Her previous appointments were as Reader in applied mathematics in King's college, London and lecturer of finance at University of Essex. So her research interests are quite diversified. She has organized at least three conferences on Non-linear dynamics and econometrics, complex behavior in economics etc. She is an Associate Editor for Journal of Economic Behavior and Organization. She has served as a referee for Physical Review Letters, Physical Review E, Physica A, European Physical Journal, Journal of Economic Behavior and Organization etc. She has been awarded as an outstanding referee reward from Journal of Economics Dynamic and Control in 2009. She has many book chapters, reviews and papers on econophysics.

Jun-ichi Inoue Inoue is working in Graduate School of Information Science and Technology, Hokkaido University, in the department of Complex System Engineering. His research interests include Statistical Mechanics: Its application to information processing, econophysics. He has published a number of papers in refereed journals such as Physical Review E, Journal of Statistical Mechanics: Theory and Experiment, Journal of Physics A and in book chapters of different books on Econophysics. His works are mainly based on 'human collective behavior at financial crisis by using stock-correlation of financial time series' etc.

Hiroshi Iyetomi Iyetomi is a professor of Econophysics at University of Tokyo, in the department of economics. He was a professor of faculty of science in physics at Nigata University, Tokyo and switched to University of Tokyo as Project Professor, Faculty of Economics, in April 2012. He adopts statistical physics approach to obtain a new insight into macroeconomic phenomena, which are regarded as outcomes (collective motion) of interactions among a number of heterogeneous agents. Such an idea is analogous to that of the many-body problem in physics. The economic issues under study include business cycles, systemic risks (contagion of failure), and group structures in stock markets. Especially, he has so far focused on empirical analyses. He takes advantage of the random matrix theory to distinguish statistically meaningful information from random noises in complicated economic data and also apply network science methodology to elucidate structural and dynamical properties of large-scale economic systems.

Sanjay Jain Sanjay Jain is a professor in Physics at Delhi University, India. He is a member of the External Faculty, Santa Fe Institute, Santa Fe, NM, USA (2000–2006, 2007–2010). He was a visiting professor to Santa Fe Institute, USA, during

1999 to 2000. He is basically a theoretical physicist having interest in diversified fields. He has also contribution to the field of 'Econophysics'. His areas of work are of complex networks, including chemical, biological, and socio-economic networks, nonlinear dynamics, random matrix models and quantum chaos besides others. He has many book chapters published by Wiley-VCH, Weinheim, 2006, The Application of Econophysics, edited by H. Takayasu (Springer-Verlag, Tokyo, 2003) etc.

Neil Johnson Neil Fraser Johnson is a Professor at Miami University and heads the team of Complexity in department of Physics. His research interests include complexity theory and complex systems, spanning quantum information, econophysics, and condensed matter physics. Current projects within his group are Social Complexity, Biological Complexity, and Physical Complexity. He has two books to his credit such as "Financial Market Complexity" (Oxford University Press, 2003) and "Simply Complexity: A Clear Guide to Complexity Theory". He has about 200 papers published in different journals. He is joint Series Editor for the book series "Complex Systems and Inter-disciplinary Science" by World Scientific Press. He is the editor of the Physics Section for the journal "Advances in Complex Systems" and associate Editor for "Journal of Economic Interaction and Coordination". He was one of the founders of CABDS (Complex Agent-Based Dynamical Systems) which is Oxford University's interdisciplinary research center in Complexity Science.

Taisei Kaizoji Taisei Kaizoji is a professor of Economics at Department of Economics and Business, International Christian University, Tokyo. He was also a visiting Professor to University of Kiel, Max-Planck Institute of Economics and ETH Zurich, Department of Management, Technology, and Economics. He has received numorous awards like 'Fellowship of the Japan Society for the Promotion of Science for Japanese Junior Scientists, 1993–1994', '1999 Award for a original paper in Japan Society for Simulation Technology' etc. He is a member of the editorial board for Journal of Economic Interaction and Coordination, Springer. He has a number of publications in physics journals like Physica A, European Physical Journal B, Progress of Theoretical Physics Supplement etc. besides Economic journals. His works are based on stock market studies. Many of his papers are co-authored by physicists.

Kimmo Kaski is a Professor in Computational Engineering and Academy Professor in Computational Science and Engineering. He is also the Director of the Centre of Excellence in Computational Complex Systems Research—COSY, Aalto University/Helsinki. His research includes Computational Science, Complex Systems,Complex Networks, Computational Systems Biology, High Performance Computing. Besides many awards and honors, he has been awarded as Fellow of the Institute of Physics, UK and granted the title of Chartered Physicist, 1998. He is in editorial board of the International Journal of Modern Physics: Computers in Physics and referee for APS journals, IOP journals, IEEE (USA) journals, and IEE (UK) journals. His research has been covered by many news agencies including his works on econophysics, social science like 'Story on computational social science' in nature news. He has more than 240 publications and more than 3000 citations.

János Kertész is one of the pioneers of econophysics, complex networks and application of fractal geometry in physical problems. He is the director of the Institute of Physics in Budapest University of Technology and Economics, Budapest, Hungary. The inaugural meeting on Econophysics was organized in 1998 in Budapest by János Kertész and Imre Kondor. He has a large number of publications in the field of Econophysics in refereed journals. His main research interest are on Networks, Econophysics, Traffic models, Pattern formation, Percolation theory, etc. . He got many fellowship including Humboldt-Fellowship and DAAD-Stipendium. He was Secretary of the Committee for Statistical Physics of the Hungarian Physical Society. He was a Board Member of many scientific communities like Computational Physics Group of EPS (1996), Statistical and Nonlinear Physics Division of EPS (1996–). Also he was a Member of the Editorial Boards of Physica A (1990–1994), Fractals (1993–2003, 2008–), Fluctuation and Noise Letters (2001–2003). He was awarded Albert Szent-Gyorgyi Award for teaching and scientific achievements by the Ministry of Culture and Education, "Santa Chiara Prize" for multidisciplinary teaching (University of Siena), etc.

Alan Kirman Kirman is an economist and was Professor of economics, Warwick University. He has received Alexander von Humboldt Stiftung prize and Fondation Urrutia Prize for Economic Diversity. He is referee of many journals like American Economic Review; Economic Journal; Econometrica; Quarterly Journal of Economics; International Economic Review; etc. He has written and edited around seventeen books like 'Introduction to equilibrium analysis', 'Complex Economics: Individual and Collective Rationality' etc. Kirman is critical of the representative agent approach in economics as representative agent models simply ignore valid aggregation concerns; they sometimes commit the so-called fallacy of composition. According to him the representative agent disagrees with all individuals in the economy. In his view, the representative agent "deserves a decent burial, as an approach to economic analysis that is not only primitive, but fundamentally erroneous." A possible alternative to the representative agent approach to economics could be agent-based simulation models which are capable of dealing with many heterogeneous agents. [Wikipedia, September 2012]

Imre Kondor Imre Kondor is professor of physics at Eötvös Loránd University, Budapest. His present research field is the application of the methods of statistical physics to problems in quantitative finance, and the theoretical aspects of risk management and of financial regulation. Since 2000 he is the chairman of the Hungarian Association of Risk Managers. Imre Kondor authored above 90 publications with more than 700 independent citations. He is an external faculty of Ecole Centrale in Paris, and core team member of the Parmenides Center for the Study of Thinking in Munich. From 1998 to 2002, he was the head of the Market Risk Research department of Raiffeisen Bank in Budapest. His research areas are condensed Bose systems, phase transitions and critical phenomena, statistical physics of disordered systems, spin glasses, application of the methods of statistical physics to problems in economics and finance, the theory of financial risk. He is a member of the editorial boards of Fractals and Journal of Statistical Mechanics and formerly review editor of Journal of Banking and Finance.

Reiner Kümmel He Studied mathematics and physics at the Technical University of Darmstadt during 1959–1964 and received degree in physics in 1964. In 1968 he was awarded Ph.D. degree at the University of Frankfurt. He retired from service in 2004. Since then he is working as a lecturer in "Thermodynamics and Economics" at the University of Wuerzburg. He is noted for his 2011 book *The Second Law of Economics: Energy, Entropy, and the Origins of Wealth*, in which he argues that we need to begin to incorporate energy and entropy thinking into economics. Kummel's central thesis, which he calls the second law of economics, is that wealth creation by energy conversion is accompanied and limited by polluting emissions that are coupled to entropy production. He argues that we need to begin teaching students about the basics of economic thermodynamics, if we are to avoid, the shrinking of natural resources, environmental degradation, and increasing social tensions. Kummel's first paper on thermodynamics and economics was the 1977 "Energy and Economic Growth" ("Energie und Wirtschaftswachstum"), followed up by the 1980 book Growth Dynamics of the Energy-dependent Economy and the 1984 book Energy and Justice (Energie und Gerechtigkeit). In circa 2005, Kummel began teaching a course on "Economics and Thermodynamics", at University of Wurzburg, with notes about the energy slave concept, among other topics, soon thereafter.

Fabrizio Lillo Lillo is a research professor in Mathematical Finance at the Scuola Normale Superiore di Pisa since January 2011. He is also Assistant Professor of Physics at Palermo University, (Italy) and Professor at the Santa Fe Institute (USA, since 2009). He has been awarded the Young Scientist Award for Socio- and Econophysics of the German Physical Society in 2007. He is a member of the editorial board of the physics journal JSTAT. His research activity focuses on financial market micro-structure, high frequency finance, portfolio optimization, and application of network theory to Finance. More recently his research activity includes topics of systemic risk and application of Complex System Theory to air traffic management.

Thomas Lux Lux is an economist, who has a lot of contributions in the field of 'Econophysics'. He is ranked among the 1,000 most often quoted economist within the decade 1990–2000. He has a number of publications in physics journals like Nature, Physica A, European Journal of Physics B, International Journal of Modern Physics, Physical Review E etc. He has served as a referee to many journals related to physics besides economics journals: such as, European Physical Journal B, Europhysics Letters, Journal of Statistical Mechanics, Journal of Wavelets, Nature, Physica A, Physica D, Physical Review E, Physical Review Letters, Proceedings of the National Academy of Sciences of the U.S.A., Quantitative Finance, Statistical Papers, Review of Modern Physics, Science etc.

Subhrangsu Sekhar Manna Manna did his M.Sc. (1978–1980) from Science College, Calcutta University and Ph.D. from Saha Institute of Nuclear Physics in 1987. He had joined IIT, Mumbai in 1992 and worked up to 1997, from 1998 he is a faculty at S.N. Bose National Centre for Basic Sciences, Kolkata. His Main areas of work are Critical Phenomena in Complex Systems, Static and Dynamic Properties of Complex Networks, Self-organized Non-equilibrium critical Systems. He is the originator of the "Manna Model" of stochastic Self-Organized Criticality & also of the "Manna Universality Class" for dynamic critical behavior and is known for the "Chatterjee-Chakrabarti-Manna model of wealth distributions". He has published a large number of papers basically on complex networks.

Rosario Mantegna Rosario N. Mantegna is, today, recognized as one of the leading pioneer in the field of econophysics. He started to work in the area of the analysis and modeling of social and economic systems with tools and concepts of statistical physics as early as in 1990. He published the first econophysics paper in a physics journal in 1991. He also co-authored the first econophysics paper in Nature, in 1995. In 1999 he published the first book on econophysics. Just after Mantegna earned his tenured position in 1999, he founded the Observatory of Complex Systems (http://ocs.unipa.it), a research group of the Dipartimento di Fisica of Palermo University. Within econophysics he has investigated a wide range of topics. Examples are the following: (i) the statistical regularities of univariate time dynamics of high frequency price returns, (ii) the hierarchical structure and correlation based networks of a portfolio of stocks, (iii) the cross sectional analysis of price returns, (iv) the presence of an Omori law during the periods of time just after a financial crash, (v) the micro-structure aspect of the price impact and of the order book dynamics and (vi) the empirical detection of resulting strategies in the trading activity of market members and individual investors acting in a financial market.

Matteo Marsili Matteo Marasili, Abdus Salam ICTP, Triste is basically a statistical physicist at Abdus Salam International Centre for Theoretical Physics. His research interest includes Statistical physics, non-equilibrium critical phenomena, disordered systems, probability Interests theory and stochastic processes, complex networks, interdisciplinary applications of statistical physics, including modeling socio-economic phenomena and financial markets, game theory, and biological networks. His work (with D. Challet & Y. C. Zhang) on Minority Games (Oxford Univ. Press, 2004) has been a classic in the field.

Sergei S. Maslov Sergei S. Maslov is a group leader at Biology Department, Brookhaven National Laboratory located on Long Island in New York state. Earlier, he was in Condensed Matter Physics and Materials Science Department. He has empirically studied the cross-correlations of stock indices in a diverse set of 37 countries all over the world and found that the more globalized is the economy of a given country, the stronger it is coupled to the world stock index. He came up with a simple model of a limit-order driven market, where agents with equal probability trade stock at the market price or place limit orders, i.e. instructions to sell (buy) if stock price rises above (falls below) a predetermined price level. In spite of a minimalistic nature of this model (no strategies, or trader psychology etc.). He is in editorial board of Biology Direct and reviewer in the NIH Panel for Technology Centers for Networks and Pathways. He has many publications with more than 2600 citations.

Jaume Masoliver is a Professor in the Department of Physics at Barcelona University, Spain. His interest of research is mainly application of statistical physics in stock market studies. He has a number of papers published in many refereed journals like Applied Mathematical Finance, Taylor and Francis Journals, The European Physical Journal B, Quantitative Finance, Journal of Economic Behavior & Organization etc. His papers are mainly based on Scaling properties and universality of first-passage time probabilities in financial markets, First-passage and risk evaluation under stochastic volatility, multifractality in financial markets, financial time series etc. Five papers by this author were announced in NEP (New Economic Papers) sponsored by the School of Economics and Finance of Victoria University of Wellington. He has more than 104 publications with 211 citations.

Tiziana Di Matteo Tiziana is a Reader in Financial Mathematics in King's College, London. She did her Ph.D. in Physics in 1999 from Dipartimento di Fisica, Università di Salerno, Italy. She was Associate Professor (Level D), Applied Mathematics at Research School of Physical Sciences, Australian National University, Canberra, Australia. Her research interests are Econophysics, Application of methods from Statistical Physics to Finance, Complex Systems, Science of Networks etc. In Econophysics she has publications in refereed journals like New Journal of Physics, Physica A, Journal of Banking & Finance, European Physical Journal B, Quantitative Finance etc. basically on financial market studies.

Joseph McCauley McCauley is a Professor of Physics in department of Physics at Houston University and at the same time he also teaches econophysics. His present research interests are on Econophysics and complexity: economics and finance like Empirically based modeling of normal, liquid finance markets. He has written many books like Dynamics of Markets: the new financial economics (2009), Dynamics of Markets, Cambridge (2004) etc. Besides he has many papers in econophysics published in different journals like Physica A.

Jürgen Mimkes Mimkes is a professor at the physics department at the University of Paderborn, Germany. He has been dealing with the subject of the statistics and thermodynamics of social and economic systems since 1990. He calls his present field as "physical economics", especially macro and micro economics, and finance. He has many works related to sociological thermodynamics and economic thermodynamics like "Society as a Many Particle System" (1997), "Chemistry of the Social Bond" (2005), "Econophysics and Sociophysics" (multi authored, 2006), and many others. In 1994, Mimkes started to bring out an analogy of segregation in populations to the miscibility gap in solutions and alloys. His first paper was the "Binary

Alloys as a Model for a Multicultural Society" in the Journal of Thermal Analysis in 1995 in this area. In 1996, Mimkes authored a 110-page article on "Politics and Thermodynamics" also. According to him economic properties may be calculated in physical terms such as capital to energy, production to physical work, and GDP per capita to temperature, and production function to entropy.

Manipuspak Mitra Manipuspak Mitra is a professor of Economics at Indian Institute of Statistics, Kolkata. Besides economics he works on econophysics and published some papers in physics journals and book chapters on econophysics. His paper 'Statistics of the Kolkata Paise Restaurant problem' jointly with Asim Ghosh, Arnab Chatterjee and Bikas K. Chakrabarti with focus on Statistical Physics Modelling in Economics and Finance, was published in New Journal of Physics 12(075033) (2010). 'The Kolkata Paise Restaurant problem and resource utilization' jointly with Anindya Sundar Chakrabarti, Bikas K. Chakrabarti and Arnab Chatterjee was published in Physica A 388(12), 2420–2426 (2009).

Jean-Pierre Nadal Nadal is Director of Research at CNRS and also works at CAMS(Centre of Analysis of Mathematical Sociology) and at Ecole Normale Superier, France. His main research interests are Statistical Physics of Information Processing in Biological and Social Networks and Computational neuroscience—Cognitive science—Complex systems in social sciences. He has many papers on econophysics published in physics journals like Phys. Rev. E, The European Physical Journal B, Condensed Matter and Complex Systems (EPJB), Physical Review Letters and other journals like quantitative finance. His papers are based on housing market dynamics, demand and supply in markets, electoral behavior among urban voters, Computation in Neural Systems network etc. His papers have 3706 number of citations. He has edited many books published by springer.

Prasanta K. Panigrahi Prasanta K. Panigrahi completed his M.Sc. from Ravenshaw College, Cuttack and Ph.D. from University of Rochester, 1988. He is now working as Professor of physics at Indian Institute of Science Education and Research, Kolkata, since 2007. He is a referee for Physical Review Letters, Physical Review B, Journal of Physics, Pattern Recognition Letters, Physics Letter A, Pramana, European Physics Letters. Besides Field Theory he has also many publications in Econophysics basically on wavelet transforms and analysis of stock markets. He has published such papers in J. Quantitative Economics, Resonance, Phys. Rev. E, Pramana, Physica A, J. Phys. A: Math. Theory etc.

Lorenzo Pareschi Lorenzo Pareschi is a professor and chair of Department of Mathematics and Computer Science, University of Ferrara. He did his Ph.D. in Mathematics [1991–1995] from University of Bologna, Italy. His research interest includes traffic flows, econophysics, sociology, bio-mathematics, besides mathematics and and plasma physics. He has published many papers basically on Kinetic models for socio-economic dynamics of speculative markets and game theory in different refereed journals.

Marco Patriarca Partriarca did his Ph.D. in Physics at University of Perugia in 1993. Now he is a Professor at Instituto de Física Interdisciplinary Sistemas Complejos IFISC (CSIC-UIB) Campus Universitat de les Illes Balears. His current research activity mainly concerns Complex Systems and Statistical Mechanics, in particular neuronal systems, social dynamics, diffusion problems etc. His research activity focuses on some topics in the fields like Brownian motion and Complex Systems. He has published many papers in refereed journals like Physica A, Quantitative Finance, EPJ B etc. Besides he has diversified research interests such as Natural Sciences and Engineering, Physics and Technical Physics (diffusional processes, complex systems, material science, computational physics).

Josep Perello Josep Perello is a faculty member of University of Barcelona. His research interests include econophysics besides science communication and others. He has worked basically on stochastic volatility, risk analysis in hedge funds, stylised fact, Random walk formalism in financial markets, applications of thermodynamics to financial markets etc. He has published 47 papers in refereed journals.

Vasiliki Plerou Vasiliki Plerou was graduated from Boston University in 1996 and also received her Ph.D. in Physics in 2001 (thesis advisor H. Eugene Stanley). She is the recipient of 2003 young scientist award for socio- and Econophysics from DGP, AKSOE. Dr. Plerou has worked on a wide range of phenomenological problems in Econophysics. She contributed to the discovery of the inverse cubic power-law for the stock return distribution and the half-cubic law of volumes, which have found to be universal patterns of financial fluctuations. Dr. Plerou was one of the first researchers who applied random matrix theory, a powerful tool in diverse fields of theoretical physics, to investigating the correlation between stock returns. She was among the first to recognize that the spectrum of the correlation matrix can be separated into two categories describing random fluctuations and collective modes. Furthermore, Dr. Plerou made successful attempts to apply these ideas and results to the more pragmatic problem of portfolio optimization. Her empirical findings have added to the touchstones of modeling the stock market and this way have initiated a lot of new research during the past years, to the advantage of the field of Socio- and Econophysics. She has 92 publications many refereed journals and 1952 citations.

Tobias Preis Tobias Preis did his M.Sc. and Ph.D. in physics from Johannes Gutenberg University of Mainz. Tobias Preis is an Associate Professor of Behavioral Science and Finance at Warwick Business School. He is a computational social scientist focusing on analysis and prediction of social and financial complexity. He is the founder of Artemis Capital Asset Management GmbH, a proprietary trading firm based in Germany in 2007. Preis has quantified and modeled financial market fluctuations. His team has discovered a link between the number of Internet searches for company names and transaction volumes of the corresponding stocks on a weekly time scale. They have found that a relationship potentially exists between the economic success of a country and the information-seeking behavior of its citizens online. He has many published papers in refereed journals on econophysics and his works have a lot of media attentions. He is an academic editor of

PLoS ONE. He was the general secretary of "Physics of Socio-Economic Systems" Division of the German Physical Society (DPG) from 2009–2011.

Sidney Redner Sidney Redner is a professor of physics and department chair of physics at Boston University. His research interests are Condensed matter and statistical physics; stochastic processes; first-passage processes; chemical kinetics; transport in disordered media; percolation theory and disordered systems; dynamics of social systems; structure of complex networks. He has 237 publications in refereed journals and book chapters, one monograph, "A Guide to First-Passage Processes" (Cambridge University Press, 2001). He was/is in the editorial board of Journal of Statistical Physics, 2000–2002, 2008–, J. Informetrics, 2006–, European Physical Journal B, 2009–2011, J. Stat. Mech., 2004–2011, Journal of Physics A, 2005–2009, American Journal of Physics, 2005–2007, Physical Review E, 1992–1997. Also he was in the advisory Panel of Institute of Physics Publishing during 2001–2009 and Chair of APS Topical Group in Statistical and Nonlinear Physics during 2008–2009.

Peter Richmond Peter is a theoretical physicist and has wide experience of research, technology, innovation and management within both academic and commercial environments. He has been Visiting Professor in Trinity College Dublin, Ireland since 1998. During the 1980's he built a new department concerned with the strategic science underpinning food processing at the Institute of Food Research at Norwich. He has published over 120 papers in the technical literature. He is an Executive Editor of the Journal for the Science of Food and Agriculture. He has turned his attention to the new area of econophysics and has published a number of papers on Pareto-Zipf Law in Non-stationary Economies, Heterogeneous Interacting Agents, Power Laws are disguised Boltzmann Laws etc.

Bertrand M. Roehner Roehner is a professor of Physics at LPTHE, University of Paris. His Research interests are Speculative trading, comparative history, market integration, social bonds, suicide, intermarriage etc. He is the author of several books like *Theory of Markets* (published by Springer-Verlag), *Hidden collective factors by speculative trading* (published by Springer), *Patterns of speculation* (published by Cambridge University Press) etc. and about 60 papers in various scientific journals in the fields of economics, sociology and physics. He is a guest faculty at many Universities like the Institute of Economics (Copenhagen, September 1996), the Harvard Department of Economics (Summer 1994, Fall 1998), the Santa Fe Institute (2002), the Harvard Department of Sociology (Spring 2002) etc. He is the member of the editorial board of one of the main websites devoted to econophysics: www.unifr.ch/econophysics, referee for various economic and physical journals; e.g. Explorations in Economic History, Journal of Development Economics, International Regional Science Review, Journal of Economic Behavior and Physica A.

Barkley Rosser Barkley Rosser is a Professor of Economics at James Madison University in Harrisonburg, Virginia since 1988. His areas of interest are nonlinear economic dynamics including applications in economics of catastrophe theory, chaos theory and complexity theory. He introduced the concepts of chaotic bubbles, chaotic hysteresis, and econochemistry into economic discourse. He is the pioneer of providing a mathematical model of the period of financial distress in a speculative bubble (1991, op. cit., Chap. 5). He has published many books and about 150 journal articles, book chapters etc. He was an Editor of the Journal of Economic Behavior and Organization, from 2001–2010 and become Co-editor since 1/1/11.

Robert Savit Robert Savit is a Professor of Physics at Michigan University. His current research interests are in the area of nonlinear and adaptive systems. His group examines two types of issues. First, they study the nature of adaptive competition for scarce resources. They examine the deep phase structure that such systems exhibit, and explore the application of their insights to problems in group decision making in business, in social systems, resource allocation problems, and evolutionary systems. Professor Savit also engages in a series of experiments with groups of human subjects that study the problem of resource allocation, competition and emergent coordination. Second, his research group considers data analysis problems of nonlinear systems. Using new techniques developed in the context of dynamical systems, Professor Savit's group studies a number of intrinsically nonlinear systems.

Enrico Scalas Scalas is an External Scientific Member of BCAM—Basque Center for Applied Mathematics, Spain. According to him the evidence-based financial mathematics is the core of contemporary financial mathematics consisting of Black-Scholes theory for derivative valuation and of Markowitz method for portfolio selection. In both cases, theoretical assumptions were falsified by empirical analyses. The problem of finding suitable stochastic models for price fluctuations in financial markets is still open and the object of active research. It can be partially solved by means of interdisciplinary work involving experimental and behavioral economics as well as probability theory and statistics. He has published many papers on econophysics basically on financial time series and stylized facts.

Parongama Sen Sen, a Kolkata based statistical physicist did her M.Sc. from Calcutta University in 1986 and Ph.D. from Saha Institute of Nuclear Physics in 1993. Now she is working in Calcutta University as an Associate professor. She has got the APS-IUSSTF Professorship Award (2012–2013) for complex system research. Her research interests include Phase transitions and critical phenomena in magnetic systems, networks, quantum systems, percolation etc. Dynamical phenomena in complex physical and social systems. She has a number of publications in refereed journals mainly based on Ising model and kinetic model of wealth exchange. She has published a book 'Quantum Ising Phases and Transitions in Transverse Ising models, Lecture Notes in Physics with B. K. Chakrabarti and A. Dutta, Springer-Verlag, 1996 and a book entitled 'Sociophysics: An introduction', with B. K. Chakrabarti, Oxford Univ. Press (2013; in press).

Ingve Simonsen Simonsen is a professor of physics at NTNU in Trondheim, Norway. His research interest includes topics of statistical physics, like disorder systems, stochastic optics, and fractals; computational physics; complex systems, including econo-physics and the study of complex networks. Within complex systems, his interests have mainly been focused on random networks (e.g. the power transmission grid) and the study and modeling of economical systems (econo-physics). A primary concern has been how the local properties of a complex network influence the global structural and dynamical transport properties of the system. In the field of econophysics and sociophysics he has experience from both the analysis of empirical data and the construction and study of (toy) models used to identify and analyze specific mechanisms.

Sitabhra Sinha is a Professor in the Physics group of the Institute of Mathematical Sciences (IMSc) at Chennai (formerly known as Madras), and adjunct faculty of the National Institute of Advanced Studies (NIAS), Bangalore. His areas of research fall broadly under complex systems, nonlinear dynamics and theoretical & computational biophysics. He is working on Physics of social and economic phenomena—genesis of scaling behavior (e.g., Pareto law of wealth and income distribution, "inverse cubic law" of stock price fluctuations in financial markets, etc.) in economics—emergence of popularity through self-organization in a population of agents—phase transitions in collective (or social) choice relevant to financial markets, movie revenue distribution and electoral behavior. He organizes Econophysics Conferences at IMSc also. He is the co-author of many books like 'Econophysics: An Introduction (Wiley, 2010)'.

Didier Sornette Didier Sornette is Professor on the Chair of Entrepreneurial Risks at Swiss Federal Institute of Technology Zurich (ETH Zurich). He is also a professor of the Swiss Finance Institute, a professor associated with both the department of Physics and the department of Earth Sciences at ETH Zurich, an Adjunct Professor of Geophysics at IGPP and ESS at UCLA. He was previously jointly a Professor of Geophysics at UCLA, Los Angeles California and a Research Director on the theory and prediction of complex systems at the French National Centre for Scientific Research. His present Fields of research interest include Social sciences, finance and economics: decision theory, behavioral decision making, societal risks, bubbles and crashes, large and extreme risks, theory of derivatives, portfolio optimization, trading strategies, insurance, macro-economics, agent-based models, and market microstructures. Physics of complex systems and pattern formation in spatio-temporal structures, dynamical system theory, pattern recognition, self-organized criticality, prediction of complex systems, time series analysis; Rupture in random media, theory of earthquakes and of tectonic deformations, rupture and earthquake prediction.

Wataru Souma Wataru is an associate Professor at Nihon University. Earlier he was also a faculty of integrated human studies of Kyoto University, Japan. His research interest includes Chemical Physics & Material Physics, Statistical Physics, Mathematical & Quantitative Methods. He has a number of publications on econophysics basically on wealth distribution in societies, business cycle, distribution of labour productivity etc. He has more than 60 publications with 331 citations. His papers have been published in journals like Quantitative Finance, Journal of The Japanese and International Economies, Physical Review E, Applied Financial Economics etc. He has written some reviews in Econophysics, agent based models and a book 'Econophysics and Companies: Statistical Life and Death in Complex Business Networks'.

H. Eugene Stanley Gene Stanley is University Professor and Director of the Center for Polymer Studies at Boston University. Stanley works in collaboration with students and colleagues attempting to understand puzzles of interdisciplinary science. His main current focus is to understand the anomalous behavior of liquid water in bulk, nano-confined, and biological environments. He has also worked on a range of other topics in complex systems, such as quantifying correlations among the constituents of the Alzheimer brain, and quantifying fluctuations in non-coding and coding DNA sequences, inter beat intervals of the healthy and diseased heart. He has published the first book on Econophysics co-authorized with R.N. Mantegna 'Introduction to Econophysics: Correlations & Complexity in Finance' (Cambridge University Press, Cambridge, 2000). This book has also been translated into Japanese, Polish, Chinese, Indonesian and Russian languages and it has more than 1221 citations. He has a large number of publications and citations. Besides he has many books and published many papers on econophysics in refereed physics journals with Mantegna, Gopikrishanan and others. He is an editor of many journals like Physica A, Graduate Texts in Physics, Springer-Verlag (Editor in Chief), Quantitative Finance, Int. J. Theor. & Appl. Finance, Finance and Economics etc. Two of his research articles are reproduced in The Physical Review: The First Hundred Years. A Selection of Seminal Papers and Commentaries, edited by H.H. Stroke (AIP Press, New York, 1995).

Dietrich Stauffer is a German theoretical physicist noted for his application of statistical physics and in particular computational physics in the areas of econophysics (since 1998) and sociophysics (since 2000). In circa 1995, Stauffer read American physicist Eugene Stanley's papers on econophysics, and when in 1997 Cont and Bouchaud applied percolation theory (Stauffer's specialty) to stock market fluctuations, he jumped onto that subject. His 1999 book Evolution, Money, War and Computers, co-written with S. Moss de Oliveira and P.M.C. de Oliveira, outlines the non-traditional applications (evolution, money and war, etc.) of computational statistical physics. In 2003, international conference Unconventional Applications of Statistical Physics: Physics of Random Networks, Econophysics, and Models of Biophysics and Sociophysics was organized in honor of the 60th birthday of Stauffer. Stauffer's 2011 article "Statistical Physics for Humanities: A Tutorial" gives a bit of historical overview, contains sections such as "humans are neither spins nor atoms", "schelling model for social segregation", on the work of Thomas Schelling, extols on an Ising model of the physics of human behavior or choice, and concludes with a suggested outline to use Fortran to build computer simulations for a type of human statistical physics.

Attilio Stella Attilio Stella is a professor of Physics at University of Padova. He did his Ph.D. at the Katholieke Universiteit Leuven, Belgium. Besides he was in many other Universities in different positions. He is the author of about 110 papers in international refereed journals, and about 20 contributions to proceedings and books. His research activity is in various fields of statistical physics.

Hideki Takayusu Econophysics was introduced to Japanese physics community by Hideki Takayasu (Sony CSL), whose pioneering work may be, the very first mathematical modeling and simulation of stock market in 1992, and application of the Langevin equation to the stochastic process and the derivation of the power law, among others. He also publicised the approach and fruits of econophysics widely in physics community through lectures and organization of conferences. Prof. Aoyama, Prof. H. Iyetomi etc., who studied various areas of physics, met econophysics under his influence. Since then, Kyoto Econophysics Group has been conducting research with emphasis on real economy, and has published two books in Japanese , and a book in English.

Giuseppe Toscani Toscani is a full Professor of Mathematical Physics Faculty of Sciences, University of Pavia. His research interests include Mathematical and numerical methods in kinetic theory of rarefied gases, Granular gases, Statistical Mechanics, Diffusion equations, Hyperbolic systems and applications, Mathematical modeling in socio-economic and related problems. He has published two books and edited some others. One of his edited book is 'Mathematical Modeling of Collective Behavior in Socio-Economic and Life Sciences, Birkhauser, Boston (2010)'. He has published many papers and book chapters on econophysics in refereed journals and books.

Victor Yakovenko Yakovenko is one of the pioneers of 'Econophysics'. He is a Professor of Physics, University of Maryland, USA. Besides his research interest on unconventional superconductors and others, he has also worked on Econophysics. He has published invited review articles in the book Encyclopedia of Complexity and System Science, edited by R.A. Meyers, Springer (2009) and many articles on 'Econophysics' in many other edited books. He has written a book entitled "Classical Econophysics" with co-authors published by Routledge (2009). He has published many papers in refereed journals on 'Econophysics', basically application of statistics to economics and finance. Many of his articles have attracted media and have been published in news papers and periodicals written by reporters in American Scientist, Australian Financial Review, New Scientist, The Newyork Times Magazine etc. He has refereed more than 292 articles in many physics and economics journals.

Yi-Cheng Zhang Cheng is a Professor of Physics in Fribourg University. His research interests are Physics, Information Science, and Complex Systems Science. He has published a large number of papers and his papers have more than 12456 numbers of citations. His paper, Dynamic scaling of growing interfaces, published in Physical Review Letters 56 (9), 889–892 in 1986 has as many as 3451 number of citations. He has works on multidisciplinary statistical mechanics, Minority games: interacting agents in financial market, Interacting individuals leading to Zipf's law etc. in journals like Physical Review E, Physica A, Physical Review Letters, Physics reports, EPL (Europhysics Letters) etc.

Other Contributors in the Field of Econophysics There are many other Physicists, Mathematicians and Economists, who have their valuable contributions in this emerging field of 'Econophysics'. Their contributions are no way less important than the scientists listed in previous pages. They have published a number of papers in refereed journals and some of them have written books on Econophysics. They are P.L. Krapivsky, Gopikrishnan from USA, Paul Ormerod, D. Brockmann from England, Sonia R. Bentes from Portugal, Yuichi Ikeda, Naoya Sazuka, Aki Hiro Sato from Japan, Marc Potters from France, R. Kitt, J. Kalda from Estonia, E. Heinsalu from Spain, Tongkui Yu, Honggang Li from China, Janusz Miskiewicz from Poland, Stefano Balietti from Hungary, Giorgio Israel G.D. Maldarella, Paolo Laureti, Franz Silvio Frank, Damien Challet from Italy, Juan C. Ferrero from Argentina, Raj Kumar Pan, Kausik Gangopadhyaya, Y. Sudhakar, Anita Mehta, P.K. Mohanty, Amit Bhaduri, Abhirup Sarkar, Indrani Bose from India etc. Thus people from different parts of the world have contributed in this field and it is difficult to say when and where econophysics took birth. Some people say it started from Osborne, some people say it is Mandelbrot, some say it is Takayusu and so on. But it is true that after 1995 it was institutionalised when the term 'Econophysics' was neolised by Stanley at Kolkata and many papers started publishing by physics journals. Other torch bearers are Hideki Takayusu in Japan, Y.-C. Zhang in China, Bikas Chakrabarti in India, Peter Richmond in Ireland, Paul Ormerod in England, Enrico Scalas in Spain, Bertrand M. Roehner in France, Jürgen Mimkes in Germany, Victor Yakovenko in USA, János Kertész in Hungary, Didier Sornette in Switzerland etc.

14.2 Bridging the Gap

To bridge the gap between natural science and social science and to adopt the principles and laws of natural science into social science, many people have tried in different ways. Some institutes have been specially established, books of interdisciplinary nature were published. Besides research in interdisciplinary subjects, courses have also been opened. Similarly Workshops, Conferences and Seminars were also organized in different parts of the world. To such events speakers from both natural sciences and social sciences were invited to deliver their talks, which helped in interaction among the two communities. Moreover, Journals publishing papers on natural science started publishing papers on econophysics. In this section we will deal with those, which have helped in bridging the gap between the natural science and social science. Such as Institutes, Books, Workshops, Conferences and Seminars, Journal publishing papers on econophysics, Opening of courses, Awards in Econophysics etc.

14.2.1 Institutes

Indian Statistical Institute Indian Statistical Institute (ISI) was first set up by Prasanta Chandra Mahalanobis in the Presidency College in Kolkata in 1932. The Institute is now considered as one of the foremost centres in the world for training and research in statistics and the related sciences. Under the leadership of Professor P.C. Mahalanobis, the Institute also initiated and promoted the interaction of statistics with natural and social sciences to unfold the role of statistics as a key technology which explicated the twin aspects of statistics—its general applicability and its dependence on other disciplines for its own development. ISI is probably the first such institute in the world, which brought closeness between natural and social science with the help of statistics. The major objectives of the Institute, as given in its Memorandum, are

(a) To promote the study and dissemination of knowledge of statistics, to develop statistical theory and methods, and their use in research and practical applications generally, with special reference to problems of planning of national development and social welfare;
(b) to undertake research in various fields of natural and social sciences with a view to the mutual development of statistics and these sciences;
(c) To provide for, and undertake, the collection of information, investigation, projects and operational research for purposes of planning and the improvement of efficiency of management and production.

Since 1950 the institute is engaged in collection and analysis of information on social, economic and demographic characteristics in India through the National sample surveys. The Institute has acquired a special distinction in India for its activities. During 1950s the interdisciplinary nature of teaching in the Institute was evolved through the guidance of stalwarts such as Sir Ronald A. Fisher, Professor P.C. Mahalanobis and Professor J.B.S. Haldane, with the encouragement of Professor Satyendra Nath Bose (Physics) who was the President of the Institute for a long time. The Indian Statistical Institute Act was amended in 1995 empowering it to grant degrees and diplomas in statistics, mathematics, quantitative economics, computer science and such other subjects related to statistics as may be determined by the Institute from time to time. The B. Stat. (Hons.) and the M.Stat. degree programs in statistics were introduced in the Institute in the year 1960 with the philosophy that the academic training of a statistician should encompass the basic principles of statistics along with its theoretical and methodological development, not merely in abstract formulation, but also in relation to concrete problems arising from natural and social sciences. Master of Science degree in quantitative economics [M.S.(QE)] was introduced in 1996. The Institute is engaged in significant research activity in many other disciplines, such as, population studies, physics, agricultural & ecological sciences, geology, biological anthropology, human genetics, linguistics, psychometry, sociology and information Science. In all these disciplines, much emphasis is given on interdisciplinary research and collaborative work with the statisticians of the Institute. The Institute thus conjures up a symbiosis of pure, applied and interdisciplinary research involving various areas of statistics, mathematics, quantitative economics, computer science, other natural and social sciences, statistical quality control and managerial decision making. This symbiosis has been systematically reflected in the teaching and training programs of the Institute [11]. Indian Statistical Institute thus can be considered as a pioneering institute in the world introducing interdisciplinary subjects probably on experimental basis and proved to be successful.

The Santa Fe Institute Several distinguished researchers in natural sciences and economics, like George Cowan, David Pines, Stirling Colgate, Murray Gell-Mann, Nick Metropolis, Herb Anderson, Peter A. Carruthers, and Richard Slansky started the Santa Fe Institute in USA in 1984. Researchers are trying to discover the common, fundamental principles in complex adaptive systems as varied as global climate, financial markets, ecosystems, the immune system, and human culture etc. Today the Institutes's researchers are approaching these problems firmly grounded in the quantitative methods of physics, chemistry, and biology. The 1990's at SFI was marked by an expanding research portfolio. The economics program remained strong and grew into nearly a dozen subsidiary pursuits. New inquiries into political behaviors and human culture further expanded SFI's work in the social sciences. Programs in biology expanded as well, exploring new territory in genomics, neurobiology, viral dynamics, biochemical networks, learning and memory in the immune system, and complex systems in ecology. Adaptive computation thrived, as did a dozen new programs in learning and cognition. Theories from physics and evolution began to play a greater roles in all of the Institute's research as its researchers sought to find parallels among fields and describe, at least metaphorically, complex behaviors they observed. True transdisciplinary research also is a norm at SFI today. The Institute supports programs that seek, for example, to understand financial market dynamics through the principles of evolution and ecology. It explores relationships among innovation in technology, genes, and human culture. Mathematics and network theory are applied to complex problems as diverse as disease propagation, terrorism, the Internet, and molecular signaling. Human culture change is studied as a biologist would examine genetic mutation. Changes in language are studied across time and space scales never before considered. A growing program in sustainability draws from archaeology, paleontology, sociology, psychology, ecology, and physics, among others. But complexity science itself is still emerging and, by its nature, continually probing the seams between mainstream disciplines and the boundaries of human knowledge. In 28 years it has transformed the scientific landscape and made possible the pursuit of new understandings of the complex, messy world in which we live. Today it is asking the questions humankind will need answers to 25 years from now [12].

Institute for New Economic Thinking (INET) It is a New York-based nonprofit foundation started by George Soros in 2009 to broaden and accelerate the development of a new field of economic thought that will lead to real world solutions to the great challenges of the 21st century. The havoc wrought by our recent global financial crisis has vividly demonstrated the deficiencies in our outdated economic theories and shown the need for new economic thinking—right now. In the wake of the 2008 financial crisis and the ongoing Euro crisis, a number of leading economists, policymakers, and business leaders have called for a fundamental re-think in economics. INET is a global community of thousands of new economic thinkers, ranging from Nobel Prize-winning economists to teachers and students, all attracted by the promise of a free and open economic discourse. On 12th April 2012 in Berlin the Oxford Martin School and the Institute for New Economic Thinking (INET) announced that they have joined forces to create INET@Oxford, a major new interdisciplinary research centre at the University of Oxford. INET@Oxford builds on an existing INET research programme at Oxford. The centre will have over 40 leading academics involved and will aim to stimulate innovation and debate in economics, support visionary interdisciplinary research, and contribute to the education of the next generation of economists as well as business and government leaders. INET@Oxford will seek to bring together thinking from across academic disciplines in its approach to economics. In addition to economists, the centre will work with physicists, biologists, psychologists, anthropologists and others across the physical and social sciences. An important part of the centre's mission will be to facilitate the application of its research to critical economic policy problems. The centre will engage with leaders from government and business. This new centre will focus on addressing some of the greatest economic challenges we face ranging from avoiding future financial crises to ensuring that the positive potential of globalization is realized and its risks mitigated. INET@Oxford will host several complementary research streams and will be able to provide expert comment and analysis to journalists and policy-makers in the research areas: Economic Modeling, Complexity Economics, Ethics and Economics, Global Economic Development, Employment and Equity, New Models of Economic Growth, Curriculum and Teaching Development [13].

14.2.2 Journals Publishing Econophysics Papers

The papers on 'Econophysics' started to be published mainly after it was institutionalized. The journals, which have started publishing the papers are Science, Nature, Physical Review Letters, Physical Review, Physica A, The European Physical Journal, International Journal of Modern Physics C, The European Physics Communications. Again journals publishing economics and financial mathematics papers also started publishing papers on 'Econophysics' although their numbers are low. Such journals are Quantitative Finance, Journal of Economic Behavior & Organization, Journal of Economic Interaction & Co-ordination etc. We can now find many papers in arXiv also.

14.2.3 Some of the Books Published on Econophysics

Many books have been published on Econophysics after it was institutionalized in 1995. The books are written by the writers from both disciplines; economics and physics and have helped bringing both subjects closure to each other. Some of these books are

- An introduction to econophysics by H. You et al. Publisher: China Renmin University Press (1991)
- Introduction to Econophysics by Rosario Mantegna, H. Eugene Stanley, Publisher: Cambridge University Press (1999)
- Dynamics of Markets: Econophysics and Finance by Joseph L. McCauley, Publisher: Cambridge University Press (2004)
- Quantitative Finance for Physicists: An Introduction (Academic Press Advanced Finance) by Anatoly B. Schmidt (2004)
- Why Stock Markets Crash: Critical Events in Complex Financial Systems by Didier Sornette and D. Sornette, Publisher: Princeton University Press (2004)
- Patterns of Speculation: A Study in Observational Econophysics by Bertrand M. Roehner, Publisher: Cambridge University Press (2005)
- Econophysics and Sociophysics: Trends and Perspectives by Bikas K. Chakrabarti, Anirban Chakraborti and Arnab Chatterjee, Publisher: Wiley (2006)
- Introduction to Econophysics: Correlations and Complexity in Finance by Rosario N. Mantegna and H. Eugene Stanley, Publisher: Cambridge University Press (2007)
- Consumer, Firm, and Price Dynamics: An Econophysics Approach: Modeling by Economic Forces by Matti Estola, Publisher: VDM Verlag (2008)
- Theory of Financial Risk and Derivative Pricing: From Statistical Physics to Risk Management by Jean-Philippe Bouchaud, Publisher: Cambridge University Press (2009)
- Econophysics: An Introduction (Physics Textbook) by Sitabhra Sinha, Anirban Chakraborti, Bikas K. Chakrabarti, Publisher: John Wiley & Sons (2010)
- Finitary Probabilistic Methods in Econophysics by Ubaldo Garibaldi, Publisher: Cambridge University Press (2010)
- Stochastic Processes: From Physics to Finance by Wolfgang Paul, Publisher: Springer (2010)
- The Second Law of Economics: Energy, Entropy, and the Origins of Wealth (The Frontiers Collection) by Reiner Kümmel, Publisher: Springer (2011)
- Econophysics and Companies: Statistical Life and Death in Complex Business Networks by Hideaki Aoyama, Yoshi Fujiwara, Yuichi Ikeda and Hiroshi Iyetomi, Publisher: Cambridge University Press (2011)
- Essentials of Econophysics Modelling by Frantisek Slanina, Publisher: Oxford Higher Education (Jan 2011)
- Classical Econophysics by Allin F. Cottrell, Paul Cockshott, Gregory John Michaelson and Ian P. Wright, Publisher: Taylor & Francis (2011)
- Econophysics of Income and Wealth Distributions by Bikas K. Chakrabarti, Anirban Chakraborti, Satya R. Chakravarty, Arnab Chatterjee, Publisher: Cambridge University Press (2013)

14.2.4 Special Issues on Econophysics [14]

- Physica A 269(1) [1999]
- International Journal of Theoretical and Applied Finance 3(1) [2000]
- European Physical Journal B 20(4) [2001]
- European Physical Journal B 27(2) [2002]
- Physica A 344(1) [2004]
- Physica A 382 [2007]
- European Physical Journal B 55(1) [2007]
- Journal of Economic Dynamics and Control 32(1) [2008]
- Complexity 14(3) [2009]
- Science and Culture 76(9–10) [2010]
- AUCO Czech Economic Review 4(3) [2010]

14.2.5 Courses Offered in Different Universities

Leiden University, Netherlands Physics department of Leiden University has started a course on Econophysics, an optional course for third-year bachelor's students from May, 2011. Dr Diego Garlaschelli has been appointed to become Leiden's first Professor of Econophysics, on the marriage between physics and the financial world [15]. The course structure is Introduction to Econophysics (historical background, interaction between Physics and Economics, past and present aims of the field). It comprises of the subjects like Stochastic processes and time series,Stylized facts of single financial time series, Cross-correlations among multiple time series, Complex networks and interactions among economic agents, Network models of wealth distribution and market behaviour and International economic interactions: the World Trade Web. The Reference book for the course is: 'Econophysics: An Introduction' (Physics Textbook) by S. Sinha, A. Chatterjee, A. Chakraborti, B.K. Chakrabarti (Publisher: Wiley-VCH, 2010).

University of Houston Econophysics is being taught from Second Year of BS-MS course at University of Houston with 'Introduction of Econophysics', in third year courses like 'Simulation of Economic Systems', Financial Engineering and Derivatives etc. Econophysics is being taught by Prof. Kevin E. Bassler, Gemunu H. Gunaratne, Joe McCauley, Donald J. Kouri, Lawrence S. Pinsky, David R. Criswell, Miguel Castro and Valery A. Kholodnyi from physics and other departments.

MMEF Course at University of Paris Mathematical Models in Economics and Finance (MMEF) is a one-year, full-time, international programme taught in English. The degree is devoted to the training of students in the use of mathematical models in economics and finance. The training is built around a large spectrum of courses, which aim to give the knowledge for the modeling of economic and financial problems, for the mathematical formalization and the tools for the numerical solutions. The course, Theory of finance 2 has the objective of introducing to theories of financial markets inspired by complex-system science (especially Statistical Physics). The aim of this course is to give an introduction to the dynamics of financial markets via the powerful tools of Mathematical Physics. It comprises of topics like Brownian motion, critical phenomena, self-organized criticality, market phase transitions etc. Reference books for this course are Dynamics of Markets—Econophysics and Finance by J.L. McCauley, Cambridge University Press, Theory of Financial Risk and Derivative Pricing—From Statistical Physics to Risk Management by J.P. Bouchaud, M. Potters, Cambridge University Press and Econophysics: An Introduction, by S. Sinha, A. Chatterjee, A. Chakraborti, B.K. Chakrabarti, Wiley.

University of Silesia The undergraduate (6 semesters) and postgraduate (4 semesters) programme in Econophysics was launched in the academic year 2009/2010 in University of Silesia as the only programme of this kind in Poland. The undergraduate programme allows students to learn the foundations of physics, mathematics, computer science and economics, and acquire specialist knowledge in the field of financial engineering, risk management and quality engineering. The postgraduate programme covers unique selection of courses in modern physics, mathematics, economics and statistics, with special emphasis on the applications and practical aspects thereof. To complete the programme, students write and defend a Master's thesis, which enables them to receive a Master's degree in Econophysics.

Econophysics teaching at the University of Wrocław A new econophysics B.Sc. course established at the University of Wrocław and organized as a result of cooperation between two university departments: Department of Physics and Department of Economic Sciences. Stress is given to interdisciplinary aspects of this course and its continuation to M.Sc. degree in Physics or M.E. (Master of Economics) degree.

Some Important Centres of Econophysics Research:

- **Boston University, USA**
- **Santa Fe Institute, USA**
- **Saha Institute of Nuclear Physics, India**
- **Ecole Centrale Paris, France**
- **University of Maryland, UK**
- **University of Palermo, Italy**
- **Kyoto University, Japan**
- **Leiden University, Netherlands**

14.2.6 Awards for Social Scientists

The Human Resource Development Ministry (India) has proposed to constitute an award for social scientists on the lines of the Bhatnagar awards for scientists. HRD

Minister Kapil Sibal said (2012) this while speaking at a conference organized by the Indian Council of Social Science & Research. As proposed there will be ten annual awards to recognize advancement in the field of social science. These awards will be known as the Amartya Sen awards.

14.2.7 Award in Econophysics

The division on "Physics of Socio-Economic Systems" (SOE, previously AK-SOE/AGSOE) is committed to support the scientific exchange between the scientific disciplines involved. It has currently more than 200 members and organizes a spring conference with more than 50 contributions each year. During this conference, it features outstanding international speakers not only from socio- and econophysics, but also from sociology and economics. Furthermore, young scientists are awarded for outstanding innovative work in this field. Names of the awardees are: 2002—Damien Challet, 2003—Vasiliki Plerou, 2004—Illes Farkas, 2005—Reuven Cohen, 2006—Xavier Gabaix, 2007—Dr. Katarzyna Sznajd-Weron, 2008—Dr. Frabrizio Lillo, 2009—Duncan Watts, 2010—Dirk Brockmann, 2011—Santo Fortunato, 2012—Arne Traulsen, and 2013—Vittoria Colizza.

14.2.8 Workshops, Conferences and Seminars

The inaugural meeting on Econophysics was organized in 1998 in Budapest. Currently, the almost regular meeting series on the topic include: Econophysics Colloquium, in Kiel, Germany, ESHIA/WEHIA, ECONOPHYS-KOLKATA, APFA, Dublin econophysics conference, Asia-Pacific Econophysics Conference, world econophysics conference in Canberra, Australia. The International Conference on Econophysics (ICE), China etc. Econophysics Colloquium was started from 2004 and still continuing as an annual event. ENEC, "Econophysics, New Economics and Complexity" International Conference is organized by the Hyperion University and the Hyperion Research and Development Institute in Bucharest, Romania. The main goal of the Conference is to provide an opportunity for scientists, researchers and professionals from econophysics, new economy and science of complexity to come together and present original papers and new ideas in the mentioned topic areas. Econophysics—Kolkata series started from 2005 by Prof. Bikas K. Chakrabarti, is an annual event. There was also a dedicated *Computing for Finance* conference held at CERN on November 21, 2007, which was specifically aimed at how to use the Grid for financial computing.

14.3 Major Research Topics

In this section, I will deal with some breakthrough research topics, which became instrumental and cleared the path for neolizing econophysics and also mention some important research topics after the term 'Econophysics' came into existence. I have already discussed at length how econophysics was very much present in preclassical and classical era and I do not want to repeat those. So here I want to discuss mainly the developments just prior to the modern era and the happenings after econophysics was institutionalised. I feel it right to start from Osborne's log normal distribution, which, in the opinion of some scientist can be considered as the origin of econophysics.

Brownian Motion in the Stock Market '*The logarithms of common-stock prices, and the value of money, can be regarded as an ensemble of decisions in statistical equilibrium, and that this ensemble of logarithms of prices, each varying with time, has a close analogy with the ensemble of coordinates of a large number of molecules*'—suggested by Osborne [16]. He used a probability distribution function and the prices of the same random stock choice at random times and became able to derive a steady state distribution function, which is precisely the probability distribution for a particle in Brownian motion. A similar distribution holds for the value of money, measured approximately by stock market indices. Thus, it was shown in his paper that prices in the market did vary in a similar fashion to molecules in Brownian motion. According to Professor Joseph L. McCauley (University of Huoston), Osborne first introduced the lognormal stock pricing model in 1958 and opined that he should be honored as the first econophysicist.

Price Changes in financial markets do not follow a Gaussian Distribution Mandelbrot found that price changes in financial markets did not follow a Gaussian distribution, but rather Lévy stable distributions having theoretically infinite variance. He found, for example, that cotton prices followed a Lévy stable distribution with parameter α equal to 1.7 rather than 2 as in a Gaussian distribution. "Stable" distributions have the property that the sum of many instances of a random variable follows the same distribution but with a larger scale parameter [17]. Some parts of the review of his book 'The (Mis)behavior of Markets' is presented below.

The roots of the book The (Mis)behavior of Markets go back to 1961 when Mandelbrot was a new researcher at IBM. Among other things, he was working on using computers to analyze the distribution of income in a society. Mandelbrot's work echoed the work of Vilfredo Pareto and showed that many economic factors, including wealth, are distributed according to an inverse power law. While at Harvard to give a talk on his work, Mandelbrot saw a diagram on a chalk board that mirrored the distributions he was seeing for income. But in this case the diagram involved cotton prices. With access to IBM's computers (and programmers) Mandelbrot started studying cotton prices. Around the time Mandelbrot was doing his work on cotton prices the work of Louis Bachelier was being rediscovered and embraced by the academic economics community. Bachelier claimed that the change in market prices followed a Gaussian distribution. This distribution describes many natural features, like height, weight and intelligence among people. The Gaussian distribution is one of the foundations of modern statistics. If economic features followed a Gaussian distribution, a range of mathematical techniques could be applied in economics. With Bachelier's work and some modern reinforcement, a new era in economics was born. Unfortunately, as Mandelbrot points out in The (Mis)behavior of Markets, the foundation of this new era of economics was rotten. As anyone who has repeatedly put money at risk in a market over a long period of time knows, market behavior does not reflect the well behaved Gaussian models proposed by economists in the 1970s and 1980s. There are far more market bubbles and market crashes than these models suggest. The change in market prices does not follow a Gaussian distribution in a reliable fashion. Like income distribution, market statistics frequently follow a power law. When a graph is made of market returns (e.g., profit and loss), the curve will not fall toward zero as sharply as a Gaussian curve. The distribution of market returns has "fat tails". The "fat tails" of the return curve reflect risk, where large losses and profits can be realized.

Inverse Quartic Power Law In the late 1990s Paramaswaram Gopikrishnan and Vasiliki Plerou, who were then graduate students working at Boston University in the US, decided to analyse every transaction of every single stock in the major US markets. At that time, the analysis of such huge data sets was not as commonplace as it is today and required a significant upgrade to their university's computer system to complete the task. Using the extra computer resources, the two students constructed a histogram that displayed the number of times the stock market changed by a certain amount, plotted as a function of that amount. They did this by analyzing 1000 different stocks each consisting of 200,000 data points. What Gopikrishnan and Plerou found was that large transactions are more common than they had expected, with the tail of their histogram not being Gaussian but following an "inverse quartic power law". This law means that if there is a probability p of, say, a $5 price change occurring, then the probability of a $10 price change is $p/2^4$, i.e. $p/16$. This inverse quartic law excels at describing the probability of very rare events, such as those occurring once every few decades. Events corresponding to 100 standard deviations, for example, have a probability of about 10^{-350} with a Gaussian model, but a far more realistic likelihood of 10^{-8} (i.e. one in a hundred million) with the inverse quartic law [18].

Economic Inequality: Is It Natural? In this paper, Arnab Chatterjee, Sitabhra Sinha, and Bikas K. Chakrabarti, India, have studied and reported that the income and wealth distribution of various countries clearly establish a robust feature: Gamma (or log-normal) distribution for the majority (almost 90–95 %), followed by a Pareto power law (for the richest 5–10 % of the population). They have shown that this 'natural' behavior of income inequality comes from a simple 'scattering picture' of the market, when the agent in the market have got random saving propensity. Models studied in physics (in kinetic theory of gases), more than a hundred years ago, helped them in formulating and understanding these 'natural' behavior of the markets.

Exponential Distribution of Income "Evidence for the exponential distribution of income in the USA" by A.A. Dragulescu and V. M. Yakovenko [19]. In this paper the authors have demonstrated that the distribution of individual income in the USA is exponential using tax and census data. Their calculated Lorenz curve without fitting parameters and Gini coefficient $1/2 = 50\ \%$ agree well with the data. From the individual income distribution, they have derived the distribution function of income for families with two earners and have shown that it also agrees well with the data. The family data for the period 1947–1994 fit the Lorenz curve and Gini coefficient $3/8 = 37.5\ \%$ calculated for two-earners families.

'Thermal' and 'Superthermal' Income Classes "Temporal evolution of the 'thermal' and 'superthermal' income classes in the USA during 1983–2001" by A.C. Silva and V.M. Yakovenko [20]—Personal income distribution in the USA has a well-defined two-class structure. The majority of population (97–99 %) belongs to the lower class characterized by the exponential Boltzmann-Gibbs ("thermal") distribution, whereas the upper class (1–3 % of population) has a Pareto power-law ("superthermal") distribution. By analyzing income data for 1983–2001, the authors have shown in this paper that the "thermal" part is stationary in time, save for a gradual increase of the effective temperature, whereas the "superthermal" tail swells and shrinks following the stock market. They have discussed the concept of equilibrium inequality in a society, based on the principle of maximal entropy, and quantitatively show that it applies to the majority of population.

"Comparison Between the Probability Distribution of Returns in the Heston Model and Empirical Data for Stock Indexes" by A.C. Silva and V.M. Yakovenko [21]. In this paper the authors compared the probability distribution of returns for the three major stock-market indexes (Nasdaq, S&P500, and Dow-Jones) with an analytical formula recently derived by Dragulescu and Yakovenko for the Heston model with stochastic variance. For the period of 1982–1999, they found a very good agreement between the theory and the data for a wide range of time lags from 1 to 250 days. On the other hand, deviations start to appear when the data for 2000–2002 are included. They have interpreted this as a statistical evidence of the major change in the market from a positive growth rate in 1980's and 1990's to a negative rate in 2000's.

KPR Problem Ghosh and Chakrabarti analyze the 'Kolkata Paise Restaurant' problem, and show that 'naive' strategies sometimes lead to much better results than sophisticated ones in their paper "Statistics of the Kolkata Paise Restaurant problem" [22].

Increase of Income Inequality Banerjee and Yakovenko construct a model that explains both the exponential and power law regions of income distribution, and show that the increase of income inequality in the United States originates primarily in the increase of the income fraction going to the upper tail, which now exceeds 20 % of total income in their paper Universal patterns of inequality [23]. They have studied probability distributions of money, income, and energy consumption per capita for ensembles of economic agents. Following the principle of entropy maximization for partitioning of a limited resource, they found exponential distributions for the investigated variables. They also discuss fluxes of money and population between two systems with different money temperatures. For income distribution, they study a stochastic process with additive and multiplicative components. The resultant income distribution interpolates between exponential at the low end and power-law at the high end, in agreement with the empirical data for USA. Analyzing the data from the World Resources Institute, they found that the distribution of energy consumption per capita around the world is reasonably well described by the exponential function. Comparing the data for 1990, 2000, and 2005, we discuss the effects of globalization on the inequality of energy consumption.

Statistical Analysis of the Change in the Stock Correlations Aste et al. carry out a statistical analysis of the change in the stock correlations due to the 2008–2009 crisis using network tools Correlation structure and dynamics in volatile markets [24].

Complex Multi-agent Model Westerhoff introduces a complex multi-agent model, which takes into account firm-firm interactions, socio-economic opinion dynamics and sales expectations depending on individual attitudes in their paper "An agent-based macroeconomic model with interacting firms, socio-economic opinion formation and optimistic/pessimistic sales expectations" [25].

Kinetic Exchange Models of Markets Understanding the distributions of income and wealth in an economy has been a classic problem in economics for more than a hundred years. Today it is one of the main branches of Econophysics. J. Angle was the first to propose an elementary version of the stochastic exchange model. In the context of kinetic theory of gases, such an exchange model was first investigated by A. Dragulescu, V. Yakovenko. The main modeling effort has been put to introduce the concepts of savings and taxation in the setting of an ideal gas-like system. Basically, it assumes that in the short-run, an economy remains conserved in terms of income/wealth and any monetary transaction therefore, represents a redistribution of money from one agent to another. Millions of such conservative transactions lead to a steady state distribution of money (gamma function-like in the Chakraborti-Chakrabarti model with uniform savings, and a gamma-like bulk distribution ending with a Pareto tail in the Chatterjee-Chakrabarti-Manna model with distributed savings) and the economy converges to it. The distributions derived thus have close resemblance with those found in case of income/wealth distributions.

The Financial Bubble Experiment First Results D. Sornette, R. Woodard, M. Fedorovsky, S. Reimann, H. Woodard, W.-X. Zhou (The Financial Crisis Observatory), Department of Management, Technology and Economics, ETH Zurich, Kreuzplatz 5, CH-8032 Zurich, Switzerland (Dated: 3 May 2010).

On 2 November 2009, the Financial Bubble Experiment was launched within the Financial Crisis Observatory (FCO) at ETH Zurich (http://www.er.ethz.ch/fco/).

In that initial report, the authors diagnosed and announced three bubbles on three different assets (IBOVESPA Brazil Index, a Merrill Lynch corporate index, gold spot price). In the subsequent release of 23 December 2009 in the ongoing experiment, they added a diagnostic of a new bubble developing on a fourth asset (cotton futures). This report presents the four initial forecasts and analyses how they fared. They found that IBOVESPA and gold showed clear signs of changing from a bubble regime to a new one within their forecast quantile windows; that the Merrill Lynch bond index changed from a strong bubble regime to one of more moderate growth just before their publication date; and that cotton was and still is in a bubble without showing a clear change of regime.

14.4 Responses to Questionnaire

I had sent some questions relating to the present status and future perspectives of 'Econophysics' to different Professors having interest and working in this field by e-mail. I represent here their responses collectively. I had sent e-mail to about 40 number of esteemed Professors around the globe. Reply I have received is only from 13 Professors. Percentage of Response is only 32.5 %. Following Professors sent their responses and their valuable comments, which I am not reproducing here.

John Angle, Sitabhra Sinha, Bertrand Roehner, Reiner Kümmel, Jüergan Mimkes, Satya R. Chakravarty, Aoyama Hideaki, János Kertész, Tiziana Di Matteo, Hagen Kleinert, Mauru Gallegati, Thomas Lux, Victor Yakovenko and Prasanta Panigrahi (partial response).

Analysis of Response

- **Q.1** a. *Your formal field of research*?
  Physics—54 %, Economics & Sociology—38.5 %, Econophysics—7.5 %
  b. *Do you identify yourself as an Econophysicist?*
  Yes—46 %, No—31 %, Unclear—23 %
  c. *Fraction of Time you put in Econophysics Research?*
  Ans. 100 %–31 %, More than 50 %–31 %, Less than 50 %–31 %. Rest 7 % have not responded to this question.
- **Q.2** a.*Do you think, the studies made generally under this econophysics banner made any significant contribution*?
  Yes—84.6 %, No—7.7 %. Rest 7.7 % have not responded to this question.
  b. *If So, To Physics or Economics or both?*
  Physics—0 %, Economics—61.5 %, Both—7.7 %, Not in either field—15.4%, rest not responded.
  c. *If not yet at any satisfactory level, do you think these studies have any potential to succeed in the near future?*
  Yes—31 %, No—7.7 %. Others have responded to the earlier question.
- **Q.3** a. *Do you think an 'Econophysics' forum/body should be formed like other fields in Physics?*

Yes—84.6 %, No—0 %, Already existing—7.7 %, No clear idea—7.7 %
b. *International conferences/workshops are conducted under its banner in different countries by rotation?*
Yes—84.6 %, No—0 %, Already existing—7.7 %, No clear idea—7.7 %

- **Q. 4**. a. *Do you think, your university/institute should also offer econophysics courses to the students?*
Yes—84.6 %, No—15.4 %
b. *Did you write any textbook/research monograph on econophysics?*
Yes—61.5 %, No—23 %. Others about—15.5 % have not responded to this question.
c. *If not, do you intend to write one?*
Yes—15.4 %, No—15.4 %, rest not responded.
- **Q.5** a. *Do you think, like many other interdisciplinary fields like biophysics, geophysics etc., 'econophysics' will be recognized as a natural one soon? If so, by whom: physics or economics, or by both*?
Yes—53.8 %, No—23.1 %, No clear idea—23.1 %
b. *Do you think, econophysics research will be recognized by Nobel Prize ever?*
Yes—53.8 %, No—23.1 %, No Clear idea—23.1 %
c. *If so, in which discipline? Physics or Economics*?
Physics—0 %, Economics—53.8 %, No Clear idea/does not matter etc.—46.2 %
d. *When?*
Near Future—0 %, Late—53.8 %, Never—30.8 %. Others have not responded to this question.

From above discussions it is evident that, Econophysics is expanding day by day and establishing itself as a branch of science and is getting popular in a good pace. Scientists and professors are very much confident for its future as we observe from their replies. Even as many as 84.6 % are of opinion that Universities should open the course in Econophysics. People have started devoting their full time for econophysics research which is certainly encouraging and more than 50 % are hopeful that it will be recognized for Nobel prize although may not be in near future.

Acknowledgements I am grateful to the participants of the Econophysics-Kolkata VII conference for their comments, and suggestions.

This piece of work is a small part of a proposed book entitled 'Evolution of Econophysics'. Would appreciate receiving comments, suggestion and documents or materials for possible inclusion in the book.

References

1. http://education.jhu.edu/PD/newhorizons/Journals/Winter2011/Tokuhama4
2. http://en.wikipedia.org/wiki/History-of-education
3. http://thinkexist.com/quotation/india-was-the-mother-of-our-race-and-sanskrit-the/289367.html
4. All India Chinmaya Yuba Kendra, Awakening Indians to India, Central Chinmaya Mission Trust-Mumbai, 116 (2003)

5. http://en.wikiquote.org/wiki/Will-Durant#The-Case-for-India-.281931.29
6. Thussu DK (2009) Internationalizing media studies, Paperback 22
7. All India Chinmaya Yuba Kendra, Awakening Indians to India, Central Chinmaya Mission Trust-Mumbai, 252 (2003)
8. http://en.wikipedia.org/wiki/History_of_economic_thought. Feb-2012
9. http://en.wikipedia.org/wiki/Econophysics. Feb-2012
10. http://www.usc.es/cpoliticas/file.php/12/files/TheValueofStatisticalLawsinPhysicsand SocialSciences.htm
11. http://www.isical.ac.in
12. http://www.santafe.edu
13. http://www.ox.ac.uk/media/newsreleasesforjournalists/120412.html
14. Chen S-H, Li S-P (2012) Int Rev Financ Anal 23:110
15. http://www.leiden.edu
16. Osborne MFM (1959) Brownian motion in the stock market. Oper Res 7(2):145
17. http://www.newscientist.com/article/mg15420784.700-flight-over-wall-st.html
18. Gopikrishnan P, Meyer M, Nunes Amaral LA, Stanley HE (1998) Eur Phys J B 3:139
19. Dragulescu AA, Yakovenko VM (2001) Eur Phys J B 20:585
20. Silva AC, Yakovenko VM (2005) Europhys Lett 69:304
21. Silva AC, Yakovenko VM (2003) Physica A 324:303
22. Ghosh A, Chakrabarti BK (2010) New J Phys 12:075033
23. Banerjee A, Yakovenko VM (2010) New J Phys 12:075032
24. Aste T, Shaw W, Matteo TD (2010) New J Phys 12:085009
25. Westerhoff F (2010) New J Phys 12:075035

Chapter 15
Econophysics and Sociophysics: Problems and Prospects

Asim Ghosh and Anindya S. Chakrabarti

Abstract Econophysics and sociophysics have survived the first ideological battle against standard theories of economics and sociology. But a mere survival and a subsequent slow decay is not in the manifesto of these interdisciplinary subjects. While there are conceptual problems and also a constant aversion of the mainstream economics community, the basic philosophy of such a physicsoriented approach can contribute largely to the soft sciences where empirical validation is still very difficult at best, and impossible at worst. Here we give a very brief description of the evolution of mainstream mathematical economics and subsequently that of econophysics. Many excerpts from writings of distinguished econophysisicts and sociophysisicts are also given so that we can take a contemporary look at where these subjects currently are and where they can go from here.

15.1 Introduction

As the names suggest, Econophysics and Sociophysics are both of interdisciplinary type. This research field stemmed from the study of many interacting units by physicists. A pertinent question that still haunts these subjects is that 'Why do physicists believe that they are more qualified than economists to answer economic phenomenon?' As physicist J. McCauley responded, the answer lies in the idea that they are trained better to see the connections between seemingly different phenomena [1]. Given the enormous stake in providing a better guideline to harness the economic catastrophes, it is not a bit surprising that all sorts of attempts have been made to explain the deep puzzles of human behavior and that of society, at a larger scale. The results are mixed. One thing is certain that we have not reached the des-

A. Ghosh (✉)
Theoretical Condensed Matter Physics Division, Saha Institute of Nuclear Physics, 1/AF Bidhannagar, Kolkata 700 064, India
e-mail: asimghosh066@gmail.com

A.S. Chakrabarti
Economics Department, Boston University, 270 Bay State Road, Boston, MA, USA
e-mail: anindya@bu.edu

F. Abergel et al. (eds.), *Econophysics of Agent-Based models*,
New Economic Windows, DOI 10.1007/978-3-319-00023-7_15,

tiny yet. The real success is far away (in the sense of deriving a result as precise and as important as finding the structure of the DNA for example). In some cases, enormous progress has been made. In others, both theory and empirics fall behind reality. The good news is that Econophysics has already become well accepted to physicists even though it is far from being impressive to mainstream economics. Before the advent of game theory, it was really difficult to figure any difference in philosophy of economics and statistical physics. However, after the birth of game theory economics drifted further and further apart from phenomenological studies and that too, with a good reason. The failure of Keynesian economics in the late sixties gave birth to the famous 'Lucas Critique' which paved the way for rational expectation and ultimately led to a modeling philosophy where it was taken for granted what ever resource we have is finite except our computational capacity.

As have been repeated dozens of times, study of economics by physicists or by mathematicians is not really a new trend. However, as always is the case with economics, the real culprit is always the market which has little empathy for one's academic credentials. So the results are to be taken with a pinch of salt. But by and large, it is the physicists and mathematicians who have shaped the economic theories as it stood about 40 years back. Daniel Bernoulli (a mathematician and a physicist) first proposed a very refined theory of preference and thereby attempted to explain risk- aversion, a remarkable achievement in itself. In 1900, French mathematician Louis Bachelier (guided by physicist Poincare who did not quite appreciate the formulation of the legendary random walk model) gave a model "Theory of Speculation" for the stock market and thereby brought a new paradigm in economics which, unfortunately, was not noticed until 1960s. Later, slight modification of the model was made by Edward Thorp (American physicist) and the model became known as Bachelier-Thorp model. Irving Fisher, who was trained by J. Willard Gibbs, was one of the founders of neoclassical economic theory. Unfortunately he was also the one who (in-)famously told 'Stock prices have reached what looks like a permanently high plateau' just before the Great Depression. This has uncanny similarity with the statement of Olivier Blanchard (of M.I.T., the then chief economist at the International Monetary Fund) that 'the state of macro is good' just before the largest depression in the last 70 years [2]. One consolation is that at least the physicists are about 80 years ahead of economists even at making bad predictions! The first institutional acknowledgement of the highest class was that the first Nobel Prize winner in economics was Jan Tinbergen who did his Ph.D. in statistical physics with Paul Ehrenfest. The situation changed later. After 1970, a large number of physics students got involved in financial market and started fidgeting with the nitty-gritty details instead of writing down a grand theory. This has become a trend in this field since then. It is often observed that models proposed by these physicists are more useful than others. Instead plunging into showing existence of a solution like what an economist does, the physicists try to actually describe a real workable solution. This methodological difference widened day by day until a situation came where the finance and economics became distant relatives who greets each other from a safe distance on an occasional meeting.

However, as with any other thing in life, the glory of Wall Street comes with its fair share of burden as well. It is often blamed that these Wall Street physicists were responsible for the Black Monday crash in 1987, crisis at Long-Term Capital Management in 1998 and worldwide financial collapse in 2007–2008. Below, we give some instance of such blames. But that did not stop physicists from asking new and more penetrating questions about economics and sociology in general. This eventually led to birth of Econophysics and Sociophysics. We will briefly discuss some statistics regarding the development of Econophysics in last 17 years (1996–till date). These, we hope to shade some light on the current controversies.

15.2 Criticisms: Some Recent Quotes

- "The proliferation of econophysicists lining the halls of university departments coincided with a flood of physicists leaving academia to enter the financial sector. Young mathematicians and physicists, equipped with little more than a PhD and a quick wit, were recruited into an industry that was fast learning the value of minds well-versed in the language of partial differential equations. So when the world at large fell into financial crisis in 2008, physicists were quickly identified by many as one of the key reasons behind the collapse. In an October 2008 interview on US television's 60 minutes, Jim Grant—founder of Grant's Interest Rate Observer, a twice-monthly journal of financial markets—blamed the mess on 'mortgage science projects devised by these Nobeltracked physicists who came to work on Wall Street for the very purpose of creating complex instruments'.... And Grant's certainly wasn't a lone voice. Some even went as far as to argue2 that the prevalence of Asperger's syndrome in the physics community was responsible for the crisis—that physicists working in the financial sector weren't capable of feeling empathy for the lives that stood to be ruined from the inevitable failure of their complex models."
  —*Editorial Note, "Net gains", Nature Physics, vol. 9, 119 (2013)*

- "The untold riches of a career on Wall Street have loomed large in the physics community these last few decades. Most physicists know someone a friend from graduate school; a former student who left academia to pursue a career in finance. And we have heard quite a bit, too, about the damage wrought by the so-called 'quants' that these physicists morph into when they take investment jobs. Quants, along with the mathematical models and algorithmic trading strategies they helped to develop, have been blamed for three decades of market crashes, from the 1987 Black Monday crash, to the 1998 crisis at Long-Term Capital Management, to the 2007-08 worldwide financial collapse." [3]
  —*J.O. Weatherall, "The Back Page", APS News, vol. 22, 3 (2013)*

15.3 A Statistical Survey on the Development of Econophysics & Sociophysics

The term 'econophysics' was introduced by H. Eugene Stanley in a Kolkata-Conference held in 1995. To see how the subject grew in scientific community, we

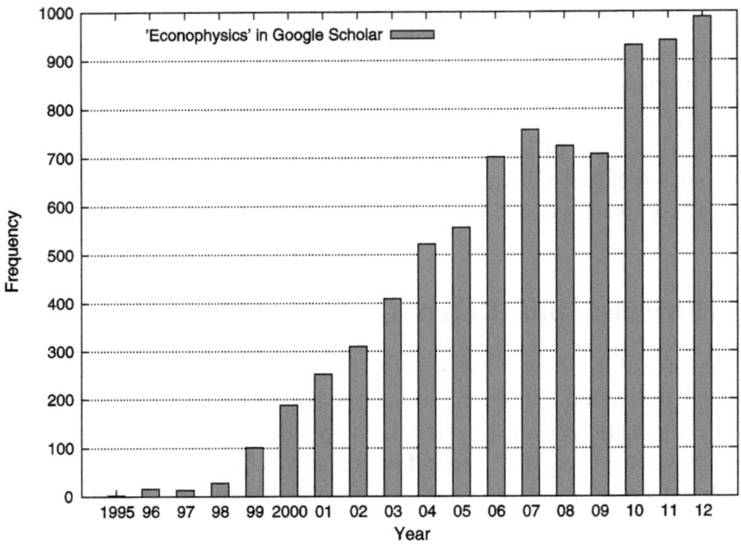

Fig. 15.1 Histogram plot of numbers of entries containing the term 'econophysics' versus the corresponding year. The data are taken from google scholar site (http://scholar.google.co.in/schhp)

have taken the statistics of the articles having 'econophysics' term any where in the article from google scholar site. Figure 15.1 shows histogram plot of the number of papers posted in google scholar over different years. From the figure it is clear that the subject is growing quite fast.

A similar study have been done for 'sociophysics'. Sociophysics is an interdisciplinary subject where statistical physics tools are applied to solve or understand the problems related to social science. To assess how the subject evolved, we have done a similar survey on it from google scholar. We have counted total number of entries having the term 'sociophysics' any where from google scholar site. Figure 15.2 shows histogram plot for every year versus corresponding total number of such articles. From the figure, it is again clear that numbers of such papers etc. are increasing. From these observations we can say that the field 'sociophysics' is also becoming popular in scientific community.

Next we give some important quotations from different sources (like books, editorial notes, reviews, etc) indicating the same point for Econophysics.

15.4 Development of Econophysics: Some Quotes

- "Econophysics is a new interdisciplinary research field applying methods of statistical physics to problems in economics and finance. The term 'econophysics' was first introduced by the theoretical physicist Eugene Stanley in 1995 at the conference Dynamics of Complex Systems, which was held in Kolkata as a satel-

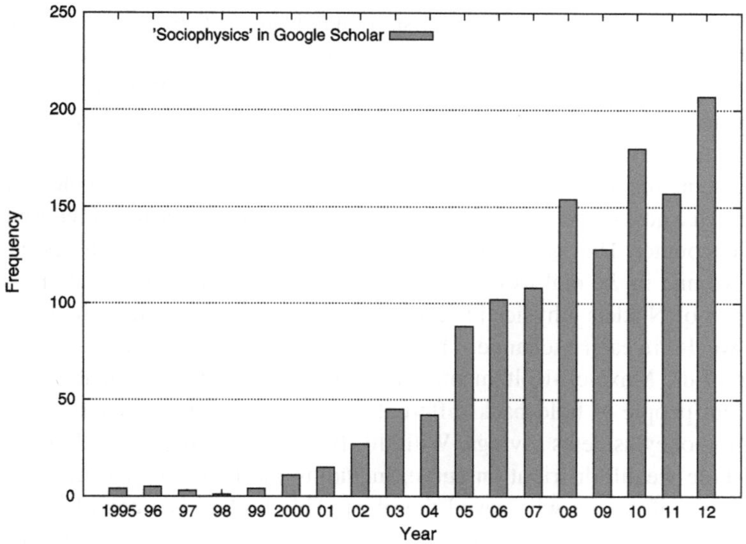

Fig. 15.2 Histogram plot of numbers of entries containing the term 'sociophysics' versus the corresponding year. The data are taken from google scholar site (http://scholar.google.co.in/schhp)

lite meeting to the STATPHYS-19 conference in China.... The term appeared first by Stanley et al. (1996) in the proceedings of the Kolkata conference. " [4]

—*V. Yakovenko & J.B. Rosser, Colloquium: "Statistical mechanics of money, wealth and income", Reviews of Modern Physics, vol. 81, 1703 (2009)*

- "Econophysics is a new research field, which makes an attempt to bring economics in the fold of natural sciences or specifically attempts for a 'physics of economics'. The term Econophysics was formally born in Kolkata in 1995. The entry on Econophysics in The New Palgrave Dictionary of Economics, 2nd Ed., Vol. 2, Macmillan, NY (2008), pp. 729–732, begins with '... the term *econophysics* was neologized in 1995 at the second Statphys-Kolkata conference in Kolkata (formerly Calcutta), India...'. The Econophysics research therefore formally completes fifteen years of research by the end of this year!" [5]

  —*Editorial Note, "Fifteen Years of Econophysics Research", Science and Culture, Kolkata, vol. 76, Sept.-Oct. (2010)*

- "When H. Eugene Stanley coined the term 'econophysics' at a Kolkata conference in 1995, the field was still in its infancy—despite having already spawned a slew of papers by statistical physicists, eager to lend their expertise to an intriguing new set of problems. This trend was born in the wake of a sudden availability of large amounts of financial data in the 1980s. But it may have had just as much to do with a frustration at the inadequacy of traditional theoretical approaches to economics, which seemed to favour model simplicity over accuracy, or agreement with empirical data."

  —*Editorial Note, "Net gains", Nature Physics, vol. 9, 119 (2013)*

- "More than 100 years ago, physicists pointed out that the broad income curve for the majority resembles the distribution of energy among molecules in a gas, a pattern called the Maxwell-Boltzmann distribution. This prompted the idea that the distribution arises because people exchange wealth when they meet, much as gas molecules exchange energy when they collide. That idea has since been tested using mathematical models that liken human beings to molecules bouncing around in a gas. In the simplest model, people risk surrendering all their wealth at each encounter. That produces a wealth curve that has far more ultra-poor people than we find in the real world. So in 2000, Bikas Chakrabarti's team at the Saha Institute of Nuclear Physics in Kolkata, India, allowed people to retain some of their wealth in each exchange. The result was a wealth curve similar to the broad hump of the Maxwell-Boltzmann distribution. The next refinement was to allow different people to hold back different percentages of their wealth—effectively setting money aside as savings. With this tweak, the model correctly reproduced the whole wealth distribution curve, including the Pareto tail, which was made up largely of people who saved the most. This finding has been backed up by other similar models, including one developed by Ferrero, in which the richest 10 per cent are once again those most inclined to save. If these simple models do capture something of the essence of real-world economics, then they offer some good news." [6]

  —*S. Battersby, "The physics of our finances", New Scientist, 28 July, issue 2875, 41 (2012)*

- "We believe, several important developments in Econophysics research have already taken place in the last one and half a decade. We now try to give a very brief (incomplete and biased!) list of some developments: (i) Empirical characterization, analyses and modeling of financial markets—in particular, the deviation from Gaussian statistics has been established, following the early observations of Mandelbrot and Fama (1960s)—beginning with the studies in 1990s by the groups of Stanley, Mantegna, Bouchaud, Farmer and others. (ii) Network models and characterization of market correlations among different stocks/sectors by the groups of Mantegna, Marsili, Kertesz, Kaski, Iori, Sinha and others. (iii) Determination of the income or wealth distributions in societies, and the development of statistical physics models by the groups of Redner, Souma, Yakovenko, Chakrabarti, Chakraborti, Richmond, Patriarca, Toscani and others. The kinetic exchange models of markets have now been firmly established; this gained a stronger footing with the equivalence of the maximization principles of entropy (physics) and utility (economics) shown by the group of Chakrabarti. (iv) Development of behavioral models, and analyses of market bubbles and crashes by the groups of Bouchaud, Lux, Stauffer, Gallegati, Sornette, Kaizoji and others. (v) Learning in multi-agent game models and the development of Minority Game models by the groups of Zhang, Marsili, Savit, Kaski and others, and the optimal resource utilization 'Kolkata Paise Restaurant' (KPR) model by the group of Chakrabarti. These have given important insights in such multi-agent collective parallel learning dynamics. In this context, it might be mentioned that to our knowledge, the KPR model might be the only model in Physics that is named after a city (namely Kolkata). Considerable literature has developed out of these and other studies [see the articles in this issue, and references therein]. We had invited all the above-mentioned groups and others to contribute their perspectives or views on these developments, and we are happy that most of them could. Unfortunately due to time constraints, several others could not contribute. The importance and proliferation of the interdisciplinary research of Econophysics is highlighted in this special issue of Science & Culture, which presents a collection of twenty nine papers (giving country wise perspectives, reviews of the recent developments and original research communications), written by more than forty renowned experts in physics, mathematics or economics, from all over the world." [5]

—*Editorial Note, "Fifteen Years of Econophysics Research", Science and Culture, Kolkata, vol. 76, Sept.-Oct. (2010)*

- "Contemporary mainstream economics has become concerned less with describing reality than with an idealised version of the world. However, reality refuses to bend to the desire for theoretical elegance that an economist demands from his model. Modelling itself on mathematics, mainstream economics is primarily deductive and based on axiomatic foundations. Econophysics seeks to be inductive, to be an empirically founded science based on observations, with the tools of mathematics and logic used to identify and establish relations among these observations. Econophysics does not strive to reinterpret empirical data to conform to a theorist's expectations, but describes the mechanisms by which economic systems actually evolve over time." [7]

  —*S. Sinha & B. K. Chakrabarti, "Econophysics: An emerging discipline", survey article in Economic & Political Weekly, vol. 46, 44 (2012)*

- "Despite the misgivings of the popular media, and the relative downturn in the employment prospects of the financial sector, the trend for physicists to enter the industry has not abated—nor, it must be said, has momentum for physicists in academia to turn their attention to problems associated with market dynamics. Indeed, the quantitative finance archive was launched in December 2008, largely to service a wealth of submissions that was being distributed amongst the existing fields. In recent years, however, the focus of these efforts has shifted towards the realm of network science. This new endeavour addresses a need to understand the structure and dynamics underlying financial markets, to explain—and anticipate—the effects that interactions between many agents are capable of inducing. An optimistic view is that, equipped with such knowledge of systemic behaviour, we might even be able to influence market dynamics using the tools of complex networks science."

  —*Editorial Note, "Net gains", Nature Physics, vol. 9, 119 (2013)*

15.5 Development of Sociophysics: Some Quotes

- "Physics has always influenced other fields.... More recently, (computational) statistical physics has been applied to biology, economics, world politics etc." [8]
  —*S. Moss de Oliveira, P.M.C. Oliveira & D. Stauffer, in Preface of Evolution, Money, War, and Computers, Teubner, Stuttgart (1999)*

- "It may be surprising but the idea of a physical modeling of social phenomena is in some sense older than the idea of statistical modeling of physical phenomena. The discovery of quantitative laws in the collective properties of a large number of people, as revealed, for example, by birth and death rates or crime statistics, was one of the catalysts in the development of statistics, and it led many scientists and philosophers to call for some quantitative understanding of how such precise regularities arise out of the apparently erratic behavior of single individuals. Hobbes, Laplace, Comte, Stuart Mill, and many others shared, to a different extent, this line of thought.... This point of view was well known to Maxwell and Boltzmann and probably played a role when they abandoned the idea of describing the trajectory of single particles and introduced a statistical description for gases, laying the foundations of modern statistical physics. The value of statistical laws for social sciences was foreseen also by Majorana (1942, 2005). But it is only in the past few years that the idea of approaching society within the framework of statistical physics has transformed from a philosophical declaration of principles to a concrete research effort involving a critical mass of physicists. The availability of new large databases as well as the appearance of brand new social phenomena (mostly related to the Internet), and the tendency of social scientists to move toward the formulation of simplified models and their quantitative analysis, have been instrumental in this change." [9]
  —*C. Castellano, S. Fortunato & V. Loreto, "Statistical physics of social dynamics", Reviews of modern physics, vol. 81, 591 (2009)*

- "Sociophysics started to become a popular topic of research in the last quarter of the twentieth century.... While physics deals mostly with nonliving systems, society is meaningful only in the presence of life. The terminology 'sociophysics' may sound strange but fact is, the term 'social physics' was introduced by the Belgian statistician Adolphe Quetelet (Quetelet 1835, 1842) long back. The concept of the 'average man' (l'homme moyen), who is characterized by the mean values of measured variables that follow a normal distribution, was outlined and data about many such variables were collected. The French social philosopher Auguste Comte also used the term social physics in his 1842 work. Comte defined social physics as the study of the laws of society or the science of civilization...." [10]
  —*P. Sen & B.K. Chakrabarti, Sociophysics: An introduction, Oxford University Press (2013)*

- "Do humans behave much like atoms? Sociophysics, which uses tools and concepts from the physics of disordered matter to describe some aspects of social and political behavior, answers in the affirmative. But advocating the use of models from the physical sciences to understand human behavior could be perceived as tantamount to dismissing the existence of human free will and also enabling those seeking manipulative skills. This thought-provoking book argues it is just the contrary. Indeed, future developments and evaluation will either show sociophyics to be inadequate, thus supporting the hypothesis that people can primarily be considered to be free agents, or valid, thus opening the path to a radically different vision of society and personal responsibility. This book attempts to explain why and how humans behave much like atoms, at least in some aspects of their collective lives, and then proposes how this knowledge can serve as a unique key to a dramatic leap forwards in achieving more social freedom in the real world. At heart, sociophysics and this book are about better comprehending the richness and potential of our social interaction, and so distancing ourselves from inanimate atoms." [11]

—*S. Galam, back cover of Sociophysics: A Physicist's Modeling of Psycho-political Phenomena, Springer (2012)*

- "In biology, people are accustomed to think that from simple animals up to dinosaurs, all species originated by Darwinian evolution and selection of the fittest. Once the principle is applied to human beings, some dislike it and rely instead on creationism. Similarly for sociophysics, not much emotion is aroused if ants are simulated by mathematically defined probabilities. But to apply the same type of modelling to humans is disliked by some: We are not just atoms. Of course we are not; neither is the planet earth a point mass. Nevertheless, for Kepler's law of how the earth rotates around our sun, a point mass is a good approximation."

  —*Hmolpedia: An encyclopedia of human thermodynamics, entry on D. Stauffer (http://www.eoht.info/page/Dietrich+Stauffer)*

- "There is indeed major scope for developing sociophysics much further, following the developments in physics of many-body nonlinear and frustrated dynamical systems and statistical physics. Though the size of any typical social system is much less compared to the Avogadro number ($\mathcal{O}(10^{23})$; and present world population is $\mathcal{O}(10^{10})$), they are sufficiently large to allow the laws of large numbers to induce and stabilize very precise statistics in most cases. Indeed, as described in various chapters of this book, the dynamics of many social phenomena are similar to those of many-body aggregates, where each constituent follows some quantifiable physical dynamics. The individual choices or decisions induce only some noise (mostly uncorrelated) on this average dynamics. Even in some special cases, like those in the cases of minority and similar adaptive games, the individual decision dynamics can be modeled after the physical models of brain or neural network having frustrated dynamics! All these inspire and allow physical modeling of every social dynamical phenomenon." [10]

  —*P. Sen & B.K. Chakrabarti, Sociophysics: An introduction, Oxford University Press (2013)*

15.6 Discussions

We have sketched how the attempts of physicists to explain economic and social phenomena have attracted both good and bad criticisms from all sides. Econophysics and Sociophysics have survived the initial battle. However, the real challenge lies in near future where they have to prove their usefulness. There are some criticisms that have been leveled against economics like lack of heterogeneity in the macroeconomic models etc. These have flimsy foundations. However, over and over a couple of more methodological (or philosophical if you wish) problems have been shown to prove to be extremely effective in making economic predictions horribly bad. One important point is that we solely describe the economics literature as that has a formidable mathematical structure and makes strict predictions where falsifiability is possible, at least in principle. As far as Sociology is concerned, probably it does not have as strong an analytical framework. Hence, we focus on economics only. A couple of problems where a new point of view (or at least different and more effective than the current economic idea) can really be of some use are as follows. 1. **Systemic risk**: even though economists are turning their attentions to the possibility of a global breakdown of market mechanism, it requires more new tools that probably physicists have. 2. **Equilibrium**: the idea of equilibrium is so deep-rooted in economics that almost every economist suffers from an obsessive-compulsive disorder about it. But to be fair to them, it is not entirely clear what else can one do? 3. **Limitation of computational facilities**: it is rarely acknowledged in the economics literature that most of us are horribly bad at mental calculations. Only in the recent years economists have started talking about 'rational Inattention' [12]. But even in those frameworks, people have to do so much of calculation on whether to spend effort thinking about something or not, the purpose of the whole exercise is probably self-defeating! But it may be too early to predict given the age of the almost newly-born idea. This leads to the fourth and last point. 4. **Economizing on models**: the usual standard neo-Keynesian DSGE (dynamic stochastic general equilibrium that the central banks use) models have about three/four exogenous shocks (TFP shock, monetary shock, possibly disaster risk), about ten endogenous variables (wage, rental rate, consumption, capital holding, prices of the intermediate goods, price of the final goods, inflation, output, employment etc.) and a huge number of exogenous variables (preference parameters, shock parameters, money growth rate etc.). Of course, the exact number of variables considered depends on the exact model specification. But in all likelihood getting a meaningful prediction from such a complicated object becomes extremely difficult. Moreover, by neglecting coordination failure on the demand side, that provides little room for a demand-side explanation to economic crisis and in the current world, crises are more demand-driven than supply-driven (we really do not have any crisis from flood or war and even if we have them we have a good idea about how to tackle them). This led to the oft-quoted comment by Robert Solow (U.S. Congress hearing on July 20, 2010), a Nobel-laureate economist that '*I do not think that the currently popular DSGE models pass the smell test. They take it for granted that the whole economy can be thought about as if it were a single, consistent person or dynasty carrying out a rationally designed, long-term plan, occasionally disturbed by unexpected shocks, but* *adapting to them in a rational, consistent way... The protagonists of this idea make a claim to respectability by asserting that it is founded on what we know about microeconomic behavior, but I think that this claim is generally phony. The advocates no doubt believe what they say, but they seem to have stopped sniffing or to have lost their sense of smell altogether*'.

Acknowledgements We would like to thank Prof. B.K. Chakrabarti, Dr. A. Chakraborti and Dr. A. Chatterjee for their valuable comments.

References

1. McCauley J (2004) Dynamics of markets. Cambridge Univ. Press, Cambridge
2. Krugman P (2009) How did economists get it so wrong?, NY Times, Sept 2
3. Weatherall JO (2013) The physics of wall street: a brief history of predicting the unpredictable. Houghton Mifflin Harcourt, London
4. Yakovenko VM, Rosser JB Jr. (2009) Colloquium: statistical mechanics of money, wealth and income. Rev Mod Phys 81:1703
5. Chakraborti A, Chakrabarti BK (eds) (2010) Fifteen years of econophysics research. Science and culture, Kolkata, vol 76
6. Battersby S (2012) The physics of our finances. New Sci 2875:41
7. Sinha S, Chakrabarti BK (2012) Econophysics: an emerging discipline. Econ Polit Wkly 46:44
8. Moss de Oliveira S, de Oliveira PMC, Stauffer D (1999) Evolution, money, war, and computers. Teubner, Stuttgart
9. Castellano C, Fortunato S, Loreto V (2009) Statistical physics of social dynamics. Rev Mod Phys 81:591
10. Sen P, Chakrabarti BK (2013) Sociophysics: an introduction. Oxford University Press, Oxford
11. Galam S (2012) Sociophysics: a physicist's modeling of psycho-political phenomena. Springer, Berlin
12. Sims C (2003) Implications of rational inattention. J Monet Econ 50:665

Chapter 16
A Discussion on Econophysics

Hideaki Aoyama

16.1 Dialog with My Friends[1]

Every year, I (**A**) and my friends **B** and **H** get together on the riverside of the Hooghly river for great catfish curry and a chat on our life, universe and econophysics.

A: Greetings, B and H, my good friends! It's so nice meeting you again here in Kolkata.

B and **C:** Greetings, A! We are glad to see you too. Let's get some food and sit down and discuss our issues.

A: Of course. No one can beat curries on Camac street. Anyhow, I am worried about the future of the word "Econophysics", as some tries to avoid this word these days.

H: They are not unreasonable. After all, all we do is to study economics. Just because some of us have backgrounds in Physics, it does not mean that we have to name all of our research topics so they end with "physics".

A: Well, even though I published papers on Linguistics [1, 2] and named it "Lingophysics" to write about it on Physics journals, I kind of agree with you. The name "econophysics" may in a sense deter good economists away from us, as it sounds as if we are in total denial of economics.

B: Wait a minute. Here we are in Kolkata, where this term "Econophysics" was born in 1995, and it was for a good reason. We are studying economics with physics

---

[1] Disclaimer: All characters appearing in this article are fictitious. Any resemblance to real persons or their opinions is purely a miracle.

H. Aoyama (✉)
Kyoto University, Kyoto 606-8502, Japan
e-mail: hideaki.aoyama@scphys.kyoto-u.ac.jp

F. Abergel et al. (eds.), *Econophysics of Agent-Based models*,
New Economic Windows, DOI 10.1007/978-3-319-00023-7_16,

ideas and philosophy. And we are proud that this true spirit of natural science is different from mathematics and engineering. We are here to find truth hidden behind all the phenomena and ...

B and **C:** (in unison) I know. I know, you don't need to tell ...

A: Have you read this "THE BACK PAGE" of the APS News by someone in *Logic and Philosophy of Science*?

> Physicists have a distinctive way of thinking ... They are experts in approximative thinking,, in building toy models and effective theories. This sort of reasoning is just what is needed to take a problem that appears hopelessly complex and find the simplifying assumptions and idealizations necessary to make it tractable.
>
> (J.O. Weatherall, March 8th, 2013)

B: Well said. So, that is why I say that what we are doing IS physics. This is true physics, only the subject is now complex phenomena of economic activity of people, firms, etc, etc. This IS physics.

A: Yes, you are indeed convincing. There is this famous textbook on statistical physics by Ryogo Kubo in Japanese, where he starts his introduction to basic concepts of statistical physics using the wealth distribution as an example. He even mentions distribution of workers to industrial sectors in the section for Bose-Einstein statistics. He was well aware of importance of statistical physics for economics, even though he himself did not venture into this area.

H: Now, that being said, let me put forward another issue before we start on Chai. It's about network: you know, every economic phenomena occurs on networks, be it trading network, credit network, ...

A: And you must know that we are among the first ones to do agent-modeling on economic networks as in the paper of Souma et al. [3].

H: I do a lot of network analysis, community identifications and all that with my slaves ..., whoops, graduate students, rather, and they are often asked "so what?" by other people in the physics department.

B: That is why "econophysics" is important: Without "physics" approaches, always trying to identify basic laws and explanations hidden under phenomena and to seek ways to predict consequences, network "analysis" as well as any other analysis are just neat way of description and pretty graphics and would never become "science".

A: That rounds up the whole discussion. We have gathered at Saha Institute of Nuclear Physics again this year, with the stress on "agent-based modeling". We are building science of economics with economic network as its base-space.

B: I am so glad that all of us agree with the value of science. Now, let me invite you to an arboretum across the river, where the largest tree in the world is. India is a great country beyond your imagination in many ways.

B and **C:** Thanks, B. We are already looking forward to our next meeting in March 2014!

References

1. Aoyama H, Constable J (1999) Lit Linguist Comput 14:339, 507
2. Aoyama H, Constable J (2001) Belg J Linguist 15:35
3. Souma W, Fujiwara Y, Aoyama H (2001) Small-world effects in wealth distribution. arXiv: cond-mat/0108482

Photographs of the participants of the Econophys-Kolkata VII meeting

MIX
Papier aus verantwortungsvollen Quellen
Paper from responsible sources
FSC® C105338

If you have any concerns about our products,
you can contact us on
ProductSafety@springernature.com

In case Publisher is established outside the EU,
the EU authorized representative is:
Springer Nature Customer Service Center GmbH
Europaplatz 3, 69115 Heidelberg, Germany

Printed by Libri Plureos GmbH
in Hamburg, Germany